Crucibles of hazard: Mega-cities and disasters in transition

Crucibles of hazard: Mega-cities and disasters in transition

Edited by James K. Mitchell

United Nations University Press

TOKYO · NEW YORK · PARIS

United Nations University Press
The United Nations University, 53-70, Jingumae 5-chome, Shibuya-ku, Tokyo, 150-8925, Japan
Tel: (03) 3599-2811   Fax: (03) 3406-7345
E-mail: sales@hq.unu.edu
http://www.unu.edu

United Nations University Office in North America
2 United Nations Plaza, Room DC2-1462-70, New York, NY 10017 USA
Tel: (212) 963-6387   Fax: (212) 371-9454
E-mail: unuona@igc.apc.org

United Nations University Press is the publishing division of the United Nations University.

Cover design by Jean-Marie Antenen, Geneva

Typeset by Asco Typesetters, Hong Kong

Printed in the United States of America

UNUP-987
ISBN 92-808-0987-3

Library of Congress Cataloging-in-Publication Data

Crucibles of hazard : mega-cities and disasters in transition / edited by James K. Mitchell.
    p.   cm.
  Includes bibliographical references and index.
  ISBN 9280809873 (pbk.)
  1. Disaster relief. 2. Metropolitan areas. 3. Disasters. 4. Natural disasters. 5. Emergency management. I. Mitchell, James K., 1943–
  HV553 .C78   1999
  363.34'09173'2–dc21   99-6054
                                                              CIP

Contents

v

Acknowledgements

Edited volumes are, by their nature, dependent on the contributions of many people, but this one has drawn on the work of an especially wide range of collaborators in addition to the participating authors. A large debt of gratitude is owed to the United Nations University (Tokyo), especially in the persons of Roland Fuchs, former Vice Rector, who encouraged the project's initial stages, and Juha Uitto, a young Finnish geographer who coordinated it. Members of the International Geographical Union's Study Group on the Disaster Vulnerability of Megacities also played a vital role. Several were participants in the 1994 Tokyo conference on which this book is based and later became founding members of the Study Group's steering committee.

Rutgers University graduate students enrolled in seminars on natural hazards during 1993, 1994, and 1996 deserve special mention. Many conducted their own case-studies of disaster-susceptible mega-cities, which helped to inform chapter 13. The following were particularly helpful: Roger Balm, Joe Center, Mike Craghan, Marla Emery, Ruth Gilmore, Jojo Hardoy, Ted Kilian, Lisa Lacourse, Juliana Maantay, Sudha Maheshwari, Elaine Matthews, Anastassia Mikhailova, Mariana Mossler, Karen Nichols, Karen Patterson, Lena Raberg, Bruce Ramsay, Karlene Samuels, Lisa Vandermark, and Doracie Zoleta-Nantes. Colleagues in a wide variety of organizations and venues served as audiences for early drafts and provided valuable inputs. These included: the Hazards Specialty Group of the Association of American Geographers; the annual

Hazards Workshops of the Natural Hazards Research and Applications Information Center (Boulder, Colorado); a special session of the 1995 meeting of the Institute of British Geographers; the International Symposium on Urban Growth and Natural Hazards (Clermont-Ferrand, France), and the First International Earthquakes and Megacities Workshop (Seeheim, Germany).

Maria Steppanen and C. Emdad Haque were external readers of manuscripts, supplied background materials, and added helpful local details that have been incorporated into case-study chapters. Scott Campbell supplied a key background paper and thought-provoking comments. Mike Siegel of the Rutgers University Cartography Laboratory oversaw the preparation or redrawing of maps and illustrations. All of these, and many others, have tried to keep me from error. If I have committed few, theirs is the praise; otherwise the flaws are my own.

List of tables, figures, and photographs

Tables

Figures

Photographs

1

Introduction

James K. Mitchell

The Great Hanshin earthquake of 17 January 1995 was a signal event in the history of urban disasters.[1] Not only was it Japan's most deadly and destructive natural disaster in over 70 years,[2] it also raised disturbing questions about existing hazard-management policies and programmes that had been regarded as among the most effective in the world. Despite decades of attention to the goals of hazard reduction by Japanese governments, industries, and citizens' organizations, over 6,000 residents of the country's second-largest metropolitan area[3] were killed, 10 times as many were injured, and large parts of the Kobe–Osaka urban region experienced heavy damage and disruption.[4] Fires took hold rapidly and burned out of control, structures and lifelines that had been designed and built to hazard-resistant standards gave way, emergency management operations failed to live up to expectations, and recovery programmes dragged on well beyond their anticipated termination dates.[5] Not since the massive Kanto earthquake of 1923 devastated Tokyo and Yokohama has a major Japanese urban area been so grievously stricken by natural disaster. Indeed, this was the first time that Japan's annual disaster death tolls have climbed back above double digits into the thousands since the Ise Bay typhoon of 1959 killed over 5,000 people and triggered a major restructuring of the country's hazard-management systems. Economic losses may have exceeded a staggering US$100 billion![6]

The Hanshin earthquake is just the most recent in a string of natural disasters that have inflicted unprecedented losses on very large cities

Table 1.1 Recent natural disaster losses in mega-cities

Year	Location	No. of deaths	Cost (US$ billion)
1985	Mexico City	5,000–10,000+	?
1989	San Francisco Bay, CA	62	8
1991	Oakland, CA	25	2
1992	Miami, FL	38	30
1994	Northridge, CA	57	20
1995	Kobe, Japan	6,300	100+

Sources: Official loss estimates provided by various national and international agencies.
Note: Figures are best available estimates.

(table 1.1). Often these have involved earthquakes, but hurricanes and wildfires have also led to heavy losses. Though the upward trend in *economic* losses is most striking, *death tolls* have also been substantial. These events have far-reaching implications for much of the world's population. Among others, they suggest that cherished notions about the security of cities in the face of natural extremes are no longer tenable and that disasters in large cities are likely to pose troubling new problems for society. Viewed against the emergence of a predominantly urban world where people increasingly live in giant urban agglomerations (i.e. mega-cities[7]), recent disasters also underscore the potential for even larger losses in the future.

This book focuses on *natural* hazards and disasters in mega-cities partly because of their potential for catastrophe. That does not mean that other kinds of hazards are incapable of producing urban catastrophes. Wars have frequently been associated with large-scale destruction of urban areas, especially in the twentieth century (e.g. Hiroshima, Dresden, Phnom Penh, Kabul). Political terrorism and crime are also potent agents of urban destruction. So too are hazardous industrial technologies (e.g. Bhopal, Texas City). In view of this record, why single natural hazards out for special consideration?

The answer is that natural hazards are joint products of nature and society. Unlike the other threats just mentioned, they are only partly created by humans.[8] This gives them a special role in debates about humanity's future because they are not, *ipso facto*, entirely susceptible to human will. They represent an "other" that can be modified by humans but is not ultimately reducible to a human construction, in either the material sense or the mental one. In other words, natural hazards invite humans to recognize that our knowledge of the Earth and its peoples is incomplete, uncertain, disjointed, and subject to penalizing contingencies beyond our control. It is likely to remain so in the foreseeable future. We should prepare our institutions and environmental management strategies for the twenty-first century with this firmly in mind, especially in the mega-cities that will likely become the pivots of global society. For mega-cities are, in effect, crucibles where new kinds of hazards are being fashioned and old ones reshaped so that existing ways of dealing with both are thrown into doubt.

Crucibles of Hazard explores the emergence, re-emergence, and transformation of environmental hazards in contemporary mega-cities. Although this process is driven primarily by changes in the location, size, structure, and functions of cities and in the composition of city populations, many other factors – including global economic restructuring and global environmental change – are also involved (International Decade for Natural Disaster Reduction, 1996). As table 1.1 shows, leading indicators of the transformation are already detectable. However, the complete results of this large-scale experiment in Nature–Society relations will not be clear for decades to come.

The thesis that natural hazards are joint products of nature and society is one of two main principles that underlie this volume. The other is that mega-cities are intensely human-constructed environments that – to varying degrees – shield their inhabitants against natural processes. When the shields are inadequate, as is often the case for poor populations in many third world cities, the human engagement with hazard can be immediate, personal, and highly unpleasant (Main and Williams 1994). But, for most people in large cities and for most of the time, extremes of nature are experienced indirectly in the form of signs and symbols that act as cues to thought and behaviour. In other words, mega-city residents are constantly bombarded with information about hazards – which is only occasionally put to the test by actual events. Unlike the simpler hazard-scapes of rural areas and small towns, evidence of mega-city hazard crowds in upon the senses, demanding consideration of many different threats and responses: direct and indirect; past, present, and future; real and imagined; stated and implied – often in startling juxtaposition.

An hour's drive along the Tijuana–San Diego urban corridor provides many examples. Here, in one of North America's fastest-growing mega-cities,[9] a plethora of natural and human-created risks is embedded in a jumble of contrasting neighbourhoods, institutions, and peoples from the first and third worlds. Evidence of hazard is plentiful – though often contradictory, ambiguous, and paradoxical. In one locality, affluent homes crowd along an eroding beach in the shadow of a massive bullring. Minutes away, new cars topped with surfboards and golf bags stream through an expressway cutting that is overhung by droop-shouldered wooden shacks clinging to a much-eroded slope. Nearby on an outdoor movie set, a large-scale replica of the ocean liner *Titanic* is poised in the act of sliding beneath the Pacific, within sight of a tollroad where wrecked vehicles have been placed in conspicuous positions to caution unwary drivers. Concrete flood-control works line the course of the Tijuana River, which usually carries only a trickle of heavily polluted water that is ultimately returned, untreated, to Mexico via a pipeline from the United States. Signs on US Interstate 5 warn motorists to beware of colliding with illegal immigrants, most of whom now prefer to avoid this well-patrolled road and to risk death from exposure in barren mountains that gird the metropolis. Further north, submarines and aircraft carriers ply navigation channels past slumping coastal bluffs and endangered migrating whales. Visitors to the San Diego Historical Museum learn about the central role of droughts and floods in the city's history, while the building – hastily concocted as a temporary exhibit for a 1915 Exposition – vibrates in the wake of large passenger jets threading down through hills and houses to land on a city-centre runway. These examples do not exhaust the range of hazards that occur in just one mega-city, but they are sufficient to illustrate a typical cross-section of threats and to hint at the complexity of contributory factors that must be taken into account by populations at risk and by urban managers. In the contemporary mega-city, knowledge and ignorance of threats help to shape the burden of hazard; so do affluence and poverty; human propensities for taking and avoiding risks; professional training and political adroitness; intended actions and unintended reactions; lessons from past realities and hopes of idealized or fantasized futures, as well as many other factors.

Global scientific, engineering, and hazard-management organizations have taken note of the increasing disaster potential of mega-cities (Ichikawa, 1995). While curiously silent about natural disasters, the global community of urban leaders, urban managers, and urban researchers has focused attention on other human crises of mega-cities such as rapid population growth, chronic unemployment and underemployment, inadequate housing, poor public services, weak pollution controls, dangerously decaying infrastructures, fractious relations among different ethnic and racial subpopulations, and insufficient citizen access to decision-making systems (see Postscript). To many observers, the crises and disasters of humankind and of Nature are interrelated; a comprehensive approach to all kinds of urban environmental hazards – natural, technological, biological, and social – is warranted. However, efforts to reduce the burgeoning natural hazards of mega-cities have been slow to develop and joint initiatives to address the full range of mega-city hazards have not yet occurred. More progress is urgently needed. This book is a contribution to the task.

Crucibles of Hazard is based on a set of invited papers delivered at the International Conference on Megacities and Disasters (Tokyo, 10–11

January 1994) and subsequently updated for publication. The conference was sponsored by the United Nations University, which is also engaged in a broad range of projects on mega-cities.[10] Initial contributed papers were later supplemented by others to provide a wider spread of case-studies drawn from different parts of the world. The Tokyo conference also functioned as a springboard for initial studies carried out by the International Geographical Union's Study Group on the Disaster Vulnerability of Megacities (Parker and Mitchell, 1995; Mitchell, 1995; see also other articles in special issues of *GeoJournal*, November 1995, and *Applied Geography*, January 1998, Vol. 18, No. 1).

The core of the book consists of 10 chapters that highlight environmental hazards in specific mega-cities on five continents. The case-studies are: Tokyo, Seoul, Dhaka, Sydney, London, Lima, Mexico City, San Francisco, Los Angeles, and Miami. The emphasis is on geological, meteorological, and hydrological hazards, but biological, technological, and social hazards are also addressed. A wide spectrum of places and communities is included (table 1.2). There are older slow-growing cities in more developed countries (e.g. London) and newer fast-expanding ones in less developed countries (e.g. Dhaka). Half of the cases are drawn from English-speaking states that share many other cultural traits (the

Table 1.2 Profiles of case-study cities

City[a]	Population (million)	Annual population growth (%)	Political function	Economic function[b]
Tokyo	26.8	1.40	National capital	Global city
Seoul	11.6	1.98	National capital	World city
Dhaka	7.8	5.74	National capital	–
London	7.3	0.00[c]	National capital	Global city
Sydney	3.6	0.47	–	World city
Lima	7.5	3.30	National capital	–
Mexico City	15.6	0.78	National capital	World city
San Francisco	3.9	1.24	–	World city
Los Angeles	12.4	1.72	–	World city
Miami	2.1	1.67	–	World city

Sources: United Nations Commission on Human Settlements (1996), pp. 21–22, 451–456; World Resources Institute (1996), pp. 10–11; Brunn and Williams (1993), p. 19.
a. Mega-cities in less developed countries are underlined.
b. Global cities are commanding nodes of the global economic system; world cities articulate large national economies into the global system.
c. Not known, but assumed to be static.

United States, the United Kingdom, and Australia) and half from culturally varied Asian and Latin American countries. Two of the world's most populous urban centres are included (Tokyo and Mexico City) together with a variety of smaller cities (e.g. Miami and Sydney). Some of the case-study cities are fulcrums of the international economic system (e.g. the "global cities" of Tokyo and London) whereas others are less strategically placed (e.g. Lima), but most can be counted among the so-called "world cities." Apart from the three US cases and Sydney, the cities are national capitals and the biggest urban agglomeration in their respective countries.

All of the case-study cities are acknowledged to be "large" or "very large," with populations ranging in size from 2 million to nearly 27 million. Precise rankings by size are unavailable and may be misleading because territorial definitions of cities are not consistent. The case-studies employ the definitions and size criteria that are used in the different mega-cities. The results may or may not be the same as those found in standardized city rankings published by the United Nations or other sources (see appendix 1). Thus, here Tokyo (i.e. the Tokyo Metropolitan Region) contains around 27 million people and London (i.e. Greater London and the Outer Metropolitan Area) houses approximately 13 million. Likewise, the urbanized area of Dhaka is 155 sq. miles whereas that of Los Angeles is 34,000 sq. miles. To some extent, gross disparities such as these are a function of choices made by the experts who establish urban classifications, but they also reflect real contrasts between sprawling automobile-based cities in more developed countries and compact animal-powered ones in some less developed countries.

The case-study authors are drawn from various countries and professions. Half (six) are geographers trained and working in the United States or the United Kingdom. The authors of the Asian and Latin American case-studies come from other fields: urban studies and urban planning (Kumagai, Nojima, Puente); anthropology (Oliver-Smith); landscape architecture (Kim); and development planning (Huq).

In chapter 2, I contend that mega-city hazards are difficult to address not just because urban natural processes are complex or because there are many competing urban issues, but also because urban areas are changing rapidly. I believe that the members of hazard interest groups and urban-management interest groups are typically mobilized by different contingencies and separated by wide gaps of experience, training, problem conceptualization, and professional outlook. Before hazard-sensitive urban sustainability can become a reality, the potential for mobilizing action in response to varied contingencies must be known and the gaps that separate the interest groups must be bridged. I liken the task to one of managing contacts between mutually unfamiliar cultures and I recommend that the goal of reducing mega-city hazards be pursued in concert with the resolution of other problems that face these places.

The case-studies begin with Tokyo, one of the premier "global" mega-cities, one of the most hazard prone, and one that possesses a highly sophisticated system of hazard management. Chapter 3 reviews the far-reaching influence of management policies for earthquakes, fires, and related hazards on the development of Tokyo during the past century. Yoshio Kumagai and Yoshiteru Nojima point out that the city is now undergoing its third major rebuilding since the Great Kanto earthquake of 1923. Whereas the once frenetic pace of national legislation on hazard management seems to have slowed, the government of Tokyo continues to plan for and implement many metropolitan hazard-reduction policies. These undergird a three-tiered strategy that combines: (1) extensive training of citizens to assume immediate responsibility for local hazard-fighting in the event of a sudden disaster; (2) land-use and building controls that permit evacuation and sheltering of exposed populations while retarding the spread of acute hazards; and (3) advanced prediction and warning systems that are intended to be linked with burgeoning technologies for the prevention of disasters. This chapter reflects the unique blend of inputs from science, technology, government, management, and citizen action that has been the hallmark of Japanese approaches to urban hazard reduction for many decades and that has been held up as a successful example to others. It was prepared before the Hanshin earthquake levelled much of Kobe and called into question widely held assumptions about the appropriateness of existing Japanese policies. Whether this system will continue unchanged in the light of Kobe's unwelcome experience remains to be seen.

Though floods, storms, and landslides are among the natural hazards of Seoul, they are neither as widely recognized nor as well documented as other environmental problems such as air pollution, traffic congestion, and shortages of affordable housing. But floods have inflicted substantial damage since the Second World War and – without proper attention – are likely to move into the front rank of Seoul's public concerns early in the twenty-first century. In chapter 4, Kwi-Gon Kim reaches this conclusion by way of a quantitative flood hazard assessment based on Geographic Information Systems that is believed to be the first of its kind employed in this city. A number of factors suggest that existing measures for reducing floods are already severely constrained. Flash-flooding challenges current prediction systems, engineering works are subject to failure, and there are no flood-control reservoirs in the surrounding hills. In addition, adequate flood insurance is unavailable in Korea, natural hazards are not taken account of in Seoul's master development plan or in local action plans, and flood-risk zoning is unheard of by those who

manage municipal lands. Without a more determined effort to incorporate flood hazards into future development strategies, the prospects are
for fast-rising losses.

From the viewpoint of effective public programmes for hazard reduction, the situation in Dhaka (chap. 5) is considerably worse than in
Seoul. Not only is the city surrounded by areas that are at grave risk from
severe storms and vast floods; it also faces multiple burdens of very rapid
population influx from rural communities, grinding poverty, under-
financed and inefficient urban services, and a disadvantageous location on
the periphery of the global market-place. Saleemul Huq offers an overview of these problems that underscores the links between rural hazards
and urban ones, and between local hazards and the global economy. He
also illustrates the importance of dykes and embankments not just as
flood-protection works but as centrepieces of planning strategies for land
reclamation in support of urban expansion and as barriers to the drainage
of polluted water that becomes trapped behind them. Increased industrialization is beginning to displace inner-city residences and to fill the last
remaining buildable spaces within existing municipal boundaries. The city
now seems poised on the edge of a major expansion phase that will push
new developments into locations that were once subject to frequent
floods and now are likely to be affected by less frequent but larger events.

John Handmer explores the growing vulnerability of Sydney to an
expanding list of environmental hazards and addresses the implications of
these for urban sustainability (chap. 6). He organizes the analysis around
several contradictions. One is the national myth of Australia as a land of
severe natural hazard, which does not match the experience of a city
where such problems have not historically been very serious. Second is
a carefully nurtured image of Sydney as a vibrant centre of hedonistic
recreation, high culture, and international investment, which sits uncomfortably with the daily realities of life in sprawling undistinguished
neighbourhoods that are chronically underserviced, subject to growing
social polarization, increasingly polluted, and at risk from environmental
extremes. Third is the ambiguity that surrounds the concept of planning
in Australia; planning is often paid a kind of lip-service that undermines
its effectiveness, so that formally "planned" communities are frequently
less well matched with their environments than those that grew fortui-
tuously. These and other contrasts throw up challenging questions about
Sydney's prospects for a sustainable future. How, for example, can
organizations whose purpose is to protect lives and property against
immediate environmental hazards contribute to the achievement of
urban sustainability when their actions usually have the effect of but-
tressing people and encouraging behaviours that are undermining longer-
term environmental stability? Handmer's assessment is that without

institutional and attitudinal reforms Sydney faces an uncertain future. Today, different types of acute hazard are generally well managed by narrowly specialized agencies. However, neither these bodies nor any foreseeable alternatives seem capable of dealing with the growing number of chronic slow-developing hazards that are emerging to threaten this mega-city.

London (chap. 7) has been a major city for almost 1,000 years and a mega-city for almost 200. During its history different types of disaster have severely damaged the city including, among others, disease epidemics, fires, floods, windstorms, and aerial bombardment. But, as revealed by Dennis Parker, the lessons of these experiences have frequently not been incorporated into improved hazard-management policies and programmes. Hazards are typically – and erroneously – viewed as problems that are separable from their social, economic, and political contexts. Instead of comprehensive responses that target the linkages among different contributory factors, the dominant approaches are usually incomplete, piecemeal, and poorly coordinated. In the coming decades, increased exposure to risk and differential vulnerability among the city's multiple interest groups are likely to be the driving forces in the hazards adjustment equation.

Chapter 8 switches the focus to Latin American mega-cities, as represented by the capital of Peru. Lima's history and hazard profile are laid out by Anthony Oliver-Smith. For almost five centuries, the city's capacity to rebound from repeated earthquakes, economic shocks, and other hazards has been seriously hampered by Peru's lack of control over the international market and political forces that permitted foreign investors to exploit its resources. During this period Lima presided over incoming flows of investment capital and metropolitan cultural influences and outgoing flows of minerals and other exports. Though repeatedly damaged by earthquakes and other extreme phenomena, the city was buffered against the worst hazards because it remained small and because it housed the most powerful Peruvian institutions and leaders who could command available resources in time of need. Unfortunately, Lima's capacity to provide for the safety of its residents has been swamped in recent decades by an unprecedented wave of poor peasants fleeing landlessness, a precarious rural economy, and the civil and military violence of rural areas in pursuit of uncertain but seductive futures in the burgeoning metropolis. What was once a medium-sized city that kept its élites well protected against most hazards, and also provided the sub-dominant classes with a modicum of security, has become a sprawling metropolis that increasingly offers a hazardous future to all its residents.

The vast metropolis of Mexico City is the focus of chapter 9. Sergio Puente describes the experience of living there as "chaos that is permanently on the verge of catastrophe" and points to continuing problems of seismic instability, flooding, fire, poor air quality, industrial explosions, and a variety of other hazards. Mexico City is viewed as the product of interactions among three groups of stakeholder: real estate developers, the state, and poverty-stricken citizens. In recent years the weakening of the state has contributed greatly to changing patterns of urban vulnerability, as evidenced by the emergence of citizens' action organizations that largely took over the task of meeting public demands for services in the wake of a catastrophic 1985 earthquake. Yet the state is one of the few third world institutions that is capable of acting against market-driven forces that have often played a large role in creating conditions of vulnerability in mega-cities. Puente indicts rapid population growth and deep social polarization as the two overriding factors that hamper state-backed attempts to address urban environmental hazards in third world cities, but he argues that the vulnerability of such places also varies widely in proportion to the weakness of their infrastructures, their degree of dependence on rural hinterlands, their positions in national urban hierarchies, and their peripherality in the global economic system. He then proposes and demonstrates a methodology for analysing vulnerability to environmental hazards in Mexico City. This employs a matrix that permits different factors of vulnerability in different urban neighbourhoods to be scored. The use of vulnerability matrices for analysing urban hazards is widely accepted among professionals, but Puente's version is more sensitive to the social dimensions of Mexico City's vulnerability than others that have preceded it. Maps of composite vulnerability based on matrix scores show that only about one-third of the mega-city is actually vulnerable to hazards, compared with over one-half as suggested by conventional analyses that rely solely on natural and physical factors.

Three chapters on North American mega-cities (San Francisco, Los Angeles, and Miami) round off the case-studies. All three cities have recently experienced major events that inflicted record losses – but none was the much-feared "Big One" of the respective communities. In other words, these are places where worse is not only possible; it is likely! Conversely, all three cities are also associated with (carefully cultivated) images of earthly paradise; they hold out the prospect that ambitious dreams of material success can be realized in an exotic location by the bold, the resourceful, or the fortunate. The tensions between these contradictions are expressed differently in each of the cities and all face daunting challenges that invite inspired responses.

In chapter 10, Rutherford Platt focuses on earthquake and wildfire problems of the San Francisco Bay Area. High levels of disaster planning and management expertise already exist among many Bay Area institutions, although coordination among different levels of government, different jurisdictions, and different service areas is often very difficult. Mutual aid agreements and other arrangements have improved the area's capacity to cope with disaster by means of greater system interconnectivity, but these adjustments are being hard pressed by changes in the broad context of hazard: increasing risks of wildfires on the urban periphery; a disturbingly long interval since the last great earthquake; the probability of renewed seismic activity along the heavily populated and poorly protected Hayward Fault; and the emergence of San Francisco as one pivot of a disaster-sensitive global economic system. Reduction of infrastructural vulnerability is perhaps the most pressing problem of the Bay Area, especially the ease with which vital lifelines can be disrupted for long periods. This is a problem that affects many other mega-cities, particularly those built near lakes, major rivers, or other physical impediments to the movement of people and resources. In light of the social and physical changes that are occurring in greater San Francisco, there are no guarantees of permanent success for the varied approaches to hazard management that have been tried there. None the less, the Bay Area's record of constructive engagement with environmental hazards has made this mega-city an important teaching laboratory for the international community of hazards scholars and managers.

Platt's guarded assessment of Bay Area hazards is followed by a vivid case-study of another metropolis that accentuates the deterioration of prospects for successful hazard management (chap. 11). This focuses on the "improbable city" of Greater Los Angeles. While acknowledging the skills of researchers, managers, and other professionals who have diligently pursued "top–down" adjustments to natural and technological hazards, Ben Wisner offers the opinion that the best efforts of such groups have been steadily undercut by a combination of post–Cold War economic shocks and widening ethnic and class divisions between rich and poor Angelenos. These have the effect of sorting and segmenting populations at risk, so that some hazards are characteristically associated with certain groups and the aggregate burden of hazard falls disproportionately on the urban poor. Alternative grass-roots (bottom–up) initiatives are recommended and examples are provided of inner-city community action groups that have incorporated hazard reduction within their agendas. From Wisner's perspective, a collaborative strategy that unites top–down and bottom–up components is called for. Though the twin forces of global economic restructuring and rampant individualism tend to work against such broad-based metropolitan policy-making, Los Angeles may – as so often in the past – once again be an exception to conventional explanations of urban development.

The final case-study examines a North American mega-city that is still recovering from unprecedented losses inflicted by a catastrophic storm (chap. 12). Greater Miami experienced relatively few deaths as a result of hurricane Andrew (1992), but property damage dwarfed the record of previous hurricanes anywhere in the world. William Solecki analyses the role of natural hazards in the development of Miami and assesses Andrew's impact on regional social ecology, land-use patterns, and demographic trends. He points out that the stereotypical image of Greater Miami as a series of more or less tranquil beach resorts is at variance with the place's dynamic and hazardous setting. Ambitious drainage and water-management projects, which expanded the originally small area available for building, have their roots in efforts to avoid a repetition of damage inflicted by a series of hurricanes that swept through the area in the 1940s. Solecki notes that this mega-city was ripe for disaster when Andrew struck, both because its vastly increased population lacked recent experience of major storms, and because the community had suffered serious economic reverses combined with the fragmentation of its political structure into fractious groups that lacked a perception of common interests. In agreement with impact studies of similar events elsewhere, Solecki reports that the disaster accelerated existing development trends but did not initiate any fundamentally new ones. However, the vulnerability gap between the poorest residents and others widened in the wake of the storm. As this study's analysis of post-disaster interest group coalitions makes clear, because of their volatility, political arrangements in large cities of the developed world pose major difficulties for the construction of long-term disaster recovery and mitigation programmes, but they also open up opportunities for creative approaches to these subjects. The challenge for scholars of hazard will be to find ways of exploiting these openings.

In chapter 13, I draw together findings and examine a series of new departures that they stimulate for urban hazards researchers and policy makers. I find that there is a strong case for ranking natural hazards and disasters among the most seriously underestimated problems of urban management and suggest that, without changes in the outlook of urban managers, they will only become worse. In recent decades, natural hazards and social hazards have been more troubling in large cities than have technological hazards and biological ones. Certain kinds of potentially valuable adjustment to hazard have been neglected in mega-cities. These include: non-expert systems, informal procedures, non-structural technologies, and a wide range of private sector initiatives suitable for families, neighbourhood groups, and small firms. Shifts in patterns of exposure and vulnerability are among the most striking changes in the parameters of hazard during recent years. A wide range of spatial indicators reflects both the dynamics of hazard within cities and the operation of causal linkages between hazard at the mega-city level and socio-political and economic processes at other scales of analysis. This chapter concludes by identifying new and potentially fruitful research areas that are indicated by the case-studies. Among these are the roles of hazards as agents of urban diversification, as catalysts of contingency in decision-making, and as sources of metaphor for urban policy-making. High priority should be accorded to the mapping of urban hazard ecologies, to exploring the influence of indirect signals of hazard in the urban environment, and to charting the role of hazards in conflicts among the major interest groups that shape contemporary urban policy.

The book ends with an extended postscript prompted by recent proceedings of the Second United Nations Conference on Human Settlements (Habitat II). Despite preparatory meetings that identified environmental hazards as an important urban problem, this bell-wether conference paid scant attention to them. Failure to recognize natural hazards as a worsening urban problem suggests a peculiarly myopic view of urban management and signals flaws in the conceptualization of sustainable development as a principle of urban management. It is to be hoped that renewed efforts will be made to correct these deficiencies in the near future.

Notes

1. The Great Hanshin earthquake is also known as the Hanshin-Awaji earthquake and the Hyogoken-Nambu earthquake. It occurred at 5.46 a.m. and registered 7.2 Ms (surface wave magnitude) and 6.9 Mw (moment magnitude) on the Richter scale. The quake's epicentre was located in Osaka Bay just south of the port city of Kobe. Major damage occurred in Kobe, with lesser effects in Osaka and adjacent communities. For details, see the special issue of the *Journal of Natural Disaster Science* 16(3), 1995.
2. It is estimated that 6,300 people were killed by the Hanshin earthquake. This surpasses the 1959 Ise Bay typhoon, which brought about over 5,000 deaths in and around the city of Nagoya, but does not compare with the Kanto earthquake of 1923, whose ground motion and fires destroyed over one-third of Tokyo and most of the adjacent port of Yokohama, killing more than 140,000.
3. The Keihanshin metropolitan region (Osaka–Kobe) contained approximately 14 million people in 1990, second only to the Tokyo region, which had almost 27 million.
4. In this book no attempt has been made to standardize estimates of city populations and city areas or quantitative estimates of disaster impacts. Varying estimates faithfully reflect the diversity of measurement criteria adopted by urban analysts and disaster assessors as well as the provisional nature of information about rapidly changing places and events.
5. Contributory factors to the disaster included, among others: a large stock of wooden houses in crowded neighbourhoods that were developed before the Second World War and never subsequently upgraded; infrastructure that was haphazardly installed during the heady post-war years of Japan's economic expansion; and transportation facilities that were seriously underdesigned for the risks that they faced (see Ichikawa, 1995).

6. American usage of the term "billion" (i.e. 10^9) is employed throughout this book.
7. As used here, "mega-city" means a metropolitan area with a population of at least 1 million people. Other analysts have employed higher population thresholds (e.g. 3 million, 8 million, or 10 million), but the term has no universally accepted definition.
8. This is a distinguishing feature of some other contemporary debates, including those about the demise of certain ancient cities and the consequences of anthropogenic atmospheric warming.
9. The combined populations of both cities are close to 4 million, of whom a slight majority live in the United States.
10. See, for example, Buendia (1990), Fuchs et al. (1994), Gilbert (1996), Lo and Yeung (1996), and Rakodi (1996).

REFERENCES

Brunn, Stanley D. and Jack F. Williams. 1993. *Cities of the world: World regional development*, 2nd edn. New York: HarperCollins.
Buendia, Hernando Gomez, ed. 1990. *Urban crime: Global trends and policies*. Tokyo: United Nations University Press.
Fuchs, Roland J., Ellen Brennan, Joseph Chamie, Fu-chen Lo, and Juha I. Uitto, eds. 1994. *Mega-city growth and the future*. Tokyo: United Nations University Press.
Gilbert, Alan, ed. 1996. *The mega-city in Latin America*. Tokyo: United Nations University Press.
Ichikawa, Atsushi. 1995. "Coping with urban disasters." *OECD Observer* 197(December 1995–January 1996): 15–16.
International Decade for Natural Disaster Reduction. 1996. *Cities at risk: Making cities safer – before disaster strikes*. Supplement to No. 28, *Stop Disasters*, Geneva: IDNDR.
Lo, Fu-chen and Yue-man Yeung, eds. 1996. *Emerging world cities in Pacific Asia*. Tokyo: United Nations University Press.
Main, Hamish and Stephen Wyn Williams, eds. 1994. *Environment and housing in third world cities*. Chichester and New York: John Wiley.
Mitchell, James K. 1995. "Coping with natural hazards and disasters in mega-cities: Perspectives on the twenty-first century." *GeoJournal* 37(3): 303–312.
Parker, Dennis and James K. Mitchell. 1995. "Disaster vulnerability of mega-cities: An expanding problem that requires rethinking and innovative responses." *GeoJournal* 37(3): 295–302.
Rakodi, Carole, ed. 1996. *The urban challenge in Africa: Growth and management of its large cities*. Tokyo: United Nations University Press.
United Nations Commission on Human Settlements. 1996. *An urbanizing world: Global report on human settlements 1996*. New York: Oxford University Press.
World Resources Institute. 1996. *World Resources 1996–97*. Special issue on "The Urban Environment". Washington, D.C.: World Resources Institute, United Nations Environment Programme, United Nations Development Programme, World Bank.

2

Natural disasters in the context of mega-cities

James K. Mitchell

"Give me a place to stand on, and I will move the earth."

(Archimedes – 287–212 B.C.)

Introduction

Among the many global transformations that are occurring at the close of the twentieth century is a shift from a predominantly rural world to a predominantly urban one – a brave new world dominated by increasingly large metropolitan complexes of 1 million or more people.[1] The societal and environmental implications of this transition are not yet fully understood, but it is already clear that the context of natural hazards management – and perhaps the very nature of hazards and human responses – will be profoundly altered as a result. At the same time, the entire universe of natural hazards issues and institutions is already changing in response to other factors. The combination of these two trends is producing a situation that is fraught with uncertainty and ripe with opportunities for new departures in mega-city hazard-reduction programmes and policies.

Background

Urbanization – especially rapid large-scale urbanization – is a major contributor to the rising global toll of disaster losses (National Research

Council, 1987; Science Council of Japan, 1989, pp. 13–14; US Office of Science and Technology Policy, 1992, p. 10). The world's big cities are rapidly becoming more exposed and more vulnerable to natural hazards and disasters (Havlick, 1986). Unfortunately, the special problems of urban disasters have not received much attention from scholars and professionals (Somma, 1991). "Too much of our efforts to assist … for too long have been focused on rural populations which are of ever diminishing relative magnitude and importance" (Jones and Kandel, 1992, p. 70). Much of the most useful information on urban disasters is part of a "grey literature" that is not readily available to potential users (Earthquake Engineering Research Institute, 1985; Organizing Committee of the International Seminar on Regional Development Planning for Disaster Prevention, 1986; IDNDR, 1990).

During the 1990s a number of researchers addressed the intersecting problems of cities and disasters. One of these was the disaster relief expert Frederick Cuny, whose strong convictions about the pivotal role of cities in international disaster operations began to appear in print just before his disappearance and probable murder in Chechnya (Cuny, 1994). Other notable contributions during this period include Sylves and Waugh's (1990) case-studies of emergency operations in American cities; Zelinsky and Kosinski's (1991) valuable compilation of data on urban evacuations; Berke and Beatley's (1992) analysis of earthquake planning in mid-sized US cities (Salt Lake City, Palo Alto, Charleston); a path-breaking edited collection of papers by geographers on housing and hazards in cities in less developed countries (Cairo, Bhopal, Mexico City, Agadir, Caracas, Hong Kong) (Main and Williams, 1994); a special issue of a European geography journal that focuses on geological hazards in mountain cities (*Revue de Géographie Alpine*, 1994); and the proceedings of a workshop on urban hunger in developing countries organized by the International Geographical Union's Commission on Vulnerable Food Systems (Bohle, 1994). The headquarters staff of the UN-sponsored International Decade for Natural Disaster Reduction (1990–2000) has also published a slim volume that reviews projects designed to improve urban disaster mitigation in 11 different countries and supplies information about organizations that are active in this field (IDNDR, 1996). Though all of these publications deal with hazards and disasters in exclusively urban settings, none focuses specifically or exclusively on large cities. Recent works with a more explicit focus on mega-city hazards include collections of papers prepared by contributors to a World Bank-sponsored conference (Kreimer and Munasinghe, 1992) and by members of the International Geographical Union's Study Group on the Disaster Vulnerability of Megacities (Parker and Mitchell, 1995). In addition, the UK Institution of Civil Engineers has published a report on mega-city vulnerability (Institution of Civil Engineers, 1995). A number of note-worthy individual papers have also appeared (e.g. Aguilar et al., 1995; Boone, 1996; Dando, 1994; Degg, 1993; Driever and Vaughn, 1988; Ezcurra and Mazari-Hiriart, 1996; Firman and Dharmapatni, 1994; Guterbock, 1990; Jairo-Cardenas, 1990; Kim, 1990; McGranahan and Songsore, 1994; Masure, 1994; Mitchell, 1993a; Reed, Tromp, and Lam, 1992; Setchell, 1995; Steedman, 1995; UNEP/WHO, 1994).

Curiously, although the published volume of mega-city hazards research is not large, certain mega-cities have attracted a dispropor-tionate amount of attention. For example, several recent books by geo-graphers, historians, and cultural analysts have dealt with the role of fire in the development of Chicago (Cronon, 1991; Miller, 1990; C. Smith, 1995) and Tokyo (Seidensticker, 1983, 1990). There has also been an unusually large number of articles on connections among rapid urban-ization, flooding, civil war, and refugees in contemporary Khartoum (Abu Sin, 1991; Agmad, 1993; Bakhit, 1994; El-Bushra and Hijazi, 1995; Han-sel, 1991; Ibrahim, 1991, 1994, 1995; Ruppert, 1991; Walsh et al., 1994; Woodruff et al., 1990; Yath, 1991, 1995). However, the bulk of mega-cities with admittedly serious hazards problems receive little or no focused coverage. Clearly, given the societal importance of mega-cities, the size of the threats, and the complexity of the associated problems, there is a need for more scholarly attention to this subject.

Because the likely impact of natural disasters on mega-cities is so great and so little understood, "the vulnerability of mega-cities" was identified as a high-priority topic for research and action during the International Decade for Natural Disaster Reduction (IDNDR). A major effort in support of that theme was mounted by the global scientific and technol-ogy community under the auspices of the International Council of Scien-tific Unions (ICSU). At ICSU's invitation, the International Association of Engineering Geologists and the World Federation of Engineering Organizations were encouraged to bring together a variety of academic and professional groups with the objective of formulating a collaborative agenda for mega-cities and disasters. This included projects by the UK Institution of Civil Engineers and the International Geographical Union as well as the World Bank and Habitat, among others. Independent but related programmes were also organized by the global community of landscape architects and other professional bodies. Most of these activ-ities involve organizations of scientists, engineers and urban managers. The conference sponsored by the United Nations University on which this book is based is distinctive because it is the only one that has adopted a broad approach to the subject of mega-cities and disasters based on human ecology and social science. The difference in approach has important philosophical and practical consequences.

The intellectual principle on which the IDNDR rests is a well-founded recognition that society already possesses a great deal of valuable information about the reduction of natural disasters but has largely failed to put that information to appropriate use. For most of the scientific and engineering community the main task is to apply existing knowledge about disaster reduction to a series of specific contexts – in this case, mega-cities. It is assumed that, like Archimedes, managers can stand on the stable base of accumulated technical knowledge and engage the lever of science to "move" the world of urban disasters. Undoubtedly, if resolutely followed, this strategy will produce important benefits, especially when it results in genuine partnerships between the producers and users of scientific information (Mitchell, 1988). By itself, however, it is an approach that incompletely engages the subject of mega-cities and natural disasters.

A basic dichotomy

Cities and disasters belong to two distinct research and management traditions that are organized by different groups of scholars and professionals around different founding assumptions, frequently in pursuit of different objectives. Often there is little communication between these groups as well as a lack of mutual understanding. The differences are readily apparent in the professional literature of both fields. Few journals about urbanization or urban issues give much prominence to environmental topics – fewer still to natural disasters – and few of the journals on natural hazards and disasters highlight urban disaster issues.

Differences in outlook and action between the two groups exist in large part because the starting points for intellectual discourse on cities and on disasters are so far apart. Natural disasters are widely regarded as Nature's ultimate sanction on human behaviour – sharp reminders that we do not live in a world that is entirely of our own making. On the other hand, cities are popularly seen as supreme human achievements – affirmations of societal power to alter Nature and sustain an environment that is, to a very large extent, divorced from Nature. When earthquakes, floods, and storms strike cities they intrude on places that are monuments to the human control of Nature. Urban disasters are, in effect, affronts to civilization![2] This helps to account for the disproportionate importance, in popular culture, of such disasters compared with similar rural events. For example, the celebrated Chicago fire of 1871 took 250 lives on the same day that 1,200 people died during a fire in rural Wisconsin. The Chicago fire has become the stuff of popular legend but only aficionados of American disasters know what happened at Peshtigo – despite the fact that it was the country's most deadly and physically destructive conflagration.

For a student of natural hazards, the city is but one among many physical settings of extreme events, albeit a setting that usually makes few concessions to the natural systems that are the agents of disaster. Conversely, from the perspective of urban managers, natural disasters are just one among many problems that must be confronted within city boundaries – and an infrequent one, at that! Hazard management has a place in the spectrum of urban issues but that place must be negotiated among competing demands for attention to other problems (e.g. new urban in-migrants, decaying infrastructure, poor housing, hazardous facilities, deindustrialization, pollution, homelessness, lack of mass transit, street crime, inadequate social services, ethnic and racial tensions).

Nor is it simply that natural disasters must compete with other urban issues for public attention; they are intimately bound up with those issues. For example, the flood and landslide disasters of Rio de Janeiro, Caracas, and Tijuana are closely connected to problems of obtaining affordable housing faced by poor migrants from rural areas. Prospects for earthquake hazard reduction in Tokyo are affected by the high cost of urban land, which encourages crowded neighbourhoods, restricts open space, and pushes new developments onto hazard-susceptible low-lying filled land in Tokyo Bay. The protection of Calcutta or Lagos against flooding is also tied up with inadequate means for disposing of solid wastes, antiquated or non-existent sewage systems, and budgetary conflicts between city and national governments. The fragmentation ("de-urbanization") of metropolitan populations and governments has a significant bearing on the design and operation of hazard-management systems in mega-cities of the United States (e.g. Los Angeles, Miami) and other developed countries. The implications are clear. If natural disasters are to be reduced in a big-city setting, the Archimedean metaphor of science-driven solutions is not an appropriate guide. Urban disasters are not merely a kind of inert problem that will yield to the direct application of scientific knowledge – they push back! In other words, not only do disasters affect cities but urbanization affects the creation of vulnerability, the scope for mitigating action, and a wide range of related topics. In brief, urbanization provides an interactive context for disaster.

There is one further twist to the relationship between disasters and mega-cities – perhaps the most important factor of all. Both elements of the problem – cities and disasters – are now changing in complex ways. (Some of these changes are treated at length later in this chapter and can be only previewed here.) The nature of disasters is changing: new kinds are emerging – some of them unprecedented. The management of disasters is changing: in many countries, existing institutions and policies are coming under increased criticism and there are calls for new ones. The ways we think about disasters are changing: for example, a post-modernist dialogue is beginning to appear in the professional literature. Equally, the nature of cities is changing: not only are they more numerous, bigger, and non-Western, but their economic and cultural functions are in flux, casting previous models of urban development into doubt. The management of cities is changing: pressed from below by burgeoning demands from their rapidly growing and diversifying populations, governments of mega-cities are also shifting to adjust to other pressures such as the changing role of national governments in the face of global economic integration and the reluctance of those governments to honour existing commitments to assist the cities. Urban mass movements and emergent neighbourhood organizations are also exerting pressure for the "reinvention" of mega-city governments. As big cities become increasingly important actors on the world stage, their populations are drawing on the experience of peer cities beyond national boundaries. As a response to such changes, it is no surprise that the intellectual basis of urban analysis is also in flux.

In these circumstances, the Archimedean metaphor requires further modification. Not only are the problem spheres of disasters and urbanization interactive (i.e. urban disasters "push back"), but given the speed and scope of changes that are now afoot in both spheres there may no longer be a truly stable base from which the leverage of scientific knowledge can be pivoted. This means that taken-for-granted assumptions about scientific information and disaster reduction need to be critically examined in light of the highly changeable contexts that bracket both disasters and mega-cities. Until these contexts are better known, the transfer of knowledge and technology among countries and cities should proceed with caution. Mega-city disaster-reduction initiatives will have to be carefully tailored to local conditions and specific settings. It is one of the tasks of this book to show how this process might proceed in different mega-cities.

A contextual model

In coming to grips with the complexities of a dynamic set of relationships between mega-cities and natural disasters it may be helpful to think in terms of a contextual model (fig. 2.1; Mitchell, Devine, and Jagger, 1989). Such a model is intended to show how a more or less discrete process relates to the broader environment of which it is a part. For the purposes of this chapter, the central process is one of hazard and the larger context is that of urbanization.[3] Both components of the model can be considered separately and in conjunction. The hazard component includes four main elements: physical processes, human populations, adjustments to hazard, and net losses. As people seek to adjust to hazardous natural environ-

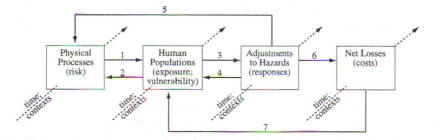

Fig. 2.1 A contextual model of hazard (Source: Mitchell, Devine, and Jagger, 1989)

ments, these four elements modify each other through seven endogenous feedback relationships. A change in one or more elements may set in motion a cascade of reflexive changes in the others. The stimulus for change may be some or all of the following: an extreme natural event such as an earthquake or a storm; community encroachment into an area of known hazard; urban technologies that inadvertently alter existing physical risks; the development of vulnerable groups, institutions, or technologies; the adoption of specific new hazard-management measures and the decay of old ones; and a broad-based commitment to raise general thresholds of public protection against hazards of various kinds.

Change can also come from outside the hazard system – in this case from the urban context in which the hazards are situated. That context includes exogenous factors or processes that interact with components of hazard, but are largely independent of them. These include variables such as the stages of city development (including the growth of mega-cities), the internal structure and functioning of urban areas under different sociocultural, political, and economic systems, and the urban implications of a wide range of contemporary changes in science, technology, environment, economics, politics, and culture. Among the latter are: the flow of new scientific information and the diffusion of new technologies; the growing recognition that humanity now has the capability to affect environmental change on a global scale; worldwide economic restructuring and integration; the collapse of an existing international geopolitical order and the development of replacement arrangements (e.g. continental trading blocs, ethnically fragmented states, "reinvented" governments, non-governmental organizations, urban mass movements); and the adoption of "post-modern" modes of thought and action both within the intellectual community and beyond.

These two components (i.e. hazard system and urban context) are a bit like the "figure" and "ground" of visual illusions. In the same way that a viewer can construct different interpretations of an illusion by focusing on either the figure or the ground, it is possible to arrive at an explanation of urban disasters that depends on disaster agents and proximate human responses or to fashion an explanation around the distal structuring characteristics of urban systems and urban lifeways. The latter approach may redefine the meaning of disaster in an urban setting and predispose cities to "disasters" that are different from those natural scientists customarily think of.[4] But urban disasters are not just amalgams of disaster and urbanization; they are the products of a set of changeable relations between both components. This feature makes it difficult to fashion effective programmes of urban disaster reduction by employing only the perspectives on hazards management or urban management. The permutations of natural hazard and urbanization produce different outcomes in different settings at different times. Moreover, the set of possible permutations is significantly larger than the set of outcomes to date; an understanding of this fact opens up a vast untapped potential for intervention to make cities safer places.

Mega-cities

Because mega-cities are a relatively recent phenomenon, the literature on mega-city disasters is sparse – though suggestive of problems to come. Much more is known about the relationship between natural disasters and urbanization in general. This section begins with a long caveat about attitudes to city hazards by urban analysts and other social commentators. It continues with a brief historical overview of mega-city disasters and then surveys the changing balance of safety in mega-cities. This is followed by a discussion of the disaster-forcing attributes of urbanization, a review of the evolution and status of mega-cities, an examination of different types of urban structure that may be reflected in mega-cities, and an exploration of the contrasting disaster issues of rich mega-cities and poor ones.

The intention is to show that, despite some tendencies toward uniformity and homogenization, there is a great deal of variety among mega-cities and considerable uncertainty about their eventual dimensions. Different types of mega-city are likely to have quite different disaster-susceptibilities and different capabilities for response. Whether this will continue is another matter. Mega-cities may grow, becoming both more complex and more alike. Or there may be perhaps two or three basic variants. For example, if the organization of the global economic and political system continues to be dominated by a relatively few countries, it may well be that the natural hazard profiles of mega-cities in the dominant (rich) countries will continue to diverge from those of the (poor) rest, thereby rendering problematic the exchange of hazard experience and hazard-reduction technologies between both. On the other hand, there is evidence that some of the characteristics of third world cities are beginning to show up in places such as Los Angeles and Miami, thus raising the possibility of eventual reconvergence.

The best of places; the worst of places

He compared Los Angeles and Mexico City (which he knew well) to volcanoes, spilling wreckage and desire in ever-widening circles over a denuded countryside. It is never wise, he averred, to live too near a volcano.

(Davis, 1992, p. 14)

Recent urban scholarship is reconceptualizing the city and the process of urbanization. But there is little consensus about what is emerging. For some analysts, cities are simply enduring features of the human landscape whose public images wax and wane (Hall, 1988; Jones, 1990). According to Peter Hall, during the past century people have passed through a great cycle of belief about cities. In the late nineteenth century, cities were widely regarded as places of physical decay and spiritual desolation – "The City of Dreadful Night" (Hall, 1988; Jones, 1990). Thereafter, especially during the period from 1920 to 1970, more positive and hopeful images appeared, partly in response to an array of efforts at urban improvement. But, towards the close of the twentieth century, the city – especially in Western societies – is again being viewed as a place of decay and despair. On the other hand, different commentators have challenged these views by suggesting that urbanization has begun to take on entirely new forms that call into question the very notion of the city. In the United States, commentators have pointed to the emergence of vast suburbias (Fishman, 1987) or "edge cities" (Chudacoff and Smith, 1994) or to "cyburbias" that are plugged into an information revolution and disconnected from particular places (Sorkin, 1992). Similar trends have been reported – at least in embryonic form – among cities of the third world such as Bangalore in India (*New York Times*, 21 December 1993). Yet other writers have identified cities whose purposes and forms are so deeply and continually contested by subgroups of inhabitants that it is difficult to form a coherent picture of the whole (Davis, 1992).

Against this background of critical scholarship, the dominant view of the city seems to be that of a place riven by problems (Brunn and Williams, 1993, pp. 32–41; Jones, 1990, pp. 163–166; McHarg, 1969; Teune, 1988). Natural hazards and disasters are usually fitted readily – all too readily – into these problem sets.

Both in rich and poor countries, the growth of such giant urban agglomerations poses major metropolitan planning problems which must be solved in a broad regional context. Food, fuel, electric power, and drinking water must be brought in from considerable distances, and massive quantities of waste materials must be disposed of. The giant metropolis needs very sophisticated transportation systems if it is to function effectively, and city dwellers may have to travel considerable distances to recreational facilities and surviving farm and wilderness areas. *Most major cities also have some special problems resulting from the peculiarities of their sites, for example subsidence, earthquakes, and air and groundwater pollution buildups in Mexico City, and flooding and typhoon damage in Dacca.* (Bromley et al., 1989, p. 372; emphasis added)

No one would contest that many cities do indeed face serious problems. However, the reality of urban living – and the role of natural hazards in mega-cities – is more complex. This is aptly recorded in Alexander Frater's book *Chasing the Monsoon* (1991). At one point in a journey around India following the monsoon, Frater asks a Bombay poet whether the sudden arrival of torrential rains at the end of the dry season is a romantic event. The reply comes back:

"Not in the big cities.... Where's the romance in mud, slush, rats and floods? Or in the anguish of watching those who live in the streets trying to cope?... For many of my fellow-citizens it is a period of misery and hardship." (Frater, 1991, p. 141)

But, a few days later in Calcutta, Frater concludes that:

despite the excess rain that kept sousing and rotting its foundations ... the city was indestructible.... For the people on the streets it was a haven. They were in Calcutta by choice, victims of starvation or persecution elsewhere, and would rather be here, living rough, than in the places from which they had fled. (Frater, 1991, p. 236)

These two, quite different, observations capture the ambiguity of living in a hazardous big city; such places may be exposed to serious physical hazards but they are also centres of human opportunity (Tuan, 1988). Both aspects of the situation need to be kept in mind by those wishing to address the hazards of mega-cities. The two contrasting characterizations of the city are freighted with policy implications for the reduction of natural hazards and disasters. A pathological perspective on cities is readily paired with an emphasis on disaster vulnerability; a more benign assessment of cities underscores their potential for resilience (Sachs, 1988). An emphasis on vulnerability tends to narrow the scope for individual actions or informal group actions by populations at risk; they are regarded as passive victims in need of protection. Conversely, an emphasis on resilience points hazard analysts in a different direction. We are reminded that laypeople are often capable of acting in their own best interests, and that this may mean finding ways to live with hazards, rather than seeking to control them or the populations they affect.

Mega-cities affected by natural disasters

Large cities of antiquity certainly experienced heavy losses as a result of natural disasters, but the data are highly incomplete.[5] Chandler and Fox (1974) show that at least a score of cities each recorded tens of thousands of deaths in the wake of earthquakes, floods, volcanic eruptions, and other events. But few of these places had populations that exceeded 200,000 and most were considerably smaller.

The Tokyo–Yokohama metropolitan area was probably the first megacity to be devastated by a natural disaster. The Great Kanto earthquake (1923) killed approximately 100,000 people in Tokyo and destroyed more than 700,000 houses in a city of over 2 million (Seidensticker, 1983, p. 259). In the half-century between 1945 and 1995, around 100 natural disasters affected large cities. The most physically destructive single events during this period took place in: Nagoya (typhoon, 1959), Tangshan (earthquake, 1976), Bucharest (earthquake, 1977), Mexico City (earthquake, 1985), Miami/Dade County (hurricane, 1992), and Kobe (earthquake, 1995). However, a focus on great catastrophes may be misplaced, because the aggregate impacts of lesser events can be much larger. This is particularly so in third world mega-cities, where small storms, floods, and seismic shocks, events that would be considered inconsequential elsewhere, often wreak disproportionately heavy losses. In such places, natural hazard is a more or less chronic everyday problem. Earthquakes, severe storms, and wildfires appear to be the most frequent catastrophic urban events.

It is worthwhile noting that urban areas are the sites of many different types of disaster: natural, technological, biological, and social. Thus far, most of these have been simple events of one type, but compound events that involve two or more of these components are beginning to appear (e.g. floods that disperse toxic materials, earthquakes that rupture oil pipelines and storage tanks, droughts that uncover flooded waste disposal sites). The potential for further "surprises" is obvious.

A shifting balance between hazard and opportunity

Evidence of increasing disaster-susceptibility in mega-cities bespeaks a change in the balance of hazard and opportunity in urban life. Of course changes in the mix of urban hazards and shifts in the hazard–opportunity balance have gone on throughout history. For several centuries the balance has tilted steadily towards higher levels of safety and opportunity for most city residents, while hydrological, meteorological, and geological hazards have declined relative to other types of threat. Urban concentrations of wealth, knowledge, talent, and power have justified – and made possible – extraordinary efforts to protect people and property against the more obvious extremes of Nature. The avoidance, prevention, and mitigation of natural hazards were a primary responsibility of those who ruled the ancient cities of Egypt, China, and Central America (Tuan, 1979; Wu Qingzhou, 1989). Dyking, drainage, and irrigation schemes were among the earliest urban public works in these places. The suppression of fire was a high priority in the crowded cities of medieval Europe and elsewhere at other periods (Rosen, 1986). A widely shared commitment to protect urban lives and property against natural extremes continues today. Contemporary Japanese cities are showplaces of hazard engineering, the metropolitan areas of Rotterdam and London are flanked by some of the world's most modern and technologically advanced flood defences, Mexico City has invested in the world's only public earthquake alarm system (*New York Times*, 19 December 1993), and the vast drainage network of greater Los Angeles has been so extensively modified by humans that it is now an almost totally artificial construct (Cooke, 1984; McPhee, 1989).

Heightened urban sensitivity to hazards and the benefits of hazard reduction do not mean that big cities have been – or are – entirely safe from extreme natural events. Nor do they mean that cities have been free of other types of hazard. (The historical record of urban epidemics, riots, fires, industrial accidents, and similar social ills is salutary.) But, over the long run, it is clear that a quite remarkable record of success has attended efforts to provide many urban dwellers with improved security against geological, meteorological, and hydrological hazards.

Now there is growing concern that cities are becoming so numerous, so large, so complex, and so volatile that the balance between risk and reward will tilt back towards conditions of increased natural hazard in the absence of determined efforts to reverse that trend. A world of megacities may well be more risky, more exposed, and more vulnerable than the world we currently inhabit. The scale of potential mega-city losses imposed by floods, storms, earthquakes, and other phenomena compels attention to the problem. So too does the timing, for we are fast becoming a predominantly urban world, a world where half the urban population will be citizens of mega-cities.

Urbanization: An agent of disaster

Before examining the links between mega-cities and disasters it is necessary to identify the general hazard-intensifying and disaster-forcing aspects of urbanization. Urban development increases disaster-susceptibility in a number of ways. First is the frequent association of cities with naturally risky locations such as seacoasts and floodplains because such places also confer important benefits (e.g. buildable land, well-appointed sites for the collection and transhipment of goods, fertile hinterlands). Initial settlements may take advantage of available safe sites but subsequent growth typically spills over into adjacent high-risk areas. Coastal metropolises of the eastern United States (Boston, New York, Miami, Houston) and the seaward-expanding cities of Japan are good examples.

Secondly, the physical process of building cities often creates or exacerbates existing environmental risks. For instance, paving over watersheds reduces infiltration, speeds runoff, and increases flood volumes; constructing coastal defences may reduce supplies of beach sand and facilitate erosion during storms. As the leading edge of urban development marches across the landscape, the incidence of natural disasters tends to keep pace (Cooke, 1984). The human role in creating conditions for disaster is clearly visible. Bangkok *klongs* (canals) that used to accommodate overflow from the Chao Phraya River have been filled in to create streets that are now chronically flood prone, while the city continues to subside owing to pumping of water from underlying aquifers.

Thirdly, cities increase disaster potential by concentrating people and investments. A disproportionate amount of material wealth is bound up with cities in the form of buildings (ceremonial, commercial, industrial, and residential) and infrastructure (i.e. the complex and expensive networks of lifelines that sustain urban populations and make it possible for them to interact with each other and the outside world). When an extreme event occurs, urban losses are often very heavy. In a matter of hours, hurricane Andrew inflicted over US$20 billion of property damage on the Miami metropolitan area, whereas it took about six weeks of heavy flooding in mostly rural sections of nine Midwestern states to produce approximately half as much material loss (Myers and White, 1993).

Fourthly, the built environment is continuously wearing out but the rate of urban replacement rarely matches the rate of urban obsolescence. As a consequence, most cities contain large concentrations of old buildings that fail to meet present standards for hazard-resistant construction. Differential ageing and uneven replacement of the physical stock typically produce a complex patchwork of disaster-susceptibilities.

Fifthly, many urban areas contain populations that are particularly vulnerable to disaster. For example, metropolitan areas often attract large numbers of in-migrants, most of them poor and all of them separated both from the familiar landscapes of home, whose risks were known, and from traditional support networks or customary behaviours that provided a modicum of security in the event of disaster.

Finally, few governments of rapidly growing cities have been able to allocate significant resources to hazard reduction when they are already stretched to breaking point by the task of providing basic support services for their expanding populations. In short, cities often contain all of the ingredients for disaster: heightened risks, concentrated exposure, and increased vulnerability. In light of the available evidence about intensified urbanization associated with mega-cities, the potential for a quantum leap in disaster-susceptibility is clear.

Changing patterns of urbanization

Some time during this decade there will be more people living in urban areas than anywhere else on Earth. In 1990 urban dwellers accounted for 47 per cent of the world's population but that number is expected to rise to around 63 per cent by 2020. Some 6,000 years separated the earliest urban centres from the first cities of more than 1 million people. Now there are about 300 such places and their numbers are growing rapidly. At the beginning of the twentieth century approximately 2 per cent of all humans lived in just 14 mega-cities.[6] Today the proportion is close to 20 per cent and it will probably rise to around 30 per cent by the year 2020 (Fellman et al., 1992, p. 371).

Moreover, the size of mega-cities is increasing rapidly. In 1900 there were no cities of more than 10 million and only three that exceeded 3 million (London, Paris, and New York). Now there are about 60 cities with populations greater than 3 million and 14 super-mega-cities of more than 10 million (table 2.1).[7] What is, by some counts, the world's largest city (Mexico City) currently contains at least 15 million people. The entire contiguous area of Mexico City (i.e. the central city plus adjacent smaller cities, towns, and suburbs) may hold more than 20 million. It is projected that there will be two or three urbanized areas of 20 million by the year 2000 (Mexico City, São Paulo, and perhaps Tokyo–Yokohama) and another 20 of between 10 and 20 million (Jones and Kandel, 1992, p. 75).

Mega-cities now occur in almost every culture and region on Earth. They are absent or scarce only in Oceania and high latitudes of the northern hemisphere. Countries that are notably rich in mega-cities include: the United States (36), China (31), India (21), Russia (15), Brazil (13), Germany (10), and Japan, the United Kingdom, and Indonesia

Table 2.1 The world's 20 largest metropolitan areas ranked by population, estimated for 1995

Rank	Location	Population (million)	Area (km²)	Density (inh./km²)
1	Tokyo-Yokohama	27.2	2,819	9,664
2	Mexico City	20.9	1,351	15,465
3	São Paulo	18.7	1,168	16,017
4	Seoul	16.8	885	18,965
5	New York	14.6	3,298	4,434
6	Osaka-Kobe-Kyoto	13.9	1,281	10,825
7	Bombay	12.1	246	49,202
8	Calcutta	11.9	541	21,990
9	Rio de Janeiro	11.7	673	17,364
10	Buenos Aires	11.6	1,385	8,416
11	Moscow	10.4	981	10,646
12	Manila	10.2	487	20,867
13	Los Angeles	10.1	2,874	3,525
14	Cairo	10.1	269	37,509
15	Jakarta	9.9	197	50,225
16	Teheran	9.8	290	33,726
17	London	9.1	2,263	4,028
18	Delhi	8.8	357	24,570
19	Paris	8.7	1,110	7,797
20	Karachi	8.0	492	16,292

Sources: United Nations (1987); Jones and Kandel (1992).

(7 each). Conversely, most African states (20) have none. Coastal locations predominate, especially in Latin America and Australia.

The spatial distribution of mega-cities has also undergone a major shift in recent decades. Most of the early twentieth-century mega-cities were located in Europe and North America (Berry, 1990, pp. 109–110). Today, by far the great majority – and most of the super-mega-cities – are in less developed countries of Asia, Africa, and Latin America (Gilbert and Gugler, 1992; Yeung, 1990). To a greater degree than ever before, mega-cities will be composed of large numbers of poor people who depend on informal economic institutions and systems for support (Perlman, 1987). As we shall see, this change has important implications for the management of urban hazards.

How far will the trend toward global mega-cities proceed? Is it essentially unilinear – towards more, bigger, and increasingly complex places? Might deconcentration and dispersal be an alternative future pattern? Are all mega-cities likely to evolve in the same way? How might expected changes affect prospects for disaster and disaster reduction? If current trends continue, and mega-city development follows the pattern of North

America and Europe, most of the signs point in the direction of increased population concentration at the global and national scales, with the possibility of regional and local deconcentrations and dispersals. In the United States, the aggregate pattern of population and investments has, for some decades, been shifting from the interior to the coasts, from the north towards the south, and from the east towards the west. Massive growth of cities such as Los Angeles and Miami follows in the wake of these trends. Similar processes are at work in many other countries. If Doxiadis's vision of the future is correct, the eventual outcome a century hence may involve the coalescence of linear mega-cities to form a more or less continuous urban "ecumenopolis" comprising a population of around 12 billion people (Jones, 1990, pp. 137–139)!

At the same time, in developed countries, there is a substantial amount of deconcentration (counter-urbanization) at the scale of the individual mega-city. Central-city growth has slowed dramatically or gone into reverse and the populations of many such cities are emptying out into a far-flung web of low-density suburbs and exurban developments. Some observers argue that there is no longer a need for urban concentration and that future large cities will be spread out over such large areas at such low densities that they will not resemble the historical city at all. Whether or not this occurs, it is clear that third world mega-cities have a long way to go before reaching this stage. In such places, the process of deconcentration is slow, indeed barely detectable in most.

Effects of city form and structure on natural disaster potential

Urban analysts recognize a number of different cultural variants of city form and structure, each with their own characteristic patterns of land use and population distribution. For example, Latin American cities tend to be made up of concentric zones, with old but well-maintained housing near a single central business district and poor shanty towns on the periphery. Industrial areas and the élite residential zone cut across these rings in the form of wedge-shaped blocks of territory. In contrast, the typical North American city more or less reverses this pattern, with a ring of decayed housing surrounding the central business district and affluent suburbs on the periphery. Also the central business district may be in decline, its functions challenged by many secondary shopping and industrial nodes scattered throughout the metropolis. West European cities often have well-marked historic cores with many preserved public buildings, surrounded by old houses – newly upgraded by middle-class gentrifiers – side by side with the cramped homes of working-class residents. Segregated urban ethnic neighbourhoods are less marked than in the United States. Suburbs may be either rich or poor and are often flanked by green belts with dormitory villages in the countryside beyond. African and south Asian cities tend to contain substantial tracts of market gardens within the built-up area. Middle Eastern cities often preserve distinct religious "quarters." The former socialist cities of Russia and Eastern Europe have a high percentage of their land given over to industrial districts and relatively few socio-economic contrasts among neighbourhoods. For much of the post–Second World War period, new investment was denied to Shanghai and other coastal cities of China, as national leaders sought both to hold down the growth of all large cities and to spread economic development more widely into the interior. The contrast between outmoded facilities and today's resurgent economic boom in the coastal mega-cities of China is striking (Gaile, 1983; Sit, 1988; *New York Times*, 22 December 1993).

These examples do not exhaust the catalogue of cultural and regional variations of urbanization but suffice to make the point that cities are functionally organized in different ways that have distinctive areal expressions. These patterns are then draped over a varied physical land-scape and adjusted to it, a process that helps to disguise the underlying organizing principles. When a natural disaster strikes, it disrupts – and may destroy – not just the lives of citizens and the city's physical fabric but also the functional organization of the metropolis. However, the same organizing principles typically reassert themselves during post-disaster reconstruction.

Different types of organization have different loss potentials, different vulnerabilities, and different prospects for recovery. Likewise, different types of city require different types of preparedness-planning and differ-ent hazard-mitigation strategies. For example, a single central business district plays an irreplaceable role in Lima (Peru), a city that contains 30 per cent of the country's population and 70 per cent of its industry; the several business districts of a typical multi-nucleated North American city share economic functions and rarely serve the entire nation. Likewise, the Lima periphery would hardly be a suitable location for the replication of suburban watershed management schemes borrowed from an American mega-city. In similar vein, the in-city agricultural lands of African and South-East Asian cities provide amounts of open space for hazard ref-uges and post-disaster relief staging areas that could only be dreamed about in the typical large Japanese city.

The preceding variants notwithstanding, there are also significant sim-ilarities among mega-cities. For example, the great majority are ports – very often the most important ports in the respective countries. Therefore protection against waterborne hazards and disasters is likely to be a shared concern, and safeguarding marine facilities, ship channels, and water-dependent land uses may receive high priority. Furthermore, if as some studies suggest, the physical fabric and organizational structure of mega-cities are becoming more uniform, then existing differences may gradually disappear in the future.

Mega-city disasters: Polycentre and periphery perspectives

A global trend towards ever-larger cities is unmistakable. So too is a trend towards increasingly similar urban forms. But it is equally clear that there remain vast differences between the mega-cities of developed countries and those of the developing world. Chief among these is the contrast between wealth and poverty. That dichotomy has important implications for the management and reduction of urban natural disasters. A brief review of the contrasting perspectives makes this clear.

Much of the concern about natural disasters in mega-cities of developed countries stems from the fact that such places contain a large proportion of the world's material investments and economic wealth. A disaster that takes many human lives is clearly possible in cities such as London, Los Angeles, or Tokyo, but such places are much more likely to suffer heavy economic and material losses in the wake of extreme natural events. As hurricane Andrew demonstrated for Greater Miami and the Hanshin earthquake for Kobe, a single metropolitan disaster in a rich country can produce economic losses that rock the global reinsurance industry.

However, the issue of mega-city disasters in rich countries is not simply one of losses that are inflicted on the affected communities themselves; it is increasingly connected with the roles that such places play in the global economy and the capacity of disasters to disrupt global economic functioning (Sassen, 1991). Several years ago the world's business community was jarred by reports that a major earthquake in Tokyo might precipitate a collapse of the global economic system (M. Lewis, 1989). This revelation followed on the heels of a record 1987 windstorm in southern England that had shut down the London Stock Exchange and may have helped to trigger the worst international stock market crisis since the Great Depression (Mitchell, Devine, and Jagger, 1989). Since then a series of other events has further underlined the disaster-vulnerability of mega-cities that anchor the global finance and trading network. They include: an underground flood that immobilized much of Chicago's financial district; riots and the possibility of a great earthquake in Los Angeles; an earthquake that disrupted the San Francisco Bay Area; and the bombing of the World Trade Center in New York. Between them these six mega-cities house the headquarters of more than 60 per cent of the world's leading private corporations (Berry, 1990).

Now there is evidence that the urban-based new information technologies on which the global economy increasingly depends are highly vulnerable to disruption by storms, floods, earthquakes, and other unexpected events. Nineteen mega-cities together make up a global "polycentre" that directs and controls the international entrepreneurial system (Berry, 1990).[8] Fifteen of these places account for 70 per cent of all electronic data flows in the contemporary world (J. Lewis, 1990). Not only are mega-cities of the polycentre full of expensive buildings and infrastructure and strategic economic functions, their sophisticated but demanding technologies that make for economic (and perhaps geopolitical) dominance may be susceptible to new kinds of vulnerability. In these cities, as well as the familiar consequences of natural disasters, there is considerable potential for disaster "surprises" in the future (Mitchell, 1996).

What do we know about the ability of mega-cities in developed countries to respond to severe natural events? The Loma Prieta earthquake and other recent mega-city disasters provided evidence that some of the measures that have proven to be effective in reducing losses in smaller cities may not work quite so well in emerging Western mega-cities (Mitchell, 1993a). Prediction, warning, and evacuation systems that depend on sophisticated technology and highly effective public bureaucracies are particularly open to question. In addition, big-city disasters possess features that have not been common in smaller communities and that may raise entirely new problems of disaster management. For example, disaster impacts in mega-cities that dominate mass media markets are likely to be extensively, continuously, and obsessively reported whereas impacts on other communities that have less access to these channels are likely to be overlooked. The consequences for skewing post-disaster relief are considerable. Secondly, the complex societal mixes of mega-cities pose new problems for the delivery of emergency response services and disaster relief; linguistic, ethnic, and other differences are often marked in such places. Thirdly, the sheer size and complexity of mega-city infrastructure networks make them particularly liable to disruption. Finally, recovery is apt to occur more slowly than in smaller places. In short, past lessons of disaster management may no longer be applicable in the mega-cities of the polycentre.

Of course, most of the world's mega-cities are not part of the polycentre. Instead they serve as primary contact points between the polycentre and regional or local markets on the global periphery. Tijuana (Mexico) is a good example. Once a small regional town, it is now the fourth-largest city in Mexico with a population of well over 1 million. Tijuana's recent growth has been fuelled by investments of multinational corporations in *maquiladora* firms near the US border. As more shanty towns go up in the steep semi-arid valleys of the city fringe, and more people crowd into the riverside lowlands, the incidence and severity of floods and landslides in Tijuana are also accelerating.

In places such as Manila, Dhaka, Ankara, or Lima there is the potential for heavy loss of life during disasters as well as catastrophic material destruction. The situation in Lima is typical. This is a city that has suffered severe earthquakes at least five times in the past 300 years. At the end of the Second World War, just over half a million people lived in the metropolitan area. Today, there are more than 5 million. Vast numbers of poor rural peasants have flooded into Lima. Not all groups are equally exposed to hazard. Indeed, the pattern of hazard-susceptibility is a complex one that has evolved in response to changes in demography, economics, land ownership, building practices, and other factors. Middle- and upper-income groups live in well-constructed houses that often comply with antiseismic codes and are sited in neighbourhoods with wide streets and ample open spaces. If affected by an earthquake there are enough resources to ensure quick recovery. The peripheral shanty towns (*pueblos jovenes*) are also low-density settlements, this time composed of light bamboo structures that do not collapse when the ground moves. People are poor, but levels of social organization are high. In contrast, seismic vulnerability is high in the inner-city slum areas. Here many poor families are crowded into old adobe brick structures, adjoining streets are narrow, and open spaces are non-existent. There are few neighbourhood organizations or other local institutions that might be called on in the event of a disaster. Here earthquake protection measures are minimal or, more often, non-existent. As summarized by one observer, the situation is full of bleak prospects:

The inhabitants of critical areas would not choose to live there if they had any alternative, nor do they neglect the maintenance of their overcrowded and deteriorated tenements. For them it is the best-of-the-worst of a number of disaster-prone scenarios such as having nowhere to live, having no way of earning a living and having nothing to eat. Given that these other risks have to be confronted on a daily basis, it is hardly surprising that people give little priority to the risk of destruction by earthquake. (Maskrey, 1989, p. 12)

In summary, there is a high degree of uncertainty about the future of mega-cities. Their growth seems assured, but at what density? New ones may spring up in unexpected places under the influence of changing geo-economic forces. Increasingly similar in outward form, mega-cities in different cultures and continents may still retain distinctively different internal structures. The divisions between rich mega-cities and poor ones may become wider and their disaster-susceptibilities may also diverge. But, at the same time, the differences between all mega-cities and their rural hinterlands may become sharper. It would be foolhardy to assume that the disaster-susceptibility of any one mega-city will be quite like that of any other. This is a period of great urban flux; it bears close watching by students of hazards and disasters.

Hazards and disasters

As the previous section shows, the process of large-scale urbanization is complex and changing. So too are the study and management of natural hazards and disasters. This section focuses on some of the policy and management-related changes of recent years. Although the US experience is highlighted, the changes noted apply to many other countries. At the same time, new ways of thinking about hazards and disasters are emerging, whose long-run implications are difficult to foresee. A brief introduction to some of these changes is provided; the reader is referred to other sources for more exhaustive coverage (Mitchell, 1993b).

Emerging crises of natural hazards management

The adequacy of existing means for managing natural hazards and other types of environmental hazards is increasingly being called into question in the United States and the global community. This is illustrated by a sampling of the issues that have recently emerged in professional and lay forums.

First are problems that are posed by new types of hazard. These come in several varieties. Some are amalgams of natural and technological hazards (Showalter and Myers, 1994). When a storm or a tsunami affects a chemicals manufacturing or storage facility it is not just the threat of high water and strong winds that is of concern; it is also the possibility that toxic materials may be dispersed throughout surrounding areas (e.g. Nagoya, 1959; Times Beach, Midwest floods, 1993). If an earthquake affects a nuclear reactor site, radioactive materials may be released. The flooding of old mines can cause surface collapses.

Given the expanding variety of technological hazards, the possibilities for new or unusual combinations of natural and technological hazards are spiralling upwards. Three classes of technological hazard pose quite different sets of problems when combined with natural hazards:
(a) Unsuspected hazards involve substances or activities that were regarded as harmless or benign until scientific evidence or human experience showed otherwise (e.g. DDT, asbestos).
(b) Improperly managed hazards involve failures of various kinds of hazard-control systems (e.g. nuclear facilities such as Windscale, Three Mile Island, Chernobyl; chemical plants such as Seveso, Basle,

Bhopal; transportation systems such as the US space shuttle Challenger and supertankers such as the Exxon Valdez; storage and disposal sites for toxic materials such as Kyshtym, Times Beach, Love Canal, Minamata).

(c) Instrumental hazards are intended to cause harm and are consciously employed towards that end; they include sabotage, arson, and warfare. Military industrial technologies belong to this group (e.g. nuclear, biological, and chemical weapons such as defoliants and nerve agents; deliberate oil-spills and oilfield conflagrations).

The UN Department of Humanitarian Affairs, created in 1992, has begun to examine a different but related set of problems that they call complex emergencies. These refer to events such as those occurring in the former Yugoslavia, Kurdistan, southern Sudan, Mozambique, and Somalia, where political conflicts, drought, famine, and other problems are intertwined (*DHA News*, January–February 1993). Hazards of global environmental change constitute a separate but related class of events that are now making their way onto the public policy agenda (Mitchell and Ericksen, 1992). It is widely accepted that a build-up of greenhouse gases in the atmosphere might trigger climate changes and other repercussions such as sea-level rise. Some of the industrial hazards are sufficiently well known to be classifiable as "routine" hazards, but others – including most of the hazards connected with global environmental change – are entirely unprecedented in the human experience. They are best considered "surprises" (Mitchell, 1996). How should public policies be changed to take account of the widening range of threats to human survival?

A second way in which natural hazards are changing grows out of the first. It is that there are now strong pressures to expand the legal definition of natural disasters. In the past, only the victims or potential victims of events triggered by natural phenomena (somewhat erroneously labelled "acts of God") were considered eligible for public assistance to upgrade preparedness or provide relief. However, in recent years there has been an unmistakable trend towards broadening the range of technological and social phenomena that are eligible for aid. In the United States this began with natural gas shortages in the cold and snowy winter of 1977 and later included the community of Times Beach, Missouri – a notorious case of contamination by the toxic chemical dioxin. More recently, the collapse of an old, disused, and deteriorating underground railroad system was treated as a "natural" disaster when water from an adjacent canal inundated the basements of high-rise buildings in downtown Chicago. In 1992, civil unrest in Los Angeles also qualified for disaster status, as did the 1993 bombing of the World Trade Center in

New York city. These events suggest that distinctions between different kinds of disasters are waning in the public policy arena. What accounts for such changes? Perhaps they reflect the growing impact of socio-technical hazards and the decline of natural phenomena in the intensely human-made environments of a rich country? Maybe they are linked to further politicization of public decisions about disasters, or to the political influence of specific interest groups that place a high premium on pre-dictability and stability (e.g. business corporations)? It is also possible that they are products of a general shift in public attitudes towards risks of all kinds.

A third type of change is swelling public dissatisfaction with hazard-management agencies. Criticism of disaster management in developing countries such as Bangladesh or the states of the African Sahel is not new. Mass media reports about the poor performance of national government organizations and international agencies are legion. Natural hazards and disasters can be volatile political issues in developed countries and a certain amount of controversy about governmental responses is the norm – as anyone who has observed the aftermath of Italian earth-quakes, or Australian wildfires, or American hurricanes can attest to. But recently there has been a sharp escalation of complaints about the effec-tiveness of hazard-management agencies in major developed countries such as the United States, the United Kingdom, Italy, Australia, and Russia.

The US Federal Emergency Management Agency (FEMA) has been a particular target. It has been accused of providing insufficient and inap-propriate relief to disaster victims. It has also been criticized for encour-aging the occupation of hazardous lands by offering low-cost insurance to rich investors; and it has drawn fire for devoting too much effort to cleaning up after past disasters and too little effort to reducing the prob-ability of future disasters. FEMA's mishandling of relief in the wake of hurricane Andrew (Florida, 1992) triggered a major investigation by the US Congress. Critics called for the nation's armed forces to replace FEMA, and large numbers of military personnel have, in fact, been deployed after recent disasters. The military is commonly in charge of disaster management in third world nations because it is often the only institution capable of providing aid during disasters and one of the few organizations that can be counted on to enforce government policies at other times. But in the United States and other Western nations pro-posals for a larger military role in civilian affairs are often controversial. Advocates of civil authority and legal due process are concerned that increased military involvement in disasters may signal an erosion of citi-zen rights and responsibilities, while others point to the reduction in international tensions and the need for more cost-effective national institutions as grounds for making creative use of military expertise in new roles.

Without going into detail, it is worthwhile to note that there is a widespread loss of faith in the ability of national public agencies to combat natural and technological hazards in many other countries. The failures of Soviet agencies in connection with the Armenian earthquake (1988) and the Chernobyl nuclear power station fire (1986) have been well documented and they are believed to have contributed to the collapse of the Soviet government. British civil defence agencies have also been roundly criticized for inadequate preparedness and lack of attention to hazard mitigation (Mitchell, 1989; Parker and Handmer, 1992).

Partly because government agencies have come under attack, there has been a determined effort to shift the burden of disaster management onto private individuals and institutions. In countries such as the United States, Germany, and the United Kingdom, this began with a conservative revolution in politics led by people such as Margaret Thatcher, Ronald Reagan, and Helmut Kohl. In the context of hazard management, policy reforms commonly took the form of insurance schemes (flood insurance, earthquake insurance, crop insurance, etc.), limitations on central government spending for disaster relief and recovery, an end to public subsidies for building in hazardous areas, and penalties for people who wilfully build – or rebuild – in such places.

Now there is a growing body of evidence that such policies may not work as intended. For example, insurance is not the panacea it was once proclaimed to be. Many potential victims are uninsured or underinsured and those who have adequate insurance often experience serious problems securing reimbursements. Not all perils are covered by insurance, and major problems arise when hazards involve several perils (e.g. hurricanes bring floods, erosion, wind damage, landslides, and other events). Cut-backs in government funding of social services have become common throughout the developed world in recent years, and spending on disasters is no exception. As a result, policies that emphasize private responsibilities for hazard management may help to widen the gap between richer and better-educated victims – that is, those who can afford to make provisions for their own security – and the poor or disadvantaged groups that lack such a capacity. In short, a hazard-protection system that relies primarily on market mechanisms may well be detrimental to broader public interests.

British experience with the great storm of 15 October 1987 illustrates several of these problems (Mitchell, Devine, and Jagger, 1989). Before the storm, local governments and private individuals in England had been encouraged to be self-reliant and not to expect the national government to provide recovery funds in the event of a disaster. But the storm –

which recorded the highest wind speeds in 250 years – blew down some 15 million trees and inflicted economic losses greater than any natural disaster in Britain since the end of the Second World War. In the process it exposed the limitations of local resources for coping with disasters and it compelled a major reversal of national policies that would have left local governments to take care of natural disasters.

If there is concern about the general effectiveness of disaster management by the private sector, there is deep consternation about the future of hazard insurance systems. Lack of insurance coverage and inadequate reimbursements are continuing problems, but the central issue is that very large disasters may bankrupt the entire international insurance system. Insurance and reinsurance companies in Germany, Japan, and the United States are all deeply troubled by this prospect. These include Lloyd's of London, which ran up losses of US$4.2 billion in 1993 as a consequence of heavy pay-outs to victims of earthquakes, hurricanes, and pollution emergencies (*New York Times*, 30 April 1993). In 1989, the Loma Prieta earthquake (northern California) and hurricane Hugo (Puerto Rico, Virgin Islands, South Carolina) inflicted severe losses on the American insurance industry and the global reinsurance market. But these losses were nothing compared with the losses sustained in 1992 as a result of three "superstorms" that occurred within a few weeks: typhoon Omar (Guam), hurricane Andrew (Florida), and hurricane Iniki (Hawaii). The total bill for insured storm losses was at least US$20 billion. Yet worse is possible – indeed likely. It is estimated that a single large hurricane could cause US$30–40 billion in insured losses and that a major earthquake in Tokyo or southern California might wreak even worse damage. These are the kinds of event that could disrupt investment markets throughout the world and trigger a collapse of the global financial system. Such circumstances underline the increasing interdependence of global society and its vulnerability to strategic loss.

Failure to make use of available hazard-reduction information and measures of known effectiveness constitutes another general policy issue. It is one that helps to stimulate the ongoing UN-sponsored International Decade for Natural Disaster Reduction (Mitchell, 1988). In many places it would be possible to mitigate losses simply by putting what is known into effect. For example, the value of warning and evacuation systems has been proven repeatedly; yet such systems are often underused. Likewise, hazard-mitigation schemes offer reliable paths toward reducing the long-term costs of disasters but they are often resisted in favour of immediate post-disaster relief, insurance, and compensation programmes. Why do individuals and governments fail to make optimal use of available knowledge? There is no single answer to this question. A large number of factors are involved.

- lack of agreement about definition and identification of problem;
- lack of awareness of hazards;
- misperception or misjudgement of risks;
- lack of awareness of appropriate responses;
- lack of expertise to make use of responses;
- lack of money or resources to pay for responses;
- lack of coordination among institutions;
- lack of attention to relationship between "disasters" and "development";
- failure to treat hazards as contextual problems whose components require simultaneous attention (i.e. reciprocity);
- lack of access by affected populations to decision-making;
- lack of public confidence in scientific knowledge;
- conflicting goals among populations at risk;
- fluctuating salience of hazards (competing priorities);
- public opposition by negatively affected individuals and groups.

Underlying all of these specific reasons is a larger problem. It is this: society fails to treat natural hazards as complex systems with many components that often require simultaneous attention. We tinker with one or another aspect of these systems when what is required are system-wide strategies. Perhaps even more important, we fail to address the direct linkage between natural hazard systems and economic investment decisions that drive the process of "development" and affect the potential for disasters. That such links exist has been known for a very long time:

If a man owes a debt, and the storm inundates his field and carries away the produce, or if the grain has not grown in the field, in that year he shall not make any return to the creditor, he shall alter his contract and he shall not pay interest for that year. (The Code of Hammurabi, King of Bablyon, c. 2250 B.C.)

But most of the decisions that are taken to build new facilities or redevelop old ones, or to adopt new production and distribution processes, or to develop new land, or to effectuate a myriad of other development goals are not currently very sensitive to considerations of natural hazards. They should become so. And that is a task that will require a great deal of effort by natural hazard scientists to go beyond the laboratory and the research office or the field study site to acquire an understanding of how best to apply their expertise in public settings. It will also require the users of scientific information about hazards (architects, engineers, planners, banks and mortgage companies, international development agencies, and investment financiers) to nurture a mutually interactive relationship with the scientists who are producers of that information.

"Development" is only one of the major public issues that overlaps with natural hazards reduction. Others include: environmental management; public health; security (personal, social, and national); and urbanization. All of them are major problem sets in their own right, each patterned by philosophical and managerial disputes and unresolved issues. Efforts to work out mutually supportive policies and programmes raise entirely new sets of contextual issues for hazards experts. How shall the different kinds of policies for these broad problem sets be articulated? Can they be organized around a single concept such as sustainable development, or will it be necessary to make trade-offs among each of them separately on an ad hoc basis? Some of the experience gained from attempts to work out sustainable development programmes in the World Bank and elsewhere might prove useful here (Cohen and Leitmann, 1994; Goode, 1996; Jones and Ward, 1994; Stren et al., 1992; White, 1994; and Wikan, 1995). So might the experience of "harmonizing" laws and regulations among countries that join together in new continental-sized free trade blocs (e.g. the European Union or the North American Free Trade Agreement).

The difficulties of merging hazards reduction in the overlapping context of sustainable development are instructive and require further comment. The rising incidence of disasters throughout the world is one of the most commonly cited indicators of non-sustainable development. (The other major villains are loss of biodiversity, impairment of natural systems, and global economic crises.) As Odd Grann, Secretary General of the Norwegian Red Cross, has said:

All major disaster problems in the Third World are essentially unresolved development problems. Disaster prevention is thus primarily an aspect of development, and this must be a development that takes place within the sustainable limits.

The evidence that disasters and development are connected is simply too overwhelming to be ignored. In the case of problems such as drought and famine in the African Sahel or the deaths of poor slum dwellers in landslides that affect the unstable slopes of Latin American cities, the connections between disasters and development are clear. But it is another matter entirely to believe that most or all natural disasters will disappear if only we adopt sustainable development policies and programmes. "Sustainable" development does not necessarily equal "safe" development.

It will probably always be necessary to support and nourish institutions that have the capability of responding to unexpected natural disasters. Much of what our existing hazard-management organizations do could be described as preparing for and coping with "routine disasters," that is, problems that were bound to occur sooner rather than later and for which we have developed reliable management responses. These are the sorts of things that it is hoped will diminish if sustainable development is properly implemented.

Clearly, the roster of hazard-management issues is packed with still unresolved old issues and emerging new ones. Among the topics that require attention are the following:

- the changing contributions of people, natural systems, and technology to the creation of hazard;
- measures to encourage improved use of available information about hazards (including scientific knowledge and folk wisdom);
- global interdependence and the vulnerability of large institutions (e.g. economies, cities) to major disruptive events;
- the relative human impacts of cumulative small-scale hazards and single large disasters;
- innovative procedures for coping with unprecedented hazards (i.e. surprises);
- attitudes toward risk and hazard;
- equity in the distribution of hazard costs and benefits;
- the illumination of polarizing debates about appropriate hazard-management strategies (e.g. "top–down" versus "bottom–up," centralization versus decentralization, rights versus responsibilities, anticipation versus reaction);
- alternative means for sustaining stakeholder involvement in decision-making beyond periods of acute crisis; and
- coalition-building between hazards interest groups and others that address overlapping problems (e.g. sustainable development).

This is by no means an exhaustive catalogue of potential research opportunities and needs. Note that urbanization or urban hazards and disasters, as such, are not explicitly identified in this list, but they are interwoven with most of these issues and topics.

Changes in the theory of hazards and disasters

The theory of research on natural hazards and disasters has been extensively reviewed elsewhere (Emel and Peet, 1989; Hewitt, 1983; Kates, 1978; Mitchell, 1984, 1987, 1989, 1990, 1991, 1993b; O'Riordan, 1986; White, 1973; Whyte, 1986; Whyte and Burton, 1980). I wish to focus only on recent (i.e. circa post-1988) changes. Two major research paradigms were in existence by the late 1980s: an older, human ecology one, which focused on decision-making by individuals and groups in the face of extreme events, and a newer, political economy paradigm, which emphasized underlying political, economic, and social structures that create conditions of hazard and prevent populations at risk from taking appropriate safety measures. Since 1988, debates between proponents of each paradigm have continued (Emel and Peet, 1989; Hewitt, 1992; K. Smith, 1992; Blaikie et al., 1994), but the two groups have tended to focus on different topics. Human ecologists have carried their ideas forward in a specialized literature on the social amplification of risk and risk communication, particularly in the context of major technological hazards (Greenberg et al., 1989; Wynne, 1989). Political economists have preferred to examine issues of vulnerability and equity, although others have also addressed those themes (Anderson and Woodrow, 1991; Dow, 1992; Kates and Weinberg, 1986; J. Lewis, 1990; Liverman, 1990; Taylor, 1990).

There have been some attempts at intellectual synthesis, including the creation of a new subfield – political ecology (Blaikie and Brookfield, 1987; Dalby, 1992; Little and Horowitz, 1987). Such syntheses incorporate elements of both paradigms and sometimes also include other perspectives (Alexander, 1993; Bryant, 1991; Palm, 1990). At the same time, a number of scholars have sought to forge links between hazards research and overlapping fields, in the hope of providing explanations for a wider range of problems. For example, medical researchers have called for attention to biological and disease hazards that were formerly neglected, and students of geopolitics have underscored the growing importance of environmental disasters as threats to national and global security (Dalby, 1992; Lewis and Mayer, 1988; Mintzer, 1992).

Overall there has been a recognition that broader interpretive frameworks are necessary – frameworks that incorporate both society and nature and a variety of contextual variables (Mitchell, Devine, and Jagger, 1989). However, whereas some have approached the task of extending hazards theory by means of synthesis among existing approaches, others have subjected long-accepted foundational concepts to critical analysis. For example, students of industrial disasters have questioned assumptions about a return to "normal" that underlie cyclical equilibrium models of disaster (Couch, 1996; Edelstein, 1988). They suggest that some disasters are essentially open-ended processes that may persist indefinitely. Other have criticized a widespread assumption among hazards researchers that public leaders want scientists to help them reach good decisions by eliminating uncertainty. Waterstone counters that the manipulation of uncertainty is a core political device that few leaders willingly surrender (Waterstone, 1993, p. 150). Wynne (1992) has also expressed doubts about the applicability of "uncertainty" as a basic concept in hazards studies. Drawing on a variety of evidence about recent surprises, he argues that many types of hazards may be better understood as indeterminate phenomena. Such risks as stratospheric ozone depletion, the build-up of greenhouse gases in the atmosphere, and the hazards of new industrial technologies are unprecedented in human experience; they represent a break with past knowledge. By invoking the concept "inde-terminacy," Wynne and others imply that hazards research is pressing against the limits of scientific methods and scientific philosophies (Wynne, 1992; P. J. Smith, 1990). Funtowicz and Ravetz (1990) have come to similar conclusions by examining the quality of scientific knowl-edge about global environmental hazards. When assessing emergent types of hazard they judge that we may need to think of several orders of scientific knowledge, each characterized by different degrees of certainty and reliability. O'Riordan and Rayner (1991) carry this analysis further and call for the creation of risk-management institutions that explicitly seek to combine science and politics. O'Riordan (1986) has also gone on record in support of creative use of knowledge from the arts and humanities to inform hazards research and education.

Thoughts like these lie at the interface between science and other forms of knowing. They are part of a general questioning of established knowledge that constitutes "postmodern discourse" in the social sciences and, especially, the humanities. In a surprising turn of events, hazards research – an interdisciplinary social science with ties to the natural sci-ences – is now beginning to be fertilized by ideas from the humanities. It is not easy to characterize either postmodernism or the special attributes of postmodernist theory (see Harvey, 1989; Hayles, 1991; Rosenau, 1992). Postmodernism is both a cultural condition and an intellectual construct. As a cultural condition, it is identified with certain architectural and artistic styles that involve the unexpected juxtaposition of forms and characters, transitory states, fluidity, and eclecticism. As an intellectual construct, postmodernism can be viewed as one outcome of a long-running border war between science and the humanities, wherein the allegiance of social scientists has been divided. The basic ideas of post-modernism have been fashioned by literary theorists and other humanists who have been concerned about the excesses and limitations of rational analysis and the neglect of alternative forms of knowing. Their thesis is as follows. The dominant global society is a product of the European Enlightenment. This "modern" society has endured for several hundred years, founded on beliefs about rationality in the organization and func-tioning of both the natural and social worlds and on the capacity of people to discover these truths and to act in light of them for their own benefit (i.e. to achieve "progress"). It is now argued that modernity has run its course and is being replaced by a fundamentally different "post-modern" era.

For social scientists, the salient characteristics of postmodern thought are: its rejection of grand theory and so-called metanarrative; its adher-ence to pluralist interpretations; its characterization of change as dis-continuity; its celebration of difference, contingency, and the primacy of local experience; its use of paradox and metaphor as investigative tools; and its non-pejorative assessment of alternative (i.e. non-scientific) forms of knowing. In the words of one commentator:

Post-modernists rearrange the whole social science enterprise. Those of a modern conviction seek to isolate elements, specify relationships, and formulate a synthesis; post-modernists do the opposite. They offer indeterminacy rather than determinism, diversity rather than unity, difference rather than synthesis, complexity rather than simplification. They look to the unique rather than the general, to intertextual relations rather than causality, and to the unrepeatable rather than the re-occurring, the habitual, or the routine. Within a post-modern perspective social science becomes a more subjective and humbling enterprise as truth gives way to tentativeness. Confidence in emotion replaces efforts at impartial observation. Relativism is preferred to objectivity, fragmentation to totalization. Attempts to apply the model of natural science inquiry in the social sciences are rejected because post-modernists consider such methods to be part of the larger techno-scientific corrupting cultural imperative, originating in the West but spreading out to encompass the planet. (Rosenau, 1992, pp. 8–9)

Few hazards researchers have self-consciously placed their work in a postmodern context, but there are clear echoes of postmodern sensibility in many recent critiques of the field (Cooke, 1992; Curry, 1988; Taylor and Buttel, 1992). The following statement by Hewitt is a good example.

[The] elements of today's dominant view of hazards ... are fully symptomatic [of] the abstracted, gender-blind, class-blind, secular and amoral, mechanistic and technocratic style of work that prevails in international "Big Science" and the preoccupations of agencies and publics in the wealthier and more powerful centers.... They not only display the Enlightenment desire to generalize and command "rational" interpretation by reducing everything to a common language and set of concepts. They are repeatedly beguiled by the most "advanced" or fashionable concepts, the worries of "fast-breaking" technologies in their own places of origin, and a belief these define and alone can solve "problems" elsewhere. (Hewitt, 1992, p. 58)

Hewitt's statement is an indictment of current theory as well as current practice in the management of hazards. He is not alone in holding such views. This is not to say that contemporary hazards research is riven by postmodern critiques. The challenges that I have reported here are not yet a dominant presence in the field. But they are beginning to tap into a distinct undercurrent of disquiet among many who are conscious of the limitations of existing theory and practice (Browning and Shetler, 1992).

Whether postmodernist critiques will fundamentally alter the nature of hazards research is an open question. At first glance there are grounds for scepticism. The existence of close links between research and praxis suggests that there will be strong counterpressures to resist any theory that calls for thoroughgoing institutional and social change – and evinces deep suspicion of the very organizations that are most involved with the making and execution of policy. Of course, current debates about hazards need not lead to postmodern theories of hazard that blend the contributions of scientists and humanists. Other purely scientific ("modern") approaches are possible, including those based on theories of evolution, chaos, and complexity (Argyros, 1991; Waldrup, 1992). (For example, some Russian colleagues are developing ideas about processes that limit the disturbance of large systems that are based on the application or extension of the well-known Le Chatelier Principle in chemistry.) But hazards researchers have been slow to link up with this more formal work, most of which is firmly in the tradition of Enlightenment thought. It is too soon to assess the advantages that a "science of chaos and complexity" might offer to hazards scholars.

Mega-cities and natural disasters reconsidered

In this chapter I have attempted to take a very broad conceptual stance towards the subject of mega-cities and natural disasters. This is deliberate because I am convinced that the task of reducing mega-city disasters is not simply one of applying existing practical knowledge that scientists, engineers, and hazards professionals – myself among them – have gained from years of working with those events. If we are to reap success, a commitment to mutual understanding and collaboration among academics, professionals, and laypersons who are hazard specialists and academics, professionals, and laypersons who are urban specialists will have to be made. Because of the changes that are now taking place in the study and management of natural hazards and the study and management of cities, and also because new forms of hazard and new types of urban phenomena (i.e. mega-cities) are emerging, the road ahead is not well signposted for either set of interest groups. So we must proceed like two well-founded cultures that are coming into contact for the first time.

This is a process with which Japanese society has had considerable experience from the late nineteenth century onwards. And the Japanese experience offers instructive parallels, both in a general sense and in a more restricted one that applies specifically to issues of hazard and disaster. After the Meiji restoration, contacts between Japan and the rest of the world were managed in such a way as to permit Japan to benefit while at the same time preventing Japanese society from being inundated by foreign cultural influences. A successful transition to a new social order was obtained. As part of that transition, Western science came to Japan and with it an interest in the scientific and technical analysis of natural hazards and disasters. Out of this arose leading institutions of disaster research and a distinctive style of hazards management that attracts visitors from all over the world. Whether one learns about the means by which Kagoshima beneficially coexists with its hyperactive backyard volcano, or the gated rivers of the Japanese coast, or Nagoya's three-tiered elevated highway, ground-level park, and underground flood reservoir scheme, or Tokyo's firewall high-rise complex, there can be no mistaking the fact that a very distinctive blend of scientific knowledge, technological innovation, and societal cooperation has been achieved in a variety of big-city contexts.[9]

In coming to grips with the burgeoning disaster potential of the world's mega-cities, the hazards community and the urban community are embarking on a somewhat similar exercise in cultural contact, this time among professional and academic cultures as well as national ones. We cannot know how the contact will turn out, but we can make sure that the best thinking from each community is incorporated into the process.

Notes

1. Most sources suggest that there are now between 300 and 400 such places, but it is not possible to provide a more precise count owing to limitations of population data and the wide range of criteria that are used to define and delimit urban areas.
2. Of course, the starting points for analysis are not necessarily reflected in contemporary theories and explanation. For example, most natural disaster specialists are uncomfortable with the label "natural" because they realize that a hazard is an interactive process that involves both people and natural systems. Some might even go so far as to assert that all so-called "natural" hazards and disasters are entirely human created. In the same way, many urbanists are well aware that natural environmental factors have affected the development of cities.
3. There is nothing to prevent the reader from reversing this order, making urbanization the process and disasters the context. Indeed, it can be argued that such an analysis is probably essential to a complete understanding of the urban disaster nexus. However, for convenience and because my own primary experience lies in the field of natural hazards research, I have placed hazard at the centre of the model. The background to this model is discussed in greater detail in Mitchell, Devine, and Jagger (1989).
4. The American humorist Garrison Keillor touched on this phenomenon in a piece about the various imagined disasters that arise in the minds of people who are trapped in malfunctioning high-rise elevators: "In the city of New York, you go out your door in the morning, you take your life in your hands. You may not get off an elevator the same person you get on. Choose your elevator wisely" (*New York Times*, 19 December 1993, p. E13). He might just as readily have said "choose your city wisely"!
5. Appendix 2 provides a list of known pre-twentieth-century urban disasters that killed more than 10,000 city residents.
6. London, Manchester, Birmingham, Glasgow, Berlin, St. Petersburg, Moscow, New York, Chicago, Philadelphia, Boston, Calcutta, Peking, and Tokyo.

7. Fellmann, Getis, and Getis (1992) identify the 10+ million cities as: Beijing, Bombay, Buenos Aires, Cairo, Calcutta, London, Los Angeles, Mexico City, Moscow, New York, Osaka–Kobe, Paris, Rhine-Ruhr, Rio de Janeiro, São Paulo, Seoul, Shanghai, and Tokyo–Yokohama.
8. New York, Chicago, Los Angeles, San Francisco, Philadelphia–Wilmington, Dallas–Fort Worth, Houston, St. Louis, Detroit, Pittsburgh, Tokyo, Osaka–Kobe, Seoul, London, Paris, Ruhrgebiet, Frankfurt, Randstadt, Rome.
9. Many critics might contend that this blend is not readily transferable elsewhere without major modifications.

REFERENCES

Abu Sin, M. E. 1991. "Migration from Eastern Gezira into Greater Khartoum – A case study in rural–urban migration and population integration processes." *GeoJournal* 25(1): 73–80.

Agmad, Adil Mustafa. 1993. "The neighborhoods of Khartoum: Reflections on their functions, forms and future." *Habitat* 16(4): 27–45.

Aguilar, Adrian Guillermo, Exequiel Ezcurra, Teresa Garcia, Marisa Mazari-Hiriat, and Irene Pisanty. 1995. "The basin of Mexico." In Jeanne X. Kasperson, Roger E. Kasperson, and B. L. Turner II (eds.), *Regions at risk*. Tokyo: United Nations University Press, pp. 304–366.

Alexander, David. 1993. *Natural disasters*. New York: Chapman & Hall.

Anderson, Mary B. and Peter J. Woodrow. 1991. "Reducing vulnerability to drought and famine: Developmental approaches to relief." *Disasters* 15(1): 43–54.

Argyros, Alexander J. 1991. *A blessed rage for order: Deconstruction, evolution, and chaos*. Ann Arbor: University of Michigan Press.

Bakhit, Abdel Hamid. 1994. "Availability, affordibility and accessibility of food in Khartoum." *GeoJournal* 34(3): 253–255.

Berke, Philip R. and Timothy Beatley. 1992. *Planning for earthquakes: Risk, politics, and policy*. Baltimore, MD: Johns Hopkins University Press.

Berry, Brian J. L. 1990. "Urbanization." In B. L. Turner II, William C. Clark, Robert W. Kates, John F. Richards, Jessica T. Mathews, and William B. Meyer (eds.), *The earth as transformed by human action: Global and regional changes in the biosphere over the past 300 years*. Cambridge: Cambridge University Press, pp. 103–119.

Blaikie, P. and H. Brookfield. 1987. *Land degradation and society*. London: Methuen.

Blaikie, Piers, Terry Cannon, Ian Davis, and Ben Wisner. 1994. *At risk: Natural hazards, people's vulnerability and disasters*. London: Routledge.

Bohle, Hans-Georg. 1994. "Cities of hunger: Urban food systems in developing countries." *GeoJournal* 34(3): 243–244 (see also additional papers, pp. 245–304).

Boone, Christopher G. 1996. "Language politics and flood control in nineteenth-century Montreal." *Environmental History* 1(3): 70–85.

Bromley, Ray, Peter Hall, Ashok Dutt, Debnath Mookherjee and John E. Benhart. 1989. "Regional development and planning." In Gary L. Gaile and Cort J. Willmott (eds.), *Geography in America*. Columbus, OH: Merrill Publishing Company, pp. 351–386.

Browning, Larry D. and Judy C. Shetler. 1992. "Communication in crisis, communication in recovery: A postmodern commentary on the Exxon Valdez disaster." *International Journal of Mass Emergencies and Disasters* 10(2).

Brunn, Stanley D. and Jack F. Williams. 1993. *Cities of the world: World regional development*, 2nd edn. New York: HarperCollins.

Bryant, E. A. 1991. *Natural hazards*. Cambridge: Cambridge University Press.

Chandler, Tertius and Gerald Fox. 1974. *3000 years of urban growth*. New York: Academic Press,

Chudacoff, Howard P. and Judith E. Smith. 1994. *The evolution of American urban society*. Englewood Cliffs, NJ: Prentice-Hall.

Cohen, Michael A. and Josef L. Leitmann. 1994. "Will the World Bank's real 'New Urban Policy' please stand up?" *Habitat International* 18(4): 117–126.

Cooke, Ronald U. 1984. *Geomorphological hazards in Los Angeles*. London: Allen & Unwin.

—— 1992. "Common ground, shared inheritance: Research imperatives for environmental geography." *Transactions of the Institute of British Geographers* 17: 131–151.

Couch, S. R. 1996. "Environmental contamination, community transformation, and the Centralia mine fire." In James K. Mitchell (ed.), *The long road to recovery: Community responses to industrial disaster*. Tokyo: United Nations University Press, pp. 60–85.

Cronon, William. 1991. *Nature's metropolis: Chicago and the Great West*. New York and London: W. W. Norton.

Cuny, Frederick C. 1994. "Cities under seige: Problems, priorities and programs." *Disasters* 18(2): 152–159.

Curry, Michael. 1988. "Commentary on 'Geographers and nuclear war: Why we lack influence on public policy'." *Annals of the Association of American Geographers* 78(4): 720–724.

Dalby, Simon. 1992. "Ecopolitical discourse: 'Environmental security' and political geography." *Progress in Human Geography* 16(4): 503–522.

Dando, W. A. 1994. *Urban famines in a future world of mega-cities: A commentary*. Professional Paper No. 20. Indiana State University, Terre Haute, Department of Geography and Geology, pp. 81–87.

Davis, Mike. 1992. *City of quartz: Excavating the future in Los Angeles*. New York: Vintage Books.

Degg, Martin. 1993. "The 1992 'Cairo Earthquake': Causes, effect and response." *Disasters* 17(3): 226–238.

Dow, Kirstin. 1992. "Exploring differences in our common future(s): The meaning of vulnerability to global environmental change." *Geoforum* 23(3): 417–436.

Driever, Steven L. and Danny M. Vaughn. 1988. "Flood hazard in Kansas City since 1880." *Geographical Review* 78(1): 1–19.

Earthquake Engineering Research Institute. 1985. *Proceedings – U.S.–Japan Workshop on Urban Earthquake Hazards Reduction.* Publication No. 85–03. Berkeley: EERI.

Edelstein, M. R. 1988. *Contaminated communities: The social and psychological impacts of residential toxic exposure.* Boulder, CO: Westview Press.

El-Bushra, El-Sayed and Naila B. Hijazi. 1995. "Two million squatters in Khartoum urban complex: The dilemma of Sudan's national capital." *GeoJournal* 35(4): 505–514.

Emel, J. and J. R. Peet. 1989. "Resource management and natural hazards." In J. R. Peet and N. Thrift (eds.), *New models in geography: The political-economy perspective.* London: Unwin Hyman.

Ezcurra, Exequiel and Marisa Mazari-Hiriart. 1996. "Are megacities viable? A cautionary tale from Mexico City." *Environment* 38(1): 15, 26–35.

Fellmann, Jerome, Arthur Getis, and Judy Getis. 1992. *Human geography: Landscapes of human activities.* Dubuque: Wm. C. Brown.

Firman, Tommy and Indira Dharmapatni. 1994. "The challenges to sustainable development in Jakarta Metropolitan Region." *Habitat International* 18(3): 74–94

Fishman, Robert. 1987. *Bourgeois utopias: The rise and fall of suburbia.* New York: Basic Books.

Frater, Alexander. 1991. *Chasing the monsoon.* New York: Alfred A. Knopf.

Funtowicz, S. O. and J. R. Ravetz. 1990. *Uncertainty and quality in science for policy.* Dordrecht: Kluwer Academic Publishers.

Gaile, Gary L. 1983. "Reanalyses of Chinese spatial inequality." *Professional Geographer* 35: 281–283.

Gilbert, Alan and Josef Gugler. 1992. *Cities, poverty and development: Urbanization in the third world.* Oxford: Oxford University Press.

Goode, Judith. 1996. "On sustainable development in the mega-city." *Current Anthropology* 36(1): 131–133.

Greenberg, Michael R., Peter M. Sandman, David B. Sachsman, and Kandice L. Salmore. 1989. "Network television news coverage of environmental risks." *Environment* 31(2): 16–20, 40–44.

Guterbock, Thomas M. 1990. "The effect of snow on urban density patterns in the United States." *Environment and Behavior* 22(3): 358–386.

Hall, Peter. 1988. *Cities of tomorrow: An intellectual history of urban planning and design in the twentieth century.* Oxford: Basil Blackwell.

Hansel, G. 1991. "Rural–urban migration and the problem of town planning – The case of Um Badda, Omdurman." *GeoJournal* 25(1): 27–30.

Harvey, D. 1989. *The condition of postmodernity: An enquiry into the origins of cultural change.* Oxford: Blackwell.

Havlick, Spencer W. 1986. "Third world cities at risk: Building for calamity." *Environment* 28(9): 6–11, 41–45.

Hayles, Katherine, ed. 1991. *Chaos and order: Complex dynamics in literature and science.* Chicago: University of Chicago Press.

Hewitt, Kenneth. 1983. *Interpretations of calamity.* Boston: Allen & Unwin.

——— 1992. "Mountain hazards." *GeoJournal* 27(1): 47–60.

Ibrahim, F. N. 1991. "The southern Sudanese migration to Khartoum and the resultant conflicts." *GeoJournal* 25(1): 13–18.

——— 1994. "Hunger-vulnerable groups within the metropolitan food system of Khartoum." *GeoJournal* 34(3): 257–261.

——— 1995. "Cultural change and the chances of reintegration after reimmigration to home areas – The case of the Southern Sudanese dislocated population in Greater Khartoum." *GeoJournal* 36(1): 103–107.

IDNDR (International Decade for Natural Disaster Reduction). 1990. *International Conference 1990 Japan: Toward a less hazardous world in the 21st century: Proceedings*. September 27–October 3, 1990, Yokohama and Kagoshima.

——— 1996. *Cities at risk: Making cities safer ... before disaster strikes*. Supplement to No. 28, *Stop Disasters*. Geneva: IDNDR.

Institution of Civil Engineers. 1995. *Megacities: Reducing vulnerability to natural disaster*. London: Thomas Telford.

Jairo-Cardenas, J. 1990. "Natural disasters and urban struggle for housing." *Architecture et Comportement/Architecture and Behaviour* 6(2): 177–191.

Jones, Barclay and William A. Kandel. 1992. "Population growth, urbanization, disaster risk, and vulnerability in metropolitan areas: A conceptual framework." In Alcira Kreimer and Mohan Munasinghe (eds.), *Environmental management and urban vulnerability*. World Bank Discussion Paper 168. Washington, D.C.: World Bank, pp. 51–76.

Jones, Emrys. 1990. *Metropolis*. New York: Oxford University Press.

Jones, Gareth A. and Peter M. Ward. 1994. "The World Bank's 'new' Urban Management Programme: Paradigm shift or policy continuity?" *Habitat International* 18(3): 33–51.

Kates, Robert W. 1978. *Risk assessment of environmental hazard*. SCOPE (Scientific Committee on Problems of the Environment) Report No. 8. New York: Wiley.

Kates, Robert W. and Alvin M. Weinberg, eds. 1986. *Hazards: Technology and fairness*. Washington, D.C.: National Academy Press.

Kim, Kwi-Gon. 1990. "Risk assessment in urban planning and management: A metropolitan example." *Habitat* 14(1): 177–190.

Kreimer, Alcira and Mohan Munasinghe, eds. 1992. *Environmental management and urban vulnerability*. World Bank Discussion Paper 168. Washington, D.C.: World Bank.

Lewis, James. 1990. "The vulnerability of small island states: The need for holistic strategies." *Disasters* 14(3): 241–249.

Lewis, Michael. 1989. "How a Tokyo earthquake could devastate Wall Street and the world economy." *Manhattan, inc.*, June: 69–79.

Lewis, Nancy D. and Jonathan D. Mayer. 1988. "Disease as natural hazard." *Progress in Human Geography* 12(1): 15–33.

Little, Peter D. and Michael M. Horowitz, eds. 1987. *Lands at risk in the Third World: Local level perspectives*. Boulder, CO: Westview Press.

Liverman, Diana M. 1990. "Drought impacts in Mexico: Climate, agriculture, technology, and land tenure in Sonora and Puebla." *Annals of the Association of American Geographers* 80(1): 44–72.

McGranahan, Gordon and Jacob Songsore. 1994. "Wealth, health, and the urban household: Weighing environmental burdens in Accra, Jakarta, and Sao Paulo." *Environment* 36(6): 4–11, 40–45.

McHarg, Ian. 1969. *Design with nature*. New York: Natural History Press (especially "The City: Health and pathology," pp. 187–195).

McPhee, John. 1989. *The control of nature*. New York: Farrar, Straus & Giroux.

Main, Hamish and Stephen Wyn Williams, eds. 1994. *Environment and housing in Third World cities*. Chichester: Wiley.

Maskrey, Andrew. 1989. *Disaster mitigation: A community based approach*. Development Guidelines No. 3. Oxford: Oxfam.

Masure, P. 1994. "Risk management and preventive planning in mega-cities: A scientific approach for action." *Regional Development Dialogue* 15(2): 183–188.

Miller, Ross. 1990. *American apocalypse: The great fire and the myth of Chicago*. Chicago: University of Chicago Press.

Mintzer, Irving M. 1992. *Confronting climate change: Risks, implications and responses*. Cambridge: Cambridge University Press.

Mitchell, James K. 1984. "Hazard perception studies: Convergent concerns and divergent approaches during the past decade." In Thomas F. Saarinen, David R. Seamon, and James L. Sell (eds.), *Environmental perception and behavior: An inventory and prospect*. Department of Geography Research Paper No. 209. Chicago: University of Chicago, pp. 33–59.

——— 1987. "Boba so stikhiynymi bedstviyami" [Human responses to natural hazards]. *Geograficheskie aspetky vzsimodeystviya khozyaystva i okruzhayushchev sredy*, akademiya nauk SSSR, Moscow, pp. 178–187.

——— 1988. "Confronting natural disasters: An international decade for natural hazard reduction." *Environment* 30(2): 25–29.

——— 1989. "Hazards research." In Gary Gaile and Cort Willmott (eds.), *Geography in America*. Columbus, OH: Merrill Publishing Company, pp. 410–424.

——— 1990. "Human dimensions of environmental hazards: Complexity, disparity and the search for guidance." In Andrew Kirby (ed.), *Nothing to fear: Risks and hazards in American society*. Tucson: University of Arizona Press, pp. 131–178.

——— 1991. "Hazards geography." In Gary S. Dunbar (ed.), *Modern geography: An encyclopedic survey*. New York: Garland, p. 76.

——— 1992. "Major natural disasters in the Pacific Basin: Status, trends and emerging issues." *International Journal of Mass Emergencies and Disasters* 10(2): 269–279.

——— 1993a. "Natural hazards predictions and responses in very large cities." In J. Nemec et al. (eds.), *Prediction and perception of natural hazards*. Dordrecht: Kluwer Academic Publishers, pp. 29–37.

——— 1993b. "Recent developments in hazards research: A geographer's perspective. In E. L. Quarantelli and K. Popov (eds.), *Proceedings of the United States–Former Soviet Union Seminar on Social Science Research on Mitigation for and Recovery from Disasters and Large Scale Hazards. Moscow, April 19–26, 1993. Vol. I: The American participation*. Newark: University of Delaware, Disaster Research Center, pp. 43–62.

————— 1996. "Improving community responses to industrial disasters: In James K. Mitchell (ed.), *The long road to recovery: Community responses to industrial disaster*. Tokyo: United Nations University Press.

Mitchell, James K. and Neil Ericksen. 1992. "Effects of climate changes on weather-related disasters." In Irving Mintzer (ed.), *Confronting climate change: Risks, implications and responses*. Cambridge: Cambridge University Press, pp. 141–152.

Mitchell, James K., Neal Devine, and Kathleen Jagger. 1989. "A contextual model of natural hazard." *Geographical Review* 89(4): 391–409.

Myers, Mary Fran and Gilbert F. White. 1993. "The challenge of the Mississippi flood." *Environment* 35(10): 6–9, 25–35.

National Research Council. 1987. *Geophysical predictions*. Washington, D.C.: Geophysics Study Committee.

Organizing Committee of the International Seminar on Regional Development Planning for Disaster Prevention. 1986. *Planning for crisis relief: Towards comprehensive resource management and planning for natural disaster prevention. Vol. 3. Planning and management for the prevention and mitigation from natural disasters in metropolis*. Nagoya: United Nations Centre for Regional Development.

O'Riordan, Timothy. 1986. "Coping with environmental hazards." In Robert W. Kates and Ian Burton (eds.), *Themes from the work of Gilbert F. White*, vol. 2 of *Geography, resources, and environment*. Chicago: University of Chicago Press, pp. 272–309.

O'Riordan, Timothy and Steve Rayner. 1991. "Risk management for global environmental change." *Global Environmental Change* 1(2): 91–108.

Palm, Risa. 1990. *Natural hazards: An integrative framework for research and planning*. Baltimore, MD: Johns Hopkins University Press.

Parker, D. J. and J. W. Handmer, eds. 1992. *Hazard management and emergency planning: Perspectives on Britain*. London: James & James.

Parker, Dennis J. and James K. Mitchell. 1995. "Disaster vulnerability of megacities: An expanding problem that requires rethinking and innovative responses." *GeoJournal* 37(3): 295–301 (see also additional papers, pp. 303–388).

Perlman, Janice E. 1987. "Megacities and innovative technologies." *Cities*, May: 128–136.

Reed, Stuart, Fred Tromp, and Alain Lam. 1992. "Major hazards – Thinking the unthinkable." *Environmental Management* 16(6): 715–722.

Revue de Géographie Alpine. 1994. "Croissance urbaine et risques naturels dans les montagnes des pays en développement," No. 4.

Rosen, Christine Meisner. 1986. *The limits of power: Great fires and the process of city growth in America*. New York: Cambridge University Press.

Rosenau, P. M. 1992. *Post-modernism and the social sciences: Insights, inroads, and intrusions*. Princeton, NJ: Princeton University Press.

Ruppert, Helmut. 1991. "The responses of different ethnic groups in the Sudan to rural–urban migration." *GeoJournal* 25(1): 7–12.

Sachs, Ignacy. 1988. "Vulnerability of giant cities and the life lottery." In Mattei Dogan and John D. Kasarda (eds.), *A world of giant cities*, vol. 1 in *The metropolis era*. Newbury Park, CA: Sage Publications, pp. 337–350.

Sassen, Saskia. 1991. *The global city: London, New York, Tokyo.* Princeton, NJ: Princeton University Press.

Science Council of Japan. 1989. *International Decade for Natural Disaster Reduction: Proposals by Japanese scientists.* Committee for Disaster Research.

Seidensticker, Edward. 1983. *Low City, High City: Tokyo from Edo to the earthquake.* New York: Alfred A. Knopf.

———— 1990. *Tokyo rising: The city since the great earthquake.* New York: Alfred A. Knopf.

Setchell, C. A. 1995. "The growing environmental crisis in the world's megacities: The case of Bangkok." *Third World Planning Review* 17(1): 1–18.

Showalter, Pamela S. and Mary F. Myers. 1994. "Natural disasters in the United States as release agents of oil, chemicals or radiological materials between 1980–1989: Analysis and recommendations." *Risk Analysis* 14(2): 169–182.

Sit, Victor F. S. 1988. *Chinese cities: The growth of the metropolis since 1949.* Oxford: Oxford University Press.

Smith, Carl. 1995. *Urban disorder and the shape of belief: The great Chicago fire, the Haymarket bomb and the model town of Pullman.* Chicago: University of Chicago Press.

Smith, Keith. 1992. *Environmental hazards: Assessing risk and reducing disaster.* London and New York: Routledge.

Smith, P. J. 1990. "Redefining decision: Implications for managing risk and uncertainty." *Disasters* 14(3): 230–240.

Somma, Mark, 1991. "Ecological flight: Explaining the move from country to city in developing nations." *Environmental History Review*, Fall: 1–26.

Sorkin, Michael, ed. 1992. *Variations on a theme park: The new American city and the end of public space.* New York: Hill & Wang.

Steedman, S. 1995. "Megacities: The unacceptable risk of natural disaster." *Built-Environment* 21(2–3): 89–93.

Stren, Richard, Rodney White, and Joseph Whitney, eds. 1992. *Sustainable cities: Urbanization and the environment in international perspective.* Boulder, CO: Westview Press.

Sylves, Richard T. and William L. Waugh, Jr., eds. 1990. *Cities and disaster: North American studies in emergency management.* Springfield, IL: Charles C. Thomas.

Taylor, Antony J. 1990. "A pattern of disasters and victims." *Disasters* 14(4): 291–300.

Taylor, Peter J. and Frederick H. Buttel. 1992. "How do we know we have global environmental problems? Science and the globalization of environmental discourse." *Geoforum* 23(3): 405–416.

Teune, Henry. 1988. "Growth and pathologies of giant cities." In Mattei Dogan and John D. Kasarda (eds.), *A world of giant cities*, vol. 1 in *The metropolis era.* Newbury Park, CA: Sage Publications, pp. 351–376.

Tuan, Yi-Fu. 1979. *Landscapes of fear.* Minneapolis: University of Minnesota Press.

———— 1988. "The city as a moral universe." *Geographical Review* 78(2): 316–324.

UNEP/WHO (UN Environment Programme/World Health Organization). 1994. "Air pollution in the world's megacities." *Environment* 36(2): 4–13, 25–37.

United Nations. 1987. *The prospects of world urbanization. Revised as of 1984–85.* ST/ESA/SER.A/101. New York: United Nations.

US Office of Science and Technology Policy. 1992. *Science and technology: A report to Congress.* Washington, D.C.: Executive Office of the President.

Waldrup, M. Mitchell. 1992. *Complexity: The emerging science at the edge of order and chaos.* New York: Simon & Schuster.

Walsh, R. P. D., H. R. J. Davies, and S. B. Musa. 1994. "Flood frequency and impacts at Khartoum since the early nineteenth century." *Geographical Journal* 160(3): 266–279.

Waterstone, Marvin. 1993. "Adrift on a sea of platitudes: Why we will not resolve the greenhouse issue." *Environmental Management* 17(2): 141–152.

White, Gilbert F. 1973. "Natural hazards research." In Richard J. Chorley (ed.), *Directions in geography.* London: Methuen.

White, Rodney R. 1994. *Urban environmental management: Environmental change and urban design.* Chichester: John Wiley.

Whyte, Anne V. 1986. "From hazard perception to human ecology." In Robert W. Kates and Ian Burton (eds.), *Themes from the work of Gilbert F. White,* vol. 2 of *Geography, resources, and environment.* Chicago: University of Chicago Press, pp. 240–271.

Whyte, Anne V. and Ian Burton. 1980. *Environmental risk assessment.* SCOPE (Scientific Committee on Problems of the Environment) No. 15. New York: Wiley.

Wikan, Unni. 1995. "Sustainable development in the megacity: Can the concept be made applicable?" *Current Anthropology* 37(1): 635–655.

Woodruff, Bradley A. et al. 1990. "Disease surveillance and control after a flood: Khartoum, Sudan, 1988." *Disasters* 14(2): 151–163.

Wu Qingzhou. 1989. "The protection of China's ancient cities from flood damage." *Disasters* 13(3): 193–226.

Wynne, Brian. 1989. "Sheep farming after Chernobyl: A case study in communicating scientific information." *Environment* 31(2): 33–39.

——— 1992. "Uncertainty and environmental learning: Reconceiving science and policy in the preventive paradigm." *Global Environmental Change* 2(2): 111–127.

Yath, A. Y. 1991. "The effect of differential access to accommodation on the Dinka migrants in Khartoum – The example of Gereif West." *GeoJournal* 25(1): 19–26.

——— 1995. "On the expulsion of rural inmigrants from Greater Khartoum – The example of the Dinka in Suq el Markazi." *GeoJournal* 36(1): 93–101.

Yeung, Y. M. 1990. *Changing cities of Pacific Asia: A scholarly interpretation.* Hong Kong: Chinese University Press.

Zelinsky, W. and L. Kosinski, L. 1991. *Emergency evacuation of cities.* London: Unwin Hyman.

3

Urbanization and disaster mitigation in Tokyo

Yoshio Kumagai and Yoshiteru Nojima

Editor's introduction

Tokyo is one of the world's largest and richest mega-cities and also one of the most hazard prone (Alden, 1984; Cybriwsky, 1993; Fujita, 1991; Geographical Review of Japan, 1990; Sassen, 1991). Since being founded over 400 years ago it has been repeatedly devastated and rebuilt after fires, earthquakes, and aerial bombings (Busch, 1962; Hewitt, 1983; Scawthorn et al., 1985; Seidensticker, 1983, 1990). Usually the urgency of reconstruction frustrated attempts to redevelop a formally planned city (Williams, 1993: 447). Tokyo has also undergone a different kind of transformation with equally profound consequences – first from provincial capital to pre-eminent national urban centre and then to co-leading global city (Douglass, 1993; Harris, 1982). Now, at the end of the twentieth century, as it awaits the possibility of another major earthquake (Katsuhiko, 1987), Tokyo has had to cope with the vicissitudes of high land costs, cramped living quarters, substandard housing, lack of open space, severe traffic congestion, air pollution, and a rash of terror gas attacks. In the face of these ills, as well as for other reasons, there are strong pressures to relocate the national capital outside of Tokyo on a safer site (http://www.keidanren. or.jp; http://www.infoweb.or.jp).

To confront its various threats the city has deployed an arsenal of engineering technologies that far surpass in scale and sophistication those of most other places. Indeed, Tokyo provides a striking example of the use of

56

high-technology means to manage high-stakes natural hazards; it also highlights the role of centralized decision-making under conditions of widespread public acceptance and cooperation. The potential appeal of this approach to other disaster-vulnerable countries is one factor that underlies Japan's leading role in the International Decade for Natural Disaster Reduction (1990–2000). Although this case-study does not take up the issue of limitations on Tokyo's methods of disaster mitigation, it will be important for others to assess the circumstances in which this city's experience might be applied elsewhere and to ask whether residents and leaders of Tokyo might in the future become resistant to heavy public expenditures for massive hazard-reduction projects.

The Tokyo Metropolis

Tokyo is one of Japan's three mega-cities (fig. 3.1). At its core is the so-called 23 Wards Area, which encompasses the city as it existed in 1943, just before it was combined with the then-existing Tokyo Prefecture to create an administrative region known as the Tokyo Metropolis.[1] The Metropolis now houses approximately 12 million people in a densely settled band of 60 municipalities that wraps around the western side of Tokyo Bay and extends into the adjacent Kanto Plain. Together with a lightly populated fringe of plateaux, hills, and mountains to the west (the Tama District) and a number of offshore oceanic islands, the built-up area is administered as one of Japan's 46 prefectures (fig. 3.2).

The Tokyo Metropolis is nested within two successively larger administrative units. The Tokyo Metropolitan Region comprises the Tokyo Metropolis and the prefectures of Chiba, Saitama, and Kanagawa. In addition to Tokyo, this region contains Yokohama – a separate city of 3 million whose history of disasters and human responses is almost as long and complex as Tokyo's. In 1990 there were 31.8 million people in the Tokyo Metropolitan Region – 25.7 per cent of the total population of Japan. The Metropolitan Region is, in turn, part of the National Capital Region, which also comprises the prefectures of Ibaraki, Tochigi, Gunma, and Yamanashi. The present chapter focuses on the mainland districts of the Tokyo Metropolitan Region.

Urban evolution

Tokyo's predecessor city (Edo) was founded in 1603 but this chapter begins with the formal birth of Tokyo as Imperial capital (1868). Thereafter the city's growth can be divided into three phases: 1868–1923; 1923–

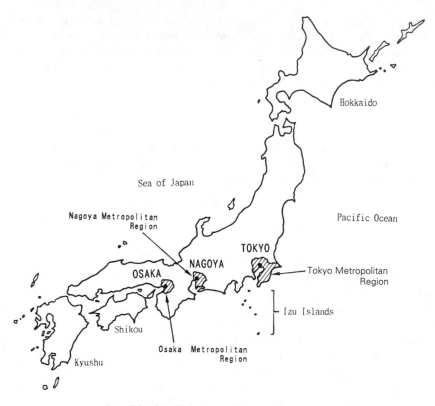

Fig. 3.1 The mega-cities of Japan

1945; and 1945–present. During the first period the city was in transition from an old provincial city to a modern national capital. Population grew steadily, from 1 million in the mid-1880s to 1.7 million in 1908 and 2 million by 1920. Urban neighbourhoods spread outwards in all directions from the new Imperial Palace near the city centre. This phase was brought to an end by the Great Kanto earthquake and fire of 1923, which destroyed much of what was already a very crowded city. The second period (1923–1945) was marked by reconstruction of the destroyed neighbourhoods and rapid suburbanization, particularly along major railroad lines that radiated north-west, west, and south from the down-town core (fig. 3.3). During this time the formerly separate built-up areas of Tokyo and Yokohama began to merge. By 1930 there were 5.5 million residents in the Metropolis and just over 7 million on the eve of the Second World War. The third period (1945–present) began after the devastation of the Second World War, when Tokyo's population had fallen to around 3 million. Post-war recovery reached a watershed in 1964

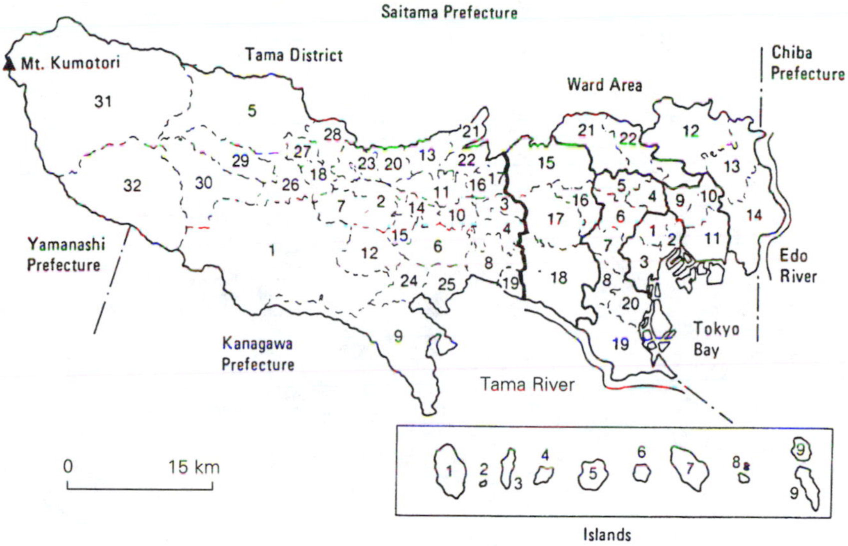

Fig. 3.2 The Tokyo Metropolis (Source: Tokyo Metropolitan Government)

during the Tokyo Olympiad; from then onwards the Japanese economy was so strong that the country became a front-rank global economic power. By 1955 there were again more than 7 million residents in Tokyo and by 1962 the total was over 10 million.

Since the end of the Second World War, population has concentrated in the Tokyo Metropolitan Region at rates that exceed those of Osaka,

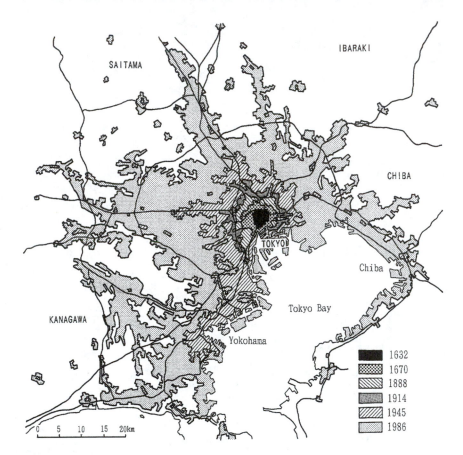

Fig. 3.3 Tokyo's expansion, 1632–1986 (Source: Tokyo Metropolitan Government)

Nagoya, and other metropolitan areas in Japan. Many of the nation's most important political, economic, and cultural activities are located in Tokyo, especially the central administration of the Japanese government, premier wholesale and finance firms, the publishing industry, and leading educational institutions. In addition, Tokyo now discharges rapidly growing international functions such as banking, investment finance, and information dissemination; these reflect Japan's recent rise to prominence in the global economy. While post-war development has brought many benefits to the city, it has also contributed to an extraordinary upsurge in land prices, severe traffic congestion, and deterioration of the physical and social environments. Significant changes in patterns of urbanization have taken place in every decade since 1945.

From the end of the Second World War (1945) to 1960: Recovery

Much of Tokyo was demolished by air raids during the Second World War. A total of 16,230 hectares (equivalent to about 40 per cent of the urbanized area) was burned and 759,000 houses were destroyed. City reconstruction projects received high priority from national and municipal governments. Passage of the Special Urban Planning Act for War Disaster Restoration (1946) ushered in a nationwide series of municipal land-use and development plans. Tokyo's War Damage Rehabilitation Plan sought to keep the city's population below 3.5 million. Additional population and industries were to be accommodated in satellite cities and outer towns all over the Kanto Plain. Post-war austerity and the weakly developed nature of municipal finance doomed this plan almost from the outset. The Korean War, which broke out in June 1950, marked a turning point in the economic fortunes of Tokyo. Private economic investment and construction activities increased rapidly and land prices began to soar. However, it proved difficult to implement large public urban development projects.

The 1960s: Rapid growth

The 1960s were marked by rapid economic growth and equally rapid urbanization of the Japanese population (Harris, 1982; Fujita and Hill, 1993). More than a quarter of a million people flowed into the Tokyo Metropolitan Region each year and buildings mushroomed everywhere. The Olympiad of 1964 was a particular catalyst for change. Planners took advantage of the Olympic Games to initiate projects that would help Tokyo achieve its new role as home to the central management functions of a vast range of private and public institutions. This included remodelling Tokyo's built-up area and building 117 km of new expressways. The Tokaido–Shinkansen (Bullet Train) railway also came into operation just before the Games opened. At the same time, the mega-city's problems were becoming acute, as evidenced by the frequency of traffic-related air pollution and a rash of disputes about sunlight deprivation owing to construction of multi-storeyed buildings. Later in the decade, Tokyo's street cars were largely abolished to improve traffic flow.

The 1970s: Budding disenchantment

Inner-city land prices were already high by the 1970s but now suburban land prices also began to soar. Urban sprawl replaced compact development as new housing sites were increasingly forced to locate further from the city centre. For example, large-scale housing projects such as the Tama New Town were started on the distant hilly lands south-east of

Tokyo. Open space declined drastically. Between 1945 and 1973 the proportion of open space fell from around 47 per cent to less than 20 per cent (Nakazawa, 1988). Although some mid-rise and high-rise housing was constructed on the sites of old factories near the central business district, almost all new housing was expensive, cramped, and far removed from workplaces. Significant clusters of new high-rise buildings emerged in the inner suburbs, especially around important junctions between radial and circumferential railroads.

Widespread signs of public disenchantment with the quality of life in Tokyo marked the 1970s. In this respect the year 1973 was particularly notable. For example, during March residents of Koto Ward forcibly blocked the transportation of garbage from Suginami Ward through their neighbourhood after people in Suginami Ward opposed construction of a new local waste incinerator. This was a highly symbolic incident in an emerging "Garbage War" brought on by the onset of a mass consumption society, subsequent pollution, and delays in the provision of waste-treatment facilities. In May, passengers damaged trains and facilities at Ageo Station in Saitama Prefecture because of accumulated discontent about overcrowded commuter trains and failures to extend the fast rail service in step with the expansion of urban settlements. In June, the Tokyo Assembly was confronted with residents' demands for the enactment of a Sunshine Ordinance. These and other incidents were a sharp indictment of the costs associated with fast economic growth, the concentration of central business management functions, and failures of urban planning in Tokyo.

During the 1970s there were also significant changes in patterns of urban migration. The inflow of new residents to the Tokyo Metropolitan Region slowed and the locus of new population growth moved further out of the city. For example, between 1970 and 1975, areas 30–40 km from the city centre experienced the largest population increases. Between 1975 and 1980 the largest population growth occurred in areas 40–50 km out. Conversely, resident population declines in the Ward Area of Tokyo (i.e. the central city) began during the 1960s and are continuing. This phenomenon is similar in appearance to the decline of inner cities in many Western countries, although some of the causes and consequences are different. For example, unlike North American cities, population out-migration to the suburbs of Tokyo has not been accompanied by increased rates of social pathology in the central city and land prices there have remained at high levels.

Since 1980

Since 1980 Tokyo has undergone yet a further transformation. According to one American observer:

Now that Japan has risen to become firmly established as a global economic power, Tokyo is undergoing yet another rebuilding – the third this century. This rebuilding is driven by money instead of by disaster, and involves expansion of the city upward to new heights of skyscrapers, outward to previously non-urbanized expanses of the Kanto Plain, to new land being made in Tokyo Bay, and to a remarkable extent, downward to even greater depths of subterranean construction. (Cybriwsky, 1993, p. 2)

The Tokyo mega-city continues to dominate Japan, and many problems of pollution, crowding, and declining quality of life that were noted during the 1970s have worsened. At the same time, deindustrialization of the metropolitan area is occurring both because of international economic forces and because of Japanese policies that encourage industries to relocate elsewhere. Between 1979 and 1986 there have been declines in the numbers of manufacturing plants, in the industrial workforce, and in the total value of manufactured goods as the city has shifted its focus to service industries.

Central-city land prices soared in the early 1980s as the yen's value accelerated relative to the US dollar. High-rise offices, hotels, and shopping centres became – and remain – the most typical buildings in the central business district (CBD) and the 22 other regional commercial centres that are scattered throughout the metropolitan region (e.g. Shinjuku). Important settlement centres are springing up at considerable distances from the old city core. They include residential communities 40 km west of central Tokyo in the Tama district, the "science city" of Tsukuba 65 km to the north-east, and new industries clustered near the city's recently expanded international airport at Narita (70 km east of the CBD). Meanwhile, high-status residential buildings are beginning to be erected on former industrial sites near the city centre and more attention is being paid to redressing the city's chronic lack of open space. For the first time in the modern era, beaches are being created along Tokyo Bay to improve recreational and aesthetic amenities for Tokyo residents.

Natural disasters in the Tokyo Metropolis

Hazard-susceptibility

Harris has pointed out that most Japanese cities are affected by serious natural and industrial hazards but that, compared with third world cities and North American cities, they are relatively free of urban social hazards (e.g. shanty towns, slums, decaying inner cities, crime, social conflicts) (Harris, 1982; Nakano, 1974). Tokyo closely fits that description. Because of its low-lying coastal location near the junction of four highly active tectonic plates, the Tokyo Metropolis is susceptible to a variety of seismic hazards, including ground shaking, liquefaction, landslides, tsunamis, and fires (Matsuda, 1980, 1990; Nakano and Matsuda, 1984; Terwindt, 1983). These are compounded by subsidence problems connected with the withdrawal of underground water and the compaction of filled land in offshore areas (Dolan and Goodel, 1986; Matsuda, 1980). About 68 km^2 of Tokyo now lie below sea level, with an additional 57 km^2 below high-tide level (Nakano et al., 1988). The city also lies in the path of typhoons that sweep north and east from the South China Sea and adjacent Pacific Ocean waters. In addition to inflicting direct damage by means of flooding, winds, and heavy rainfall, typhoons often trigger secondary hazards such as landslides and other ground failures (Tamura, 1993). Widespread wind and water damage, including riverine floods, occur occasionally; economic losses associated with heavy winter snowfalls are more frequent. Although the imposing bulk of quiescent Mount Fuji reminds many Tokyo residents of the potential for volcanic eruptions, the likelihood of a damaging event from this source is generally considered to be very low.

Concern about the susceptibility of Tokyo to another major earthquake has been growing since the mid-1960s. For example, in June 1968 the Tokyo Metropolitan Disaster Prevention Conference took up the issue of evacuation routes that would be needed in the event of a future quake. Today, much of the inner city is particularly at risk, including the main commercial and governmental districts. Industrial plants, warehouses, and port facilities that were built on filled land around the edges of Tokyo Bay and in deeper waters further offshore are also at risk. So too are residential and artisanal zones in the Koto Delta and elsewhere. Moreover, a large part of the transportation infrastructure that ties together the entire metropolis occupies areas that are susceptible to subsidence, inundation, or seismic disturbances.

Urbanization and hazard-susceptibility

The built-up area of Tokyo is sited on two kinds of land forms: uplifted and dissected river and coastal terraces that comprise modestly hilly inland districts (2–50 metres); and a series of floodplains (e.g. on the Edo, Ara, Sumida, and Tama rivers) that cut across the terraces to form deltas along the shores of Tokyo Bay (fig. 3.4). Dyking, filling, and reclamation projects were begun as early as 1590 and have progressively moved the shoreline seawards, especially during the twentieth century. These two subregions have been characterized by contrasting patterns of human occupance throughout the history of Tokyo, so much so that they are often referred to as the High City (*yamanote*) and the Low City (*shita-*

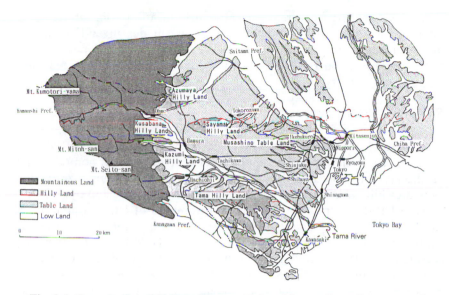

Fig. 3.4 Topography of Tokyo (Source: Tokyo Metropolitan Government)

machi) (Seidensticker, 1983). Tokyo's main middle-class and upper-class residential neighbourhoods are generally located in the hillier districts, whereas the lowlands are primarily occupied by commercial and industrial facilities as well as blue-collar housing. For example, the Keihin Industrial Zone stretches for 40 km along the Tama River and Tokyo Bay and is home to steel mills, machinery fabrication plants, and chemical works. The portion of lowland that is sandwiched between the Ara and Sumida rivers (i.e. the Koto Delta District) is now heavily occupied by housing as well as other uses.

The lowlands have historically been more susceptible to natural disasters than the upland terraces (Nakano et al., 1988). However, the relation between urbanization and hazard losses is a complex one. For example, the need quickly to provide housing for Tokyo's rapidly growing population after the Second World War led to the construction of tightly packed masses of wooden apartment buildings in western parts of the city. Many of these have been replaced by reinforced concrete structures but dense clusters of inflammable wooden buildings still remain in other places. There has also been an upsurge in the use of city gas, propane gas, petrol, and kerosene throughout Tokyo. In addition, the number of multi-storey buildings, the number of large-scale underground shopping malls, and the volume of automobile traffic have all increased. These changes have added to the disaster-vulnerability of the Tokyo Metropolis. In summary, the general level of environmental hazard in

Tokyo has tended to increase because urbanization concentrates populations and investments. Innovations of urban design and occupancy often introduce unexpected patterns of exposure, and the potential for large losses has spurred extra efforts to reduce both physical risks and human vulnerability.

Significant hazard events

Most natural hazard losses in 1963–1993 were caused by heavy rain and high winds associated with typhoons and other storms, but non-damaging earthquakes also occurred as well as an assortment of other hazards such as winter blizzards. In an average year the city was affected by 5–20 significant hazard events, but total annual numbers of affected households rarely exceed a few thousand. In an unusual year (1966) typhoons disrupted or damaged almost 200,000 households; even more damaging storms have also affected Tokyo outside this period of record (e.g. the Kanogawa typhoon, 1958). Severe earthquakes have historically occurred at intervals of several decades, although none took place between 1963 and 1993. Nevertheless, the threat of a major earthquake is treated very seriously by Tokyo's leaders and its citizens.

Earthquakes and fires

Since 1615 the area that is today encompassed by Tokyo has experienced at least six earthquakes of magnitude 5 or greater on the Japanese scale (1615, 1649, 1703, 1855, 1894, 1923) (Matsuda, 1993).[2] Two of these quakes (1703 and 1923) were centred offshore in the Sagami Trough, which bounds the Philippine Sea tectonic plate and the Pacific Ocean plate, and four occurred directly under the site of Tokyo. Heavy losses were sustained in both types of quake. For example, the 1855 event (M 7) caused the collapse of 14,000 dwellings with an estimated death toll of around 10,000 (Katsuhiko, 1987). The Great Kanto quake of 1923 and the fire that followed killed 140,000 and devastated approximately 44 per cent of the built-up area. The latter was the most destructive natural disaster to have affected any of the world's large cities during the twentieth century and one of the largest disasters to have affected any community during recorded history.

Earthquake damage in Japan is closely related to weather conditions that favour the spread of fires ignited by ruptured gas lines and the scattered embers of cooking fires. For example, the Kanto earthquake of 1923 occurred just before noon at the end of a hot and dry summer when strong winds were blowing. Non-earthquake-related fires have also been a frequent hazard. Before the Second World War, the city was overwhelmingly composed of inflammable wooden buildings. During the Edo period (1603–1868) there were at least 89 urban fires that burned large areas. It was said that new wooden buildings had an average lifespan of seven years before being consumed by fire. Since the 1960s, the incidence of large fires has declined sharply as a consequence of improved fire-fighting services and increased use of non-flammable materials in urban buildings (Scawthorn, Arnold, and Scholl, 1985).

Floods

Tokyo is subject to three types of flooding: storm surges associated with typhoons; heavy rainfall that fills low-lying areas directly; and breaches in sea walls and dykes caused by earthquakes (Nakano et al., 1988). Although there have been at least eight major surge floods since 1900 and numerous rainfall floods, there has been no significant flood damage in the Tokyo lowlands since 1966. This is primarily due to massive invest-ments in embankments, pumping stations, and water gates throughout the lowlands. In contrast, flooding has increased in the upland tributary valleys. This illustrates a well-known relationship between urbanization and increased flooding as permeable undeveloped soils are replaced by impermeable roads, parking lots, and buildings. The resulting pattern marks a dramatic shift in the location of flood risks (fig. 3.5). For example, widespread flooding in September 1949 inundated areas that were almost totally confined to the lowlands. Nine years later the flood zone included valleys in the hillier regions of the "tableland." By June of 1966 areas affected by urban flooding were showing up on the west (inland) side of the Ward Area and flooding in the lowlands was less of a problem.

Disaster mitigation

Public policies for disaster reduction in Tokyo are heavily influenced by the city's experience with devastating fires. These have occurred as a result of accidents, earthquakes (1923), and aerial bombings (Second World War). Early efforts focused on fire awareness, fire prevention, and fire fighting, and these are still central components of hazard manage-ment. For example, residents and visitors are constantly reminded about the dangers of fires by signs and information pamphlets; lay citizens are expected to provide the first line of defence against fires; specialized local stores sell emergency supplies; neighbourhood fire houses are stocked with ingenious lightweight fire-fighting devices powered by human muscle and other portable energy sources; and the Tokyo Fire Service is both highly trained and equipped with advanced machinery for fire fighting and rescue operations. These kinds of responses are already well devel-oped and widespread throughout the city. A second tier of protection is

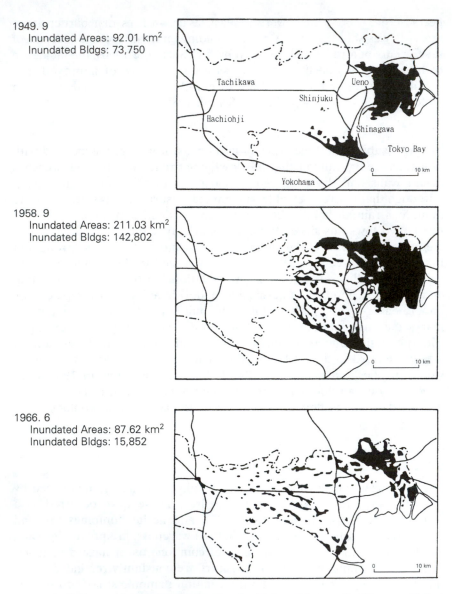

1949. 9
  Inundated Areas: 92.01 km²
  Inundated Bldgs: 73,750

1958. 9
  Inundated Areas: 211.03 km²
  Inundated Bldgs: 142,802

1966. 6
  Inundated Areas: 87.62 km²
  Inundated Bldgs: 15,852

Fig. 3.5 Changes of inundated area in the Tokyo Metropolis (Source: Tokyo Metropolitan Government)

provided by various land-use and building construction measures that are designed to: (1) retard the spread of fires; (2) provide shelters for evacuees; and (3) supply relatively secure bases from which emergency countermeasures can be mounted. In some parts of Tokyo all of these responses are already in place, but elsewhere many are still being developed. Eventually there will be a third tier of protection which is based on assumptions about the prediction of risks and the reliability of "disaster-proofing" technologies. Residents may not only be forewarned about earthquakes and other threats by sensors and monitoring devices; they will be provided with enough protection to obviate the necessity of evacuation. Many of the adjustments to fires also serve to offset other environmental hazards (e.g. earthquakes, industrial accidents) but they do not necessarily guarantee protection against flooding, windstorms, and landslides. Separate initiatives are sometimes undertaken for hazards such as these.

Japan possesses a highly centralized, hierarchical form of government, which is reflected in the disaster-mitigation process. This is complemented by an extensive system of consultation and consensus-seeking among different constituencies before policy changes are enacted. Although rancorous conflicts are not precluded (and may become more frequent), there is rarely much publicly voiced opposition to government edicts.[3] National legislation and policies set the context for local responses and it is appropriate to begin with a brief review of disaster mitigation at the national level.

The evolution of disaster countermeasures in Japan

Since the Meiji Restoration of 1868 a wide range of natural hazard laws has been enacted in Japan (table 3.1). Between 1880 and 1911 there were seven major national laws that addressed issues of flooding, erosion, and disaster relief. After an interval of 36 years, marked by numerous earthquake and typhoon disasters as well as the trauma of the Second World War, a renewed round of national disaster legislation began in 1946.[4] This continued until 1980 and involved the passage of 35 additional acts and related initiatives.

The most important of these was the Disaster Countermeasures Basic Act (1961). Other laws passed during this period addressed fire fighting, flood control, forest conservation, slope failures, coastal hazards, heavy snowfall, petrochemical complexes, and the prediction of earthquakes and volcanic eruptions. In addition, several disaster science and management agencies were established, an annual national "disaster prevention day" (1 September – the anniversary of the 1923 earthquake) was proclaimed, and legislation was enacted to address the special problems of disasters in large cities. Particular attention was directed towards the threat of future earthquakes in urban areas.

Since 1980 the pace of national disaster legislation in Japan appears to have slowed. Only two major laws were passed in almost 15 years; one of

Table 3.1 Disaster countermeasures and major natural disasters in Japan: 1868–1993

Year	Disaster measures	Major disasters
1880	Provision and Saving Law	
1888		Mt. Bandai eruption (Jul.)
1891		Nohbi earthquake (Oct.)
1896	River Law	Sanriku earthquake tsunami (Jun.)
1897	Erosion Control Law	
	Forest Law	
1899	Disaster Preparation Funds Special Account Law	
1908	Flood Prevention Association Law	
1910		Floods in central and northern Japan (Aug.)
1911	Flood Control Expenditure Funds Special Account Law	
1914		Sakurajima Is. eruption (Jan.)
1923		Great Kanto earthquake (Sep.)
1926		Mt. Tokachigake eruption (May)
1927		Kita-tango earthquake (Mar.)
1933		Sanriku earthquake tsunami (Mar.)
1934		Muroto typhoon (Sep.)
1944		Tohnankai earthquake (Dec.)
1945		Makurazaki typhoon (Sep.)
1946		Nankai earthquake (Dec.)
1947	Disaster Relief Law (Oct.)	Typhoon hits Kanto (Sep.)
	Fire Organization Law (Dec.)	
1948	Fire Service Law (Jul.)	Fukui earthquake (Jun.)
		Typhoon hits northern Japan (Sep.)
1949	Flood Control Law (Jun.)	
1950	Temporary Measures Law for Subsidizing Recovery Project for Agriculture, Forestry and Fisheries Facilities Damaged Due to Disasters (May)	

70

Year	Legislation / Measures	Disasters
1951	Law concerning National Treasury Share of Expenses for Recovery Projects for Public Civil Engineering Facilities Damaged Due to Disasters (Mar.)	Typhoon hits southern Japan (Oct.)
1952	Meteorological Service Law (Jun.)	
1954		Tohyamaru typhoon (Sep.)
1955	Temporary Measures Law for Financing Farmers, Forestrymen and Fishermen Suffering from Natural Disaster (Aug.)	
1956	Seashore Law (May)	
1957	Landslide Prevention Law (Mar.)	Torrential rains in Isahaya (Jul.)
1958		Kanogawa typhoon (Sep.)
1959		Ise Bay typhoon (Sep.)
1960	Forest Conservation and Flood Control Urgent Measures Law (Mar.) Designation of "Disaster Prevention Day: 1. Sep." (Jun.)	Chile earthquake tsunami (May)
1961	Disaster Countermeasures Basic Act (Nov.)	
1962	Special Measures Act for Countermeasures in High Snowfall Areas (Apr.) Act concerning Special Financial Support to deal with the Designated Disaster of Extreme Severity (Sep.)	
1963	Formulation of "Basic Plan for Disaster Prevention" (Jun.)	Hokuriku heavy snowfall (Jan.–Feb.)
1964	Revision of River Law (Jul.) Geodesy Council's Proposition on "Earthquake Prediction" (Jun.)	Niigata earthquake (Jun.)
1968	Establishment of Coordination Committee for Earthquake Prediction (Apr.)	Tokachi-oki earthquake (May)
1969	Law concerning Prevention of Steep Slope Collapse Disaster (Jul.)	
1971	Essentials of Earthquake Countermeasures for Larger Cities (May) Start of Earthquake Disaster Countermeasures Drill (Sep.)	San Fernando earthquake (Feb.)
1973	Formulation of "Plan for Volcanic Eruption Prediction" (Jun.) Act concerning Improvement, etc. Refuges, etc. in Vicinal Areas of Active Volcanoes (Jul.) Law for the Payment of Solatia for Disaster (Sep.)	

71

Table 3.1 (cont.)

Year	Disaster measures	Major disasters
1974	Establishment of National Land Agency (Jun.) Establishment of Coordination Committee for Prediction of Volcanic Eruption (Jun.)	Izu-hanto-oki earthquake (May)
1975	Law on Prevention of Disaster in Petroleum Industrial Complexes and Other Petroleum Facilities (Dec.)	
1976	Establishment of Hdqrs for Earthquake Prediction Promotion (Oct.)	
1977	Establishment of "Tokai Area Assessment Council" (Apr.)	Mt. Usu eruption (Aug.) Izu-Ohshima-Kinakai earthquake (Jan.)
1978	Large-scale Earthquake Countermeasures Act (Jun.) Designation of "Areas under Intensified Measures against Tokai Earthquake" (Aug.)	Miyagi-ken-oki earthquake (Jun.)
	Act on Special Measures for Active Volcanoes	
1979	Establishment of "Prediction Committee for the Area under Intensified Measures against Earthquake" (Aug.)	
1980	Special Fiscal Measures Act for Urgent Improvement Projects for Earthquake Countermeasures in Areas under Intensified Measures against Earthquake (May)	
1982		Torrential rain in Kyushu (Jul.)
1983		Nihon-kai-chubu earthquake (May) Torrential rain in western Japan (May) Miyakejima Is. eruption (Oct.) Izu-Ohshima Is. eruption (Nov.)
1986		
1987	Law concerning Dispatch of Japan Disaster Relief Team (Sep.)	
1991		Mt. Unzendake eruption (Jun.) Typhoon No. 19 hits whole of Japan (Sep.)
1992	General Essentials of Countermeasures for Earthquake directly beneath the South Kanto Area (Aug.)	
1993		Hokkaido-nansei-oki earthquake (Jul.) Cold-weather damage in northern Japan

Source: Tokyo Metropolitan Government.

these was again directed at preparations for the anticipated Kanto earth-quake and the other focused on Japan's emerging role as a supplier of disaster-management expertise to other countries. However, the Great Hanshin earthquake that devastated Kobe in January 1995 may well stimulate an upsurge in disaster-related legislation and other public policy initiatives. The pattern of disaster legislation directly affecting Tokyo is somewhat different (fig. 3.6). Disproportionately more laws and regulations were enacted during the period 1919–1924 (which encom-passes the 1923 earthquake) and again in the years between 1952 and 1980. Whereas the pace of national legislation appears to have decreased in recent decades, there has been no diminution of the legislation affect-ing hazard management in and around Tokyo.

The Disaster Countermeasures Basic Act (1961)

After the Second World War, Japan experienced a series of large dis-asters, culminating in 1959 in the Ise Bay typhoon, which took a particu-larly heavy toll (5,098 deaths) and prompted national leaders to adopt a wholesale reform of the disaster-management system. Subsequently, the Disaster Countermeasures Basic Act was promulgated in November 1961 with the goal of developing a nationwide, comprehensive, governmental system for disaster prevention.

This Act required the central government to form a Central Disaster Prevention Council headed by the Prime Minister. Members of the council are ministers and heads of designated administrative organs and designated public corporations. The Council deliberates important matters relating to disaster prevention and reports to the Prime Minister. The Council also prepared the Basic Plan for Disaster Prevention in 1963. Individual administrative organs are required to make their own Opera-tional Plans for Disaster Prevention in accordance with the Basic Plan. Local governments (prefectural governments and municipalities such as cities, towns, and villages) are also required to establish Prefectural Disaster Prevention Councils and Municipal Disaster Prevention Coun-cils. These bodies prepare Local (Prefectural and Municipal) Plans for Disaster Prevention that define specific actions to be taken in their respective areas.

Disaster countermeasures plan of the Tokyo Metropolitan Government

The Prefectural Plan for Disaster Prevention

In accordance with the national Basic Plan for Disaster Prevention, the Tokyo Metropolitan Government has prepared a Prefectural Plan for

Era	Urban development	Construction of fireproof buildings	Countermeasures against disaster	Main disasters and major theories
Meiji era	1872 Ordinance to construct brick buildings; 1881 Ordinance to use non-flammable material to build rooftops; 1888 Ordinance to improve urban area of Tokyo; 1909 Agricultural Land Arrangement Act	1872–1874 Construction of brick buildings at Ginza in Tokyo; 1881 Designation of roads along which buildings should be made fireproof; 1892–1912 Construction of a block with fireproof buildings at Marunouchi in Tokyo	Acceleration of construction of fireproof buildings by the Ordinance of 1881	1872 Big urban fire at Ginza in Tokyo; 1879 Big urban fire at Nihonbashi in Tokyo; 1894 Tokyo earthquake (M7)
Taisho era	1919 Urban Planning Act (old); 1919 Set-up of land readjustment institution using Agricultural Arrangement Act; 1923 Special Urban Planning Act for Earthquake Disaster Restoration	1919 Urban Building Act; Two Categories of fireproof district established, plus structural controls; 1924 Ordinance on Subsidies and Control over Buildings in Fireproof Districts	1924 Subsidies for building construction	1923 Great Kanto earthquake (M7)
Showa era	1937 Air Defence Act; 1946 Special Urban Planning Act for War Disaster Restoration; Fireproof district; Quasi-fireproof district; Structural controls; Building standards	1936 Ordinance on Special Buildings; 1942 Ordinance on Building Repair for fireproofing; 1950 Building Standard Act	←→ War Time Special System; 1939 Control over building construction for air defence; 1948 Extra Building Control on fireproofing; Two Categories of fireproof district; Quasi-fireproof district	1933 Sanriku offshore earthquake (M7.5); 1945 Second World War; 1948 Fukui earthquake (M7.3); 1950 Big fire in Atami; 1952 Big fire in Tottori

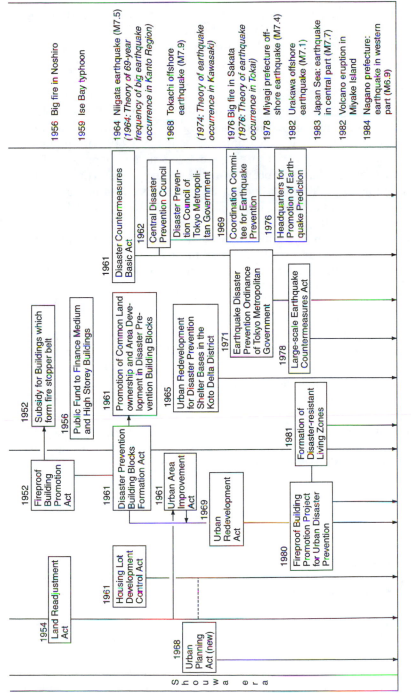

Fig. 3.6 Disaster legislation in Tokyo and Japan (Source: Tokyo Metropolitan Government)

Disaster Prevention. This plan is checked every year and amended when necessary. It is printed in book form and available for sale to citizens. The Earthquake Disaster Volume (last amended in 1986) alone consists of 803 pages of information about appropriate disaster countermeasures.

Plan for earthquake disaster prevention

There is a high level of concern about the potential effects of large earthquakes on Japan. The earliest major legislative manifestation of this is the Large-scale Earthquake Countermeasures Act (1978). Under the terms of the Act, special measures have been taken to prepare for a massive earthquake that may occur in or near Suruga Bay (the so-called Tokai earthquake). An intensive earthquake observation system has been set up to monitor much of Kanagawa, Yamanashi, and Shizuoka prefectures beyond the southern and western edges of the Tokyo Metropolis. This system is designed to provide warnings of an impending quake which will trigger various disaster prevention actions throughout the region. According to most Japanese studies it is not expected that Tokyo buildings would experience devastating effects from a repeat of the 1923 Kanto (offshore) quake. Most of the worst physical impact would be felt further south in Shizuoka Prefecture. However, there is concern about the possibility of social disruption, panic, and loss of confidence in public authorities. A Stanford University engineering researcher, Haresh Shah, has calculated that economic losses might amount to US$1.2 trillion and other commentators have speculated about the possibility that this scale of losses might trigger vast global financial repercussions (Lewis, 1989).

In August 1992, the Expert Committee of the Central Disaster Prevention Council pointed out that there was a significant possibility of a Richter scale 7 earthquake occurring just beneath the South Kanto land area (Saitama, Chiba, Tokyo, and Kanagawa prefectures) in the near future. At the same time, the Council announced a set of guidelines for addressing the consequences of such an event. These noted the difficulty of predicting this kind of earthquake and the localized pattern of damage that would result. The Metropolitan Government was charged to begin risk assessment of earthquakes in the Tokyo Metropolis during fiscal year 1994, and the National Land Agency was given responsibility for developing methods of assessing risks associated with local quakes.

In Tokyo the system for coping with earthquakes is highly detailed. The Metropolitan Government has enacted a Tokyo Metropolitan Ordinance for Earthquake Disaster Prevention (1971). A specific Plan for Earthquake Disaster Prevention has been developed and is updated every five years. It includes a provision that requires the Metropolitan Government to assess earthquake vulnerability regionally and to publi-

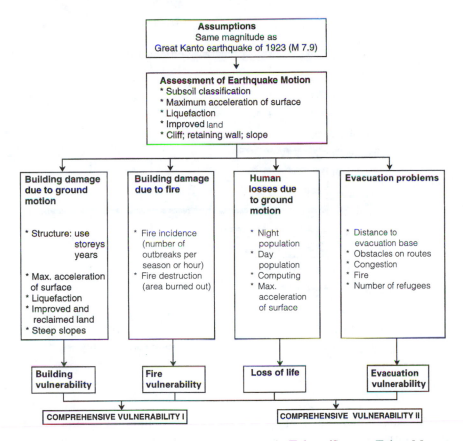

Fig. 3.7 Framework of vulnerability assessment in Tokyo (Source: Tokyo Metropolitan Government)

cize the results. The findings of the vulnerability assessment were made
public in 1975 and 1984 for the Ward Area, in 1980 for the Tama District,
and in 1992 for the whole of the Metropolis.

The assessment was based on the effects of an earthquake similar to
the one that occurred in 1923 (i.e. 7.9 Richter scale magnitude with an
epicentre located under Sagami Bay). The urbanized area of Tokyo was
divided into 5,017 blocks (500 m × 500 m) and a survey of vulnerability
conducted for each block using the following indicators on a scale of 5
(most vulnerable) to 1 (least vulnerable) (fig. 3.7):

1. collapse of buildings
2. human lives lost
3. occurrence and spread of fire
4. evacuation.

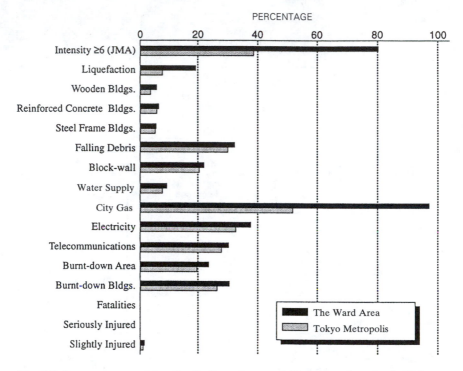

Fig. 3.8 Loss assessment for the forthcoming great Kanto earthquake in Tokyo

This survey provides a picture of the relative vulnerability among different districts of Tokyo at the time of the earthquake. It does not take account of cumulative losses incurred before, during, and after the earthquake, including chains of consequences throughout the recovery period. These omissions have been rectified by a later loss assessment that is based on an assumed repeat of the 1923 earthquake at 6 p.m. during winter when winds are blowing with a velocity of 6 metres per second. Damage has been estimated in three stages: (1) immediate damage to buildings due to ground shaking, liquefaction, and slope failure; (2) loss of life and property due to subsequent fires, floods, and other hazards; and (3) after-effects such as the prolonged loss of urban functions and services. Figure 3.8 summarizes the overall estimated losses for the Tokyo Metropolis. It suggests that approximately 80 per cent of the old built-up (Ward) district of Tokyo would be affected by ground shaking and 20 per cent by liquefaction. In this area, loss of gas supplies would be nearly total (98.5 per cent) and 30 per cent of the 1.5 million buildings would be burned down. Losses in the rest of the Tokyo Metropolis would be proportionately less but still very substantial. Loss of life is estimated at around 10,000 with total casualties around 70,000, but these figures may be revised upwards in the light of Kobe's recent earthquake experience. There a Richter scale 7.2 quake on 17 January 1995 caused much more damage than anyone had anticipated.

Urban disaster-prevention planning

The basic elements of disaster-prevention planning in Tokyo are intended to increase the fire resistance of buildings, to provide residents with safe evacuation routes to secure shelter areas, and to equip these areas with the means to combat surrounding fires.

Fireproof buildings

The Metropolitan Government has designated 56,553 ha (or 95 per cent) of the Tokyo Ward Area as special districts for purposes of fire suppression and reduction. There are two types of districts: (1) fireproof districts and (2) quasi-fireproof districts. In the former, all buildings of more than three storeys or more than 100 m^2 of floor space are required to be built of non-flammable materials such as reinforced concrete. In quasi-fireproof districts, fireproofing is required of only certain taller and larger buildings; other buildings are regulated to a lesser degree. Nearly all parts of the Ward Area are designated as quasi-fireproof districts and areas within Loop Road 6 are designated as fireproof districts. The latter area represents the "disaster-proof" core of modern Tokyo.

Government subsidies to encourage the use of fireproof materials in building construction constitute a second strand of the fireproof building strategy. Subsidies are available to owners of buildings in the designated fireproof districts of Tokyo as well as similar districts in Osaka, Nagoya, and other large cities that face significant earthquake risks. So far under this programme, 34 districts (732.3 ha) have been designated as so-called "Fireproof Building Promotion Districts." Therein the main objective is to retard fire and permit evacuation in the wake of an earthquake rather than to prevent fire occurring. Of these districts, 28 (638.8 ha) are located in the Ward Area of Tokyo. In other words, just over 1 per cent of Tokyo city is included in this programme at present. The rate of the subsidy is 50 per cent (local government 25 per cent, national government 25 per cent); in fiscal year 1990, ¥510 million were spent on 198 buildings in 29 Fireproof Acceleration Districts (i.e. ¥2.6 million or US$25,000 per building).

Open spaces for evacuees

More than 1 million people survived fires that followed the 1923 earthquake by evacuating to large open areas such as public parks. Compared with most cities in Europe and North America, however, Tokyo is chronically short of public open spaces. Creation of parks, green promenades, and similar open areas receives high priority in parts of the city where they are currently lacking.

In Tokyo, 149 evacuation sites have been officially designated. Typically they are large parks, university or high school campuses, other public facilities (e.g. Haneda Airport), non-flammable housing complexes such as the so-called Metropolitan Residents' Housing, cemeteries, and riverbanks. The largest are located in an arc close to Tokyo Bay and in a band of shrines and parks surrounding royal residences that stretches west from the Imperial Palace. Each is intended to provide space for evacuees at a density of not less than 1 m^2 per person. According to government guidelines, no person should have to travel more than 2 km to reach an official evacuation site. Non-flammable buildings, parks, and open spaces, which provide the majority of evacuation sites, are shown in figure 3.9. Other evacuation sites are also shown.

The number of evacuation sites has increased in recent years with the

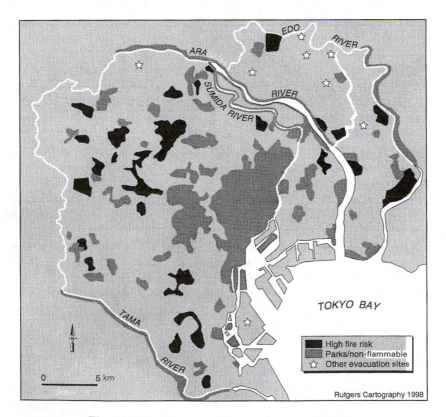

Fig. 3.9 Fire risk, open space, and evacuation sites

departure from Tokyo to Tsukuba of many national research institutes and one university. The grounds of these facilities are typically converted into public parks, which are also frequently designated as evacuation sites. In addition, the government has made use of former US military bases now returned to Japanese control. Eleven such bases cover 1,116 ha and four of these (244 ha) are now designated as evacuation sites. The Tokyo Metropolitan Government is also eagerly purchasing abandoned factory sites for use as roads, parks, and housing that incorporate the principle of disaster resistance. Finally, agricultural land also plays a role in increasing human survivability during disasters in the Tokyo Metropolis. Most such land is privately owned. Farmers can sell it to developers at any time without penalty. For a long time taxes on agricultural land were also held at very low levels but the tax laws were amended in 1991 to remove this provision; property taxes are now set at the same level as on built-up land. If an owner's land is designated as "Agricultural Green Space" it is taxed at the previous low rate provided that it remains in agriculture for at least 30 years. The new taxation system in effect divides farm land into two different groups: land that will be converted to urban uses sooner or later; and land that will continue to be farmed for at least 30 years. In the Tokyo Metropolis, 53 per cent (3,983 ha) of the farm land in the Urbanization Promotion Areas (17,520 ha) is designated as "Agricultural Green Space." Although such land is scattered in small parcels, these might perhaps be amalgamated and put to work as evacuee reception areas at some time in the future.

Evacuation routes

Successful evacuation depends in part on the availability of suitable roads, and in much of the metropolis these are lacking. A need for better roads is recognized both inside the built-up (Ward) area of Tokyo – where 1,705 km of new main roads have been included in urban planning documents – and in the (Tama) hinterland, where an additional 1,368 km are so designated. Unfortunately, owing to the high price of land and the difficulty of negotiations with prospective sellers, only 54 per cent of these roads in the Ward Area and 38 per cent in the hinterland have been completed.

Urban redevelopment and land adjustment projects

Urban redevelopment projects provide a suitable – though expensive – means of converting districts full of wooden buildings into fireproof districts. For example, Tokyo Metropolitan Government has completed the Shirahige-higasi District Urban Redevelopment Project (27.6 ha) on the eastern bank of the Sumida River; three more projects are under way (Shirahige-nishi District, 49.6 ha; Kameido-oojima-komatsugawa

District, 98.4 ha; Kouenji-eki-kita District, 4.0 ha). In each of these cases, improved disaster resistance is the primary motivation for the project. Elsewhere, disaster resistance is an important secondary goal of two completed urban redevelopment projects (Nishi-ookubo District, 2.9 ha, and Iidabashi District, 2.3 ha), and one ongoing project (Akabane-kita District, 3.5 ha). The foregoing are publicly sponsored efforts. Private sector redevelopment projects are also being undertaken.

Land readjustment projects are an additional means of making urban areas less vulnerable during disasters. These projects involve realigning roads and property boundaries so that disorderly patterns of urban sprawl can be replaced by more coherent land-use patterns. Many of Tokyo's existing roads were former agricultural tracks and as such they are often narrow and winding. Land readjustment projects were successfully adopted to achieve the rationalization of urban land uses after the Kanto earthquake of 1923 and again after the Second World War.

Disaster-prevention bases

Since 1978 the National Land Agency has been promoting a model project for the construction of disaster-prevention bases. The main purpose is to strengthen disaster-prevention systems at the neighbourhood level. In the event of a disaster such places are intended to act as bases from which disaster countermeasures would be mounted; during normal times they serve a variety of other disaster-prevention functions including public relations, education, and training. So far, six bases have been completed in different parts of the country: Kawasaki City (1980), Osaka City (1981), Nagoya City (1982), Kita Ward of the Tokyo Metropolis (1984), Amagasaki City (1986), and Shizuoka City (1989). Two additional bases are under construction.

One particular disaster-prevention base deserves special mention. The Tachikawa Regional Disaster-Prevention Base is located at an airport that was formerly a US military base. Its purpose is to provide emergency countermeasures in the event of a widespread disaster affecting the South Kanto area – including parts of the Tokyo Metropolis. Together with the adjacent Showa Memorial National Governmental Park, the disaster-prevention base encompasses 180 ha. Since 1983 there has been a build-up of investment in this facility. All related ministries, other agencies, and the Tokyo Metropolitan Government have begun to improve the base, under the coordination of the National Land Agency. It is intended to function as the substitute facility of the national government's Headquarters for Disaster Countermeasures.

Comprehensive water-control measures

Flooding in the western (Tama) district of the Tokyo Metropolis is mainly a problem of overflow from small rivers during periods of heavy rainfall, exacerbated by increasing urban occupancy of floodplains. Here the main strategy involves structural engineering measures such as embankments, underground river systems, diversion channels, and sewers. These are designed to accommodate rainfall at the rate of 75 mm per hour (a once-in-15-year event). The eastern half of Tokyo includes a number of so-called "zero metre areas" (e.g. Koto Delta) that are at or near sea level and at risk of typhoons, high tides, and earthquake subsidence. Here, engineering responses again dominate, including tidal dykes, coastal embankments, water gates, and inland water discharge facilities.

Along large rivers such as the Tama in the west and the Ara and Sumida in the east, the Ministry of Construction is promoting construction of "super banks." These are wide flat-topped berms that have shown superior ability to resist earthquakes and to protect adjacent urban areas against floods. Property developers and landowners have welcomed the super banks because they afford view lot sites for housing in areas that are otherwise flat. At present, "super banks" are found only in scattered locations but – when connected in a continuous line – it is believed that they will afford permanent protection to the areas that they enclosed. (They are similar in concept to the medieval Dutch "ring dykes" that preceded polderization and reclamation schemes, but on a more monumental scale.)

Disaster-prevention shelter bases in the Koto Delta District

One of the most earthquake-vulnerable districts in the Tokyo Ward Area is the Koto Delta District, which covers about 4,500 ha between the Sumida River and the Ara River. Most of the land here is very low-lying owing to subsidence caused by excessive use of underground water for industrial purposes since the late 1920s. It is barely at sea level or lower and the ground condition is soft. Homes and industries are primarily constructed of wood and there are a great number of petrochemical installations.

Land reclamation has dominated the history of the Koto Delta District. After a large fire in 1657 destroyed most of the existing buildings of Edo, including the shogun's castle, it was decided to reduce overcrowding (and vulnerability to fire) by expanding the amount of buildable land. Refuse and other materials were dumped along both sides of the Sumida River to create suitable sites, and the area east of the river became a residential zone, which was also home to many temples that were transferred there. By this means the Koto Delta District came into existence and continued to expand throughout the Edo era.

In the wake of the Meiji Revolution (1868), Japan became an increasingly industrialized society and the Koto Delta District was the primary location of industrial production in Tokyo. This situation persisted until

1960, when protective trade and monetary controls were abolished by the national government. That action had the effect of encouraging many private companies to concentrate production in smaller numbers of new, very large facilities located elsewhere. By 1964 the Tokyo Metropolitan Government had taken advantage of these changes to acquire former factories for badly needed residential development sites. But it was clear that the entire area was susceptible to disaster and that a plan for disaster prevention was needed both for the occupants of new residential developments and for existing residents living in dense clusters of old wooden houses.

The startling effects of earthquake-induced liquefaction and subsidence on new low-rise housing in Niigata (1964) underscored the urgency of a disaster-prevention project for the Koto Delta District. As a result, the Ministry of Construction proposed to develop 16 square-shaped disaster prevention bases, each 250,000 m² in size. This plan was modified by replacing the square bases with cross-belt (X) shaped ones to take account of the possibility that massive post-earthquake firestorms might occur. In the end this proposal was abandoned because it was too costly. Finally the present "Disaster-Prevention Shelter Base Concept" was adopted by the Tokyo Metropolitan Government in 1969 because it would address the dual goals of reducing disaster and increasing the supply of housing. Six disaster-prevention shelter bases were proposed, each having an area of 50–100 ha. The objective of this plan is to secure safe open spaces to which people can evacuate within approximately 30 minutes in the event of a large earthquake. The estimated total cost (in 1969 prices) was ¥500 billion (table 3.2).

Considerable progress has been made toward implementing this plan. Urban redevelopment projects were adopted for Shirahige-higashi District, Shirahige-nishi District, and Kameido-oujima-komatsugawa District; the first of these was completed in 1985. Park development projects were adopted for Shirahige-higashi District (Higashi-shirahige Park, 10.3 ha), Kameido-oujima-komatsugawa District (Oujima-komatsugawa Park, 24.7 ha), and Kiba District (Kiba Park, 24.2 ha). Fireproof building promotion projects for urban disaster prevention were also adopted in Shirahige-nishi District, Kiba District, Yotsugi District, Ryougoku District, Chuou-sarue District, and Chuou-sumida District.

Creation of disaster-resistant living zones

Recently a new round of more advanced countermeasures is being added to Tokyo's defences. These include so-called "disaster-resistant living zones" where residents will not be called upon to evacuate in anticipation of future threats. The area included in a disaster-resistant living zone is roughly the same as the service area of an elementary school or junior

Table 3.2 Shelter bases in the Koto Delta District

Name of base	Area of base (ha)	Area of open spaces (ha)	Population of refuge ('000)
1. Shirahige			
Shirahige-higashi	37.9	10.3	80
Shirahige-nishi	57.9	11.6	120
2. Kameido-oujima-			
komatsugawa	114.1	22.6	200
3. Kiba	75.8	24.2	150
4. Yotsugi[a]	88.1	44.1	230
5. Ryougoku	47.5	5.3	53
6. Chuou			
Chuou-sarue	53.0	17.4	110
Chuou-sumida	94.5	18.1	139
Total	568.8	153.6	1,082

Source: Tokyo Metropolitan Government.
a. Yotsugi district includes 44 ha of riverbed.

high school (about 65 ha). Disaster-resistant living zones shall be surrounded by firebreaks composed of main roads, parks, railways, and rivers. Along these firebreaks, construction of fireproof buildings is promoted. So long as fires do not break out within these zones, the residents are presumed to be completely safe; they will not suffer from fires in neighbouring blocks.

Within the Tokyo Ward Area disaster-resistant living zones, a series of additional improvements will be made. These include: widening of passageways; improvement of areas crowded with wooden rental apartment houses; development of parks, shopping malls, and community bases for disaster-prevention activities; preservation of residential amenities; and promotion of voluntary community activities for disaster prevention. The Tokyo Metropolitan Government is now experimenting with promotion schemes for disaster-resistant living zones in three areas of the city. However, even under the present plan, many of the outer western neighbourhoods of the Ward Area will not be designated as disaster-proof living zones.

Countermeasures against other aspects of disaster

Traditional problems of disaster such as building collapses and fires have now been joined by new ones that are a consequence of recent changes in urban infrastructure and associated technologies. In particular, the spread of high-rise buildings and the multiplication of supporting lifelines (i.e. electricity, city gas, telecommunications, etc.) have raised problems of complexity and interdependence. This affects the choice of disaster countermeasures in various ways. For example, gas systems require at least four types of hazard adjustment: (1) earthquake shock sensors that are attached to household gas meters and shut off supplies when strong ground motion is detected; (2) earthquake-resistant gas pipelines to generation plants; (3) shut-off valves on large gas holding tanks at these plants; and (4) technologies that permit some parts of the entire gas distribution network to be isolated from the rest when disruptions affect the system. Similar principles of hazard-resistant design and redundancy (e.g. loops, bypasses, duplicate facilities, multiple distribution channels, emergency standby generators; portable satellite communications) are appropriate for infrastructure systems that convey electricity, water, voice communications, and electronic data, among others. Bridges, tunnels, railways, and harbour facilities pose related network-failure problems.

Although many of the newer buildings of Tokyo are in compliance with hazard-resistance standards, much of the city is occupied by old buildings that do not satisfy these provisions. In addition, a variety of problems are not adequately covered by any building controls. These include standards for glass windows and cladding on exterior walls. Another problem that is peculiarly modern is the presence of large numbers of day-time shoppers in the centre of Tokyo. Many of the city's 2 million daily commuters are concentrated around stores in the vicinity of railroad stations. These people are unfamiliar with the local area and far removed from relatives, friends, and known environments, so they will require special handling in the event of a disaster. The elderly and foreigners already pose special problems, particularly with respect to providing information and assistance about appropriate evacuation routes and shelters. As Tokyo's role as a global city strengthens and its population ages, these will become increasingly serious issues. All of these problems underscore an increasing need for improved public education about disasters.

Long-range planning

Since the end of the Second World War, two long-range plans for the Tokyo Metropolis have been completed and a third is now in progress. Unlike its predecessors, which were both oriented to encouraging economic development, the 1990 plan – known as "My Town Tokyo" – is a direct attempt to redress some of the problems that have arisen as a result of rapid development.

According to the 1990 plan, Tokyo in the future should be organized to achieve the following goals:
• an expanding global role,
• a balanced polycentric layout,

- the safety and comfort of residents,
- high levels of environmental quality,
- a healthy industrial base,
- lives with a feeling of affluence,
- a lively ageing society,
- a city that is open to the world.

Clearly, improved disaster management is called for if the third of these goals is to be realized. Accordingly, the plan calls for additional emphasis to be placed on: (a) disaster-proof construction and fireproof construction; (b) measures to protect lifelines; (c) training of disaster-prevention volunteers; (d) measures to prevent earthquake disasters; and (e) development of a city-wide drainage system that can handle rainfall at rates up to 75 mm per hour. All of these efforts will take place in a city that is more international in outlook and function, more dependent on information acquisition and transfer, and occupied by an ageing population. These characteristics will have important – if not entirely foreseeable – effects on the formulation of future disaster-mitigation programmes.

Summary outlook

As in the past, the emphasis of Tokyo's future disaster policies and programmes will be on technologically based hazard-prevention and hazard-management projects, but additional attention will probably be given to integrating disaster education, voluntary private mitigation initiatives, disaster drills, and other non-structural measures into non-structural strategies.

Earthquakes are the primary concern, and long-term plans call for various improved responses. Most of these are incremental changes over existing practice, but there are significant new initiatives that involve region-wide coordination of energy, transportation, and communications services, and special provisions for vulnerable subgroups such as the aged, the handicapped, and (non-Japanese-speaking) foreigners. Planning is still largely oriented toward an offshore-centred earthquake, whereas a local onshore earthquake in the South Kanto region may be more imminent. It is urgently necessary for the Tokyo Metropolis to assess possible losses and to strengthen the emergency response system against such a local earthquake.

Future adjustments to flooding are geared to the characteristics of different areas. For lowland and waterfront districts, the preferred methods will rely on: strengthened river banks; deeper, wider, and straighter river channels; and sea walls. In the uplands, the emphasis will be on comprehensive flood-control plans for specific river basins. These would include land-use controls as well as rainfall retention and absorption measures on building sites.

Overall, integration and coordination are watchwords of the envisaged disaster-proof Tokyo Metropolis. Soft technologies will be integrated with hard ones and the various organizations that are involved in emergency response activities will be closely articulated. Finally, even though the city of Tokyo intends to pursue an aggressive programme of countermeasures against natural disasters, it must also be prepared to absorb some losses. The capacity for rapid recovery and restoration still needs to be maintained and the general quality of public facilities in many areas of the city requires upgrading – especially during future post-disaster periods.

Notes

1. The Metropolis includes 23 wards, 26 cities, 7 towns, and 8 villages in a territory of 2,156 km^2. Wards were the basic unit of Tokyo's municipal government, but much of the urbanized part of the metropolis now lies outside the "23 Wards Area" of old Tokyo, (593 km^2). This complicates the interpretation of urban data, which are often presented separately for wards and other areas.
2. On the Japanese Meteorological Agency scale, 5 is roughly equivalent to VIII on the Modified Mercalli scale of earthquake intensity.
3. For details of recent Japanese environmentally oriented protest movements, see Li (1992), Hasegawa (1991), Huddle et al. (1987), Kawamura (1994), and Maruyama (1996).
4. Curiously, the Great Kanto earthquake of 1923 did not spur many changes in national legislation, although it did stimulate governmental investigations that gave a major boost to earthquake research (Imamura, 1932).

BIBLIOGRAPHY

Alden, J. 1984. "Metropolitan planning in Japan." *Town Planning Review* 55(1): 55–74.
Busch, Noel F. 1962. *Two minutes to noon: The story of the great Tokyo earthquake and fire*. New York: Simon & Schuster.
Cybriwsky, Roman. 1993. "Tokyo." *Cities* 10(1): 2–11.
Dolan, R. and H. G. Goodel. 1986. "Sinking cities." *American Scientist* 74(1): 38–47.
Douglass, Mike. 1993. "The 'New Tokyo' story." In Kuniko Fujita and Richard Child Hill (eds.), *Japanese cities in the world economy*. Philadelphia: Temple University Press, pp. 83–119.

Fujita, K. 1991. "A world city and flexible specialization: Restructuring of the Tokyo metropolis." *International Journal of Urban and Regional Research* 15(2): 269–284.

Fujita, K. and C. Hill, eds. 1993. *Japanese cities in the world economy.* Philadelphia, PA: Temple University Press.

Geographical Review of Japan. 1990. Special issue on Tokyo. Series B, 63(1).

Harris, Chauncy D. 1982. "The urban and industrial transformation of Japan." *Geographical Review* 72(1): 50–89.

Hasegawa, Koichi. 1991. "Han Genshiryoku Undo ni okeru Jyosei no Ichi: Post-Chernobyl no Atarashii Syaki Undo" [Women of the anti-nuclear energy movement: A post-Chernobyl new social movement in Japan]. *Leviathan* (Tokyo: Bobutakusha) 8: 41–58.

Hewitt, Kenneth. 1983. "Place annihilation: Area bombing and the fate of urban places." *Annals of the Association of American Geographers* 73(2): 257–284.

Huddle, Norrie, Michael Reich, and Nahum Stiskin. 1987. *Island of dreams: Environmental crisis in Japan.* Cambridge: Schenkman Books.

Imamura, A. 1932. "The present state of seismological study in Japan." *Transactions of the Asiatic Society of Japan* (Second series) IX, December: 155–163.

Katsuhiko, Ishibashi. 1987. "Countdown to the next Tokyo earthquake." *Japan Echo* 14(4): 75–84.

Kawamura, Nozomu. 1994. "The citizens' movement against environmental destruction in Japan." In *Sociology and society of Japan.* New York: Kegan Paul International, pp. 122–144.

Lewis, Michael. 1989. "How a Tokyo earthquake could devastate Wall Street and the world economy." *Manhattan, inc.*, June: 69–79.

Li, Jun, ed. 1992. *Industrial pollution in Japan.* Tokyo: United Nations University Press.

Maruyama, Sadami. 1996. "Responses to Minamata Disease." In James K. Mitchell (ed.), *The long road back: Recovery from industrial disaster surprises.* Tokyo: United Nations University Press, pp. 41–59.

Matsuda, Iware. 1980. "Land subsidence as dynamic environment in the Tokyo lowland." *Records of Symposia on the cartography of the environment and of its dynamics*, 1–36.

——— 1990. "Natural disasters and countermeasures for the Tokyo lowland." *Geographical Review of Japan*, Series B, 63(1): 108–119.

——— 1993. "Earthquake vulnerability assessment and damage prediction for the Tokyo Metropolis." *International Congress on Geomorphological hazards in the Asia-Pacific region: Abstracts, September 6–10.* Tokyo, pp. 77–80.

Nakano, T., with H. Kadomura, T. Mizutani, M. Okuda, and T. Sekiguchi. 1974. "Natural hazards: Report from Japan." In Gilbert F. White (ed.), *Natural hazards: Local, national, global.* New York: Oxford University Press, pp. 231–242.

Nakano, Takamasa and Iware Matsuda. 1984. "Earthquake damage, damage prediction and countermeasures in Tokyo, Japan." *Ekistics* 51(308), no. 413.

Nakano, Takamasa, Toshio Mochizuki, Iware Matsuda, and Itsuki Nakabayashi. 1988. "Basic studies on earthquake disaster prevention for Tokyo district." In *Planning for crisis relief: Towards comprehensive resource management and planning for natural disaster prevention. Vol. 3: Planning and management for the prevention and mitigation from natural disasters in Metropolis.* Tokyo: United Nations Centre for Regional Development, pp. 99–125.

Nakazawa, Morimasa. 1988. "Prevention of urban disasters in Japan." In *Planning for crisis relief: Towards comprehensive resource management and planning for natural disaster prevention. Vol. 4: Planning and management for the prevention and mitigation from natural disasters – Japanese experience.* Tokyo: United Nations Centre for Regional Development, pp. 203–224.

Prime Minister's Office. 1990. *Earthquake disaster countermeasures in Japan.* National Land Agency, Government of Japan.

Sassen, Saskia. 1991. *The global city: New York, London, Tokyo.* Princeton, NJ: Princeton University Press.

Scawthorn, Charles, Christopher Arnold, and Roger E. Scholl. 1985. *Proceedings of the U.S.–Japan Workshop on Urban Earthquake Hazards Reduction.* Publication No. 85-03. Berkeley, CA: Earthquake Engineering Research Institute.

Seidensticker, Edward. 1983. *Low City, High City: Tokyo from Edo to the earthquake.* New York: Alfred A. Knopf.

———— 1990. *Tokyo rising: The city since the great earthquake.* New York: Alfred A. Knopf.

Shinomoto, Katsumi. 1988. "The planning for restructuring Tokyo Metropolis into a disaster-proof megalopolis." In *Planning for crisis relief: Towards comprehensive resource management and planning for natural disaster prevention. Vol. 4: Planning and management for the prevention and mitigation from natural disasters – Japanese experience.* Tokyo: United Nations Centre for Regional Development, pp. 225–246.

Tamura, Toshikazu. 1993. "Changing features of hillslope hazards in urban areas of Japan." *International Congress on Geomorphological hazards in the Asia-Pacific region: Abstracts, September 6–10.* Tokyo, pp. 85–88.

Tanaka, Yoshirou. 1979. *The city with no protection against earthquake disaster* (Japanese edition). Tokyo Newspaper Co.

Terwindt, J. H. J. 1983. "Prediction of earthquake damage in the Tokyo Bay." *GeoJournal* 7(3): 215–227.

Tokyo Fire Department. 1993. *Present situation of earthquake disaster countermeasures* (Japanese edition).

Tokyo Metropolitan Government. 1983. *City planning of Tokyo.*

———— 1989. *Tokyo: Yesterday, today and tomorrow.* Bureau of Citizens and Cultural Affairs.

———— 1991. *White Paper on city of Tokyo '91* (Japanese edition). City Planning Bureau.

———— 1992. *Tokyo Metropolitan Regional Disaster Prevention Plan/ Earthquake Disaster Volume* [revised in 1992] (Japanese edition). Disaster Prevention Council.

———— 1992. *The third long-term plan for the Tokyo metropolis.*

———— 1992. *One hundred years of city planning in Tokyo* (Japanese edition). City Planning Bureau.

———— 1992. *Planning of Tokyo.* City Planning Bureau.

———— 1993. *The Fifth Tokyo Metropolitan Earthquake Disaster Prevention Plan* (Japanese edition).

———— 1993. *Tokyo: The making of a metropolis.* Bureau of Citizens and Cultural Affairs.

———— 1994. *White Paper on city of Tokyo '94* (Japanese edition). City Planning Bureau.

Tokyo Metropolitan University. 1988. *TOKYO: Urban growth and planning 1868–1988.* Center for Urban Planning.

Williams, Jack F. 1993. "Cities of East Asia." In Stanley D. Brunn and Jack F. Williams, *Cities of the world: World regional development,* 2nd edn. New York: HarperCollins, pp. 431–477.

Internet sites

⟨http://www.keidanren.or.jp/english/policy/pol040.html⟩ Discussion of the recent Japanese Law on the relocation of Tokyo in the face of earthquake risks, Japanese Federation of Economic Organizations (keidanren), October 1997.

⟨http://www.infoweb.or.jp/dkb/infoecon/9605/9605-1-3.html⟩ Discussion of Tokyo relocation sponsored by the Dai-Ibchi Kangyo Bank, October 1997.

⟨http://www.tfd.metro.tokyo.jp/eng/earth.html⟩ Earthquake preparedness page of the Tokyo Fire Department.

⟨http://www.eas.slu.edu/earthquake_Center/⟩ Historic earthquake photographs of 1923 earthquake postcards. 85 views with captions. St. Louis University.

⟨http://www.iac.co.jp/ellis/quake.html⟩ "The Tokyo office market after Kobe earthquake." Information provided by Richard Ellis, International Property Consultants, Tokyo.

4

Flood hazard in Seoul: A preliminary assessment

Kwi-Gon Kim

Editor's introduction

Seoul is sometimes known as the Phoenix City because of its ability to recover from disaster (Williams, 1993, p. 450). Such disasters have usually been the result of military actions, most recently the Korean War, which completely destroyed the city. Today's rebuilt Seoul lies only 20 miles south of the demilitarized zone with the People's Republic of Korea; the city exists in an uncertain political relationship with its flood- and famine-stricken northern neighbour. There is also uncertainty about the country's ability to weather recent economic shocks that are linked with a broader crisis of East Asian economies. Although blizzards, winter freezes, and earthquakes are significant natural risks, and the city's best-known environmental hazard may be its severe air pollution, this chapter focuses on the relatively neglected – but worsening – hazard of riverine flooding.

Flooding is an increasing problem in metropolitan Seoul and the management of floods and floodplains is a neglected priority for urban planning. The causes of flooding are well known but the effects are poorly documented. Typically, floods are triggered by heavy rain, rapidly melting snow, or tropical storms, especially under conditions where soils are already saturated. Rivers overflow into surrounding built-up areas, bringing death and injury to people as well as considerable damage to buildings. During the past four decades, particularly serious floods

Table 4.1 Flood damage in Seoul, 1960–1991

Year	Population ('000)	Total precipitation (mm)	Monthly peak precipitation (mm)	WLHRB[a] (metres)	Damage (million won)	Buildings lost
1960	2,445	1,188	313	6.84	3	216
1961	2,577	1,437	361	6.25	57	687
1962	2,983	986	139	7.04	8	42
1963	3,255	1,627	514	8.30	30	2,380
1964	3,424	1,794	504	8.27	64	7,654
1965	3,471	1,216	632	10.80	442	10,770
1966	3,793	2,020	898	10.78	605	23,000
1967	3,969	1,269	284	6.60	16	1,170
1968	4,335	1,288	412	7.75	33	1,584
1969	4,777	1,737	455	9.71	67	2,196
1970	5,525	1,708	426	9.02	33	8,557
1971	5,851	1,360	530	6.75	120	1,728
1972	6,076	1,770	882	11.24	1,980	41,200
1973	6,290	928	191	4.97	15	134
1974	6,542	1,251	319	7.22	11	44
1975	6,890	1,067	385	7.45	3	198
1976	7,255	1,110	462	8.41	52	497
1977	7,526	1,148	405	4.28	713	8,158
1978	7,823	1,161	375	8.86	63	732
1979	8,114	1,269	354	7.98	57	218
1980	8,367	1,242	332	7.72	24	3
1981	8,676	1,216	464	8.12	188	122
1982	8,916	949	256	4.99	15	134
1983	9,204	1,057	399	4.23	188	121
1984	9,501	1,250	348	11.03	20,274	34,905
1985	9,639	1,545	439	4.58	173	218
1986	9,799	1,247	370	5.05	287	14
1987	9,991	1,751	651	7.00	15,595	17,603
1988	10,287	761	382	5.46	16	6
1989	10,577	1,426	344	3.98	119	17
1990	10,613	2,356	570	11.27	22,528	21,599
1991	10,905	1,158	488	6.74	993	614
Total					64,771[b]	186,521
Mean		1,353	436	7.46		

Source: Annual records, Seoul Municipal Government.
a. WLHRB = Water level at Han River Bridge.
b. Approximately US$80 million (1995).

occurred in 1966, 1972, 1977, 1984, 1987, and 1990. Between 1960 and 1991 more than 130 people died, 975 were injured, and over 185,000 buildings suffered flood damage at a total cost of around US$80 million (table 4.1). Among the residents of Seoul, flooding is most often attributed to failure of the pumping system that is the city's chief line of defence against inundation. However, other analysts believe that a major reduction in the amount of green open spaces is contributing to increased flood risks. Issues of urban poverty are also mixed in with the flood problem. This is made clear by a newspaper report:

The residents who live in the frequently inundated areas regard flood disasters as man-made disasters rather than natural disasters. Those people, who are low income class people, believe that flood disasters are due to their poverty. (*Dong-a*, 15 September 1990)

This chapter emphasizes recent changes in the relationship between flood risks and urbanization in Seoul. It also assesses the consequences of development plans for human safety and well-being and it explores possible management solutions. An overview of urban problems, processes, and issues in Seoul is followed by what is believed to be the first systematic assessment of Seoul's flood hazard. A simple procedure for identifying flood risks is described. Opportunities for making hazards reduction a priority of urban government and community life are also reviewed.

Urbanization and urban issues in the Seoul Metropolitan Area

It is important to understand how particular factors of geography, history, economics, climate, ecology, culture, and politics have combined to create a distinctive contemporary context for flood hazard in Seoul.

Physical setting

Seoul occupies a location that commands major transportation routes through Korea. It is sited on the banks of the Han River in a mountain-rimmed basin near the edge of the western coastal plain (fig. 4.1). The Han drains much of central Korea into the Yellow Sea. Nearby mountain peaks of up to 836 metres are linked together by scenic ridges, and well-vegetated smaller hills are scattered throughout the built-up area. The ocean port of Inch'on lies approximately 30 km to the west, and intervening areas are served by a network of major highways and interconnecting railroads.

Urbanization

The earliest traces of urban settlement in this vicinity date from 2,000 years ago when the town of Viryesong was chosen as capital of the

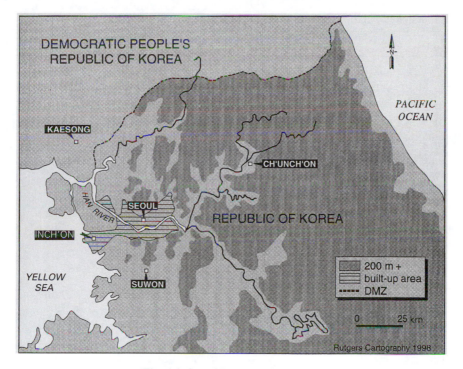

Fig. 4.1 Seoul Metropolitan Area

Paekche Kingdom (18 B.C.–A.D. 660) (Seoul Metropolitan Government, 1992). Following a number of name changes, Seoul became capital of all Korea at the beginning of the Choson Dynasty (1392–1910). The new capital was laid out in 1394 according to principles of geomancy. After 600 years of existence it now houses around 11 million people – more than a quarter of the entire population of the Republic of Korea.

In 1428, approximately one-third of a century after Seoul's founding, 103,328 people lived within the city walls and approximately 110,00 more lived outside them. Thereafter, the urban population remained close to 200,000 for over 400 years. Then, in the late nineteenth century – when the country was opened to foreign contacts – the population of Seoul began to grow. This process continued when Korea was annexed by Japan in 1910, and by 1936 about 700,000 people lived in the city. After the Second World War, municipal boundaries were expanded to include 268.35 km^2, which encompassed about 1,400,000 people.

Seoul was devastated during the Korean War (1950–1953) and had to be extensively rebuilt. By means of a special legislative measure the Seoul Metropolitan Government came under the direct control of the

Table 4.2 Population and area of Seoul, 1961–1991

Year	Population ('000)	Total area (km^2)
1961	2,577	268.35
1962	2,983	268.35
1963	3,255	613.04
1964	3,424	613.04
1965	3,471	613.04
1966	3,793	613.04
1967	3,969	613.04
1968	4,335	613.04
1969	4,777	613.04
1970	5,525	613.04
1971	5,851	613.04
1972	6,076	613.04
1973	6,290	627.06
1974	6,542	627.06
1975	6,890	627.06
1976	7,255	627.06
1977	7,526	627.06
1978	7,823	627.06
1979	8,114	627.06
1980	8,367	627.06
1981	8,676	627.06
1982	8,916	627.08
1983	9,204	627.08
1984	9,501	605.38
1985	9,639	605.43
1986	9,799	605.42
1987	9,991	605.42
1988	10,287	605.40
1989	10,577	605.43
1990	10,613	605.34
1991	10,905	605.33

Source: Seoul Metropolitan Government, *Seoul Statistical Yearbook*, 1992, p. 4.

Prime Minister in 1962 and this enabled it to develop rapidly. The municipal boundaries were further expanded in 1963 to encompass 613.04 km^2 and were redrawn yet again in 1973 to include 627.06 km^2 (table 4.2, fig. 4.2). If present trends continue, it is estimated that the city's population may reach 15 million by the year 2000, at a density of 400 persons per hectare. However, it is also government policy to disperse urban activities to less developed areas. A Capital City Construction Plan has been prepared, which includes incentives and other provisions for moving industries out of Seoul as well as increased taxes on urban in-migrants (Working Group for the Middle-Range Development Plan of Seoul, 1979).

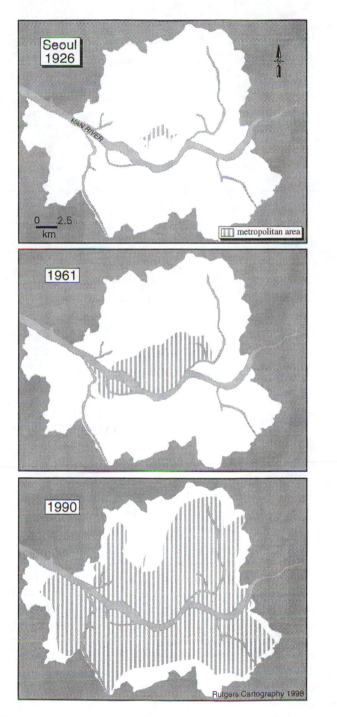

Fig. 4.2 Expansion of the Seoul Metropolitan Area, 1926–1990 (Source: Seoul Metropolitan Government)

Table 4.3 Land-use changes in Seoul City, 1979–1991

| | 1979 | | 1991 | | Change |
	km²	%	km²	%	(km²)
Built-up area	253	42	390	65	+137
Agricultural land	87	14	36	6	−51
Forest land	242	40	153	25	−89
Water	23	4	26	4	−3

Source: Seoul Metropolitan Government.

Table 4.4 Land-use changes in Seoul Region, 1979–1991

| | 1979 | | 1991 | | Change |
	km²	%	km²	%	(km²)
Built-up area	528	11	997	20	+469
Agricultural land	1,543	31	1,342	27	−201
Forest land	2,805	56	2,525	50	−280
Water	151	3	163	3	+12

Source: Seoul Metropolitan Government.

Recent demographic trends have brought about considerable land-use changes in Seoul. Based on an analysis of LANDSAT imagery carried out by the Korea Research Institute for Human Settlements, it is known that built-up areas within the municipal boundaries increased by 54.2 per cent between 1979 and 1991. During the same period, agricultural land declined by 58.6 per cent and land in forests decreased by 36.8 per cent (table 4.3). In the larger Capital Region (within 40 km of Seoul City Hall), built-up areas have increased by 8.9 per cent, while agriculture and forests have decreased by 13.1 per cent and 10.0 per cent, respectively (table 4.4).

During the years leading up to the 1960s, Seoul maintained a single central business district and grew by peripheral expansion along major access highways. Thereafter, the pattern of urban functions began to change. In the 1970s, large-scale city planning projects resulted in rapid development of selected localities such as Kangnam. Following construction of an extensive subway system in the 1980s, urban functions have become more decentralized. Now subway stations are the foci of secondary business districts. In other words, multiple economic nuclei have emerged to challenge the dominance of the downtown business core.

Table 4.5 Centralization of functions in Seoul, 1989

	Country	Seoul	% in Seoul
Area	99,237 km^2	605 km^2	0.6
Population	42,380,000	10,577,000	25.0
Central government organizations	45	31	69.0
Headquarters of large firms	1,711	851	50.0
Loans	62 trillion won	35 trillion won	56.0
Cars	2,660,000	990,000	37.0
Physicians	34,498	14,577	42.0
Colleges	474	161	34.0
Publishing companies	1,654	1,546	94.0
Newspapers (daily and weekly)	195	162	83.0

Source: Seoul Metropolitan Government.

Urban environmental problems

Seoul suffers from a variety of urban environmental problems, including overcrowding, housing shortages, transportation congestion, air pollution, building fires, and flooding. The city is the pre-eminent centre of national life in Korea and a magnet for the rural population. Political, economic, social, and cultural organizations of Korea are heavily concentrated in Seoul. For example, half of the headquarters of large businesses are here and 56 per cent of all financial loans are made for investments in Seoul (table 4.5). The city produces 40 per cent of the country's gross national product, houses 90 per cent of the nation's administrative functions, and hosts 70 per cent of Korea's college students (Koo, 1989). The country's publishing industry and most of its public institutions are also located here. Inner districts of Seoul are now centres of labour-intensive industries while suburban Seoul is occupied by capital-intensive industries. Technical and computer services are also concentrated in the city.

In-migration from rural areas is heavy, demand for housing and public services is acute, and urban population densities are among the highest in the world. Between 1980 and 1990 housing units grew from 1 million to over 1.4 million, while numbers of schools increased from around 1,100 to over 2,600. The population density is around 18,000 people per km^2. Because the city's population is growing so fast (about 4 per cent per

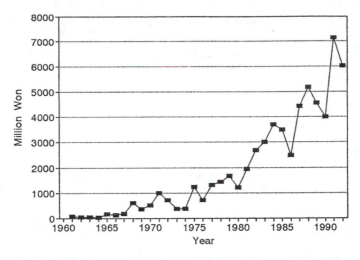

Fig. 4.3 Trends of fire damage in Seoul, 1961–1992 (Source: Seoul Metropolitan Government)

year), there is strong pressure to develop existing open spaces within the metropolitan boundaries and to permit expansion further into the countryside. High-rise construction helps to ease some of the pressure on land and the city is surrounded by an officially designated green belt, which was created in 1971. More than 27 per cent of all land within the boundaries of the Seoul administrative area is included in the green belt. It has been effective in minimizing the spread of non-conforming urban land uses (Kim, 1990).

Housing shortages are severe. In 1988 there were fewer than six dwellings for every ten families (Clifford, 1989), and squatters now make up about 20 per cent of the metropolitan population. Most squatters are poor migrants from rural areas who congregate in the peripheral districts of the city. With demand for downtown land at record levels, housing and land prices there have risen steeply. Closely packed wooden homes are also susceptible to fire, and total fire losses are accelerating. In the years between 1980 and 1990, annual losses rose by about 400 per cent and by a further 50 per cent in the two succeeding years (fig. 4.3).

During the 1980s it was estimated that 400,000 additional housing units would be required in the near future. These will consume about 34 km^2 of land. Some of the land will come from redevelopment of existing residential areas and abandoned factories, but at least 12.6 km^2 of green open space land in the outer parts of the city will also be consumed. By 1989 work was already under way to convert 8.5 km^2 of green areas to housing sites in 17 city districts. Profits from the sale of housing are routinely reinvested in the construction of roads, public utilities, and additional houses (Seoul Metropolitan Government, 1989). Conversion of green spaces inside Seoul's boundaries for new towns-in-town is believed to increase flood risks and lessen human safety.

Crowding produces other serious problems. These include traffic congestion, air pollution, water pollution, and solid waste disposal (Shin, Lee, and Song, 1989; Moon, Lee, and Yoon, 1991; Sohn, Heo, and Kang, 1989; Shin, Koo, and Kim, 1989). More than 40 per cent of Korea's vehicles are concentrated in Seoul and approximately 300 more cars are being added every day (Won, 1990). In contrast, relatively little land is devoted to streets and roads – around 14.5 per cent compared with 20–35 per cent in American and European cities of similar size. Since 1981, road space has risen from 15 per cent to 17 per cent of the city's total area. Traffic accidents have increased from 27,483 in 1981 to 49,413 in 1987 (Lee, 1988; Parsley, 1993).

The Han River provides a major barrier to surface transportation systems and is a main contributor to congestion. Roads are funnelled over nine major as well as eight lesser bridges; and there are three main railroad bridges. Only one of the seven cross-river subway lines uses a tunnel, the remainder being carried on combined road and rail bridges. Because Seoul occupies a mountain basin protected from high winds, the potential for air stagnation is high. Especially in winter there is a high frequency of inversions marked by calm atmospheric conditions and limited vertical mixing. Urbanization and industrialization have combined with a rise in living standards to produce a dramatic increase in energy consumption. As a result, the air over Seoul contains high levels of sulphur dioxide and suspended particulates. Most of these problems are traceable to the use of particular fuels: anthracite briquettes for home heating and cooking; bunker oil for industry and power stations; and gasoline for motor vehicles (UNEP/WHO, 1992).

Flood-risk assessment

Framework for the assessment of flood risk in Seoul

The following conceptual and methodological framework for flood-risk assessment is based on two premises: (1) urbanization and land-use decisions are important determinants of flood risk; and (2) urban planning and decision-making have not effectively incorporated flood risks or made provisions for their reduction. The analytic procedure used here produces estimates of likely future losses by combining floodplain location data from Geographic Information Systems with statistical analysis

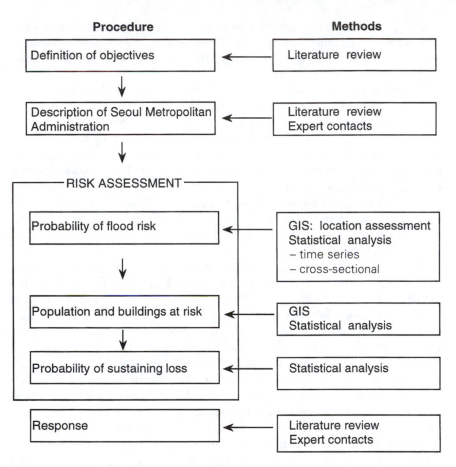

Fig. 4.4 Framework for flood-risk assessment in Seoul

of flood probabilities and populations at risk (fig. 4.4). Public perceptions of floods and appropriate responses are also examined by means of a survey of newspaper articles.

The Han River basin covers 26,200 km², or 26 per cent of Korea's land area, and contains nearly one-third of the country's population (Republic of Korea and Asian Development Bank, 1982). In Seoul, floodplains border the Han and its tributaries (fig. 4.5). Although flood risks are increasing, no systematic analysis of these risks has ever been completed.

Methods

For the purposes of this analysis, exposure to flooding in Seoul is assumed to be a function of two variables: (1) elevation; and (2) distance from a river channel. Generally, in Seoul's reach of the Han River valley, land up to 10.5 m above sea level is at risk of flooding. LANDSAT images were analysed to identify three levels of elevation corresponding to three different probabilities of exposure to floods: less than 10 m above mean sea level; 10–15 m above mean sea level; and more than 15 m above mean sea level. The three elevation land classes are shown in table 4.6 (areas below 15 m appear in fig. 4.5). Areas under 10 m include 31.37 per cent of the total area of Seoul. Areas between 10 and 15 m include 7.2 per cent of Seoul. And areas above 15 m include 61.43 per cent of the city. Thus, about one-third of Seoul is below the threshold elevation for serious flood risk. There are many dwelling units and other buildings in this area. Of the 22 wards (*gu*) of Seoul, Yongdungp'o-gu has the greatest percentage of land under 10 m (84.21 per cent).

The study area was also divided into three zones based on distance from the boundary of the Han River channel: within 500 m; 500–1,000 m (1 km); beyond 1 km. Table 4.7 illustrates these three distance classes. Just under 6 per cent of the area of Seoul is within 500 m of the Han River channel. An additional 5.24 per cent lies between 500 m and 1,000 m, but most of the city is located further away from the river (88.79 per cent). Map'o-gu contains the largest amount of land closest to the river (19.83 per cent within 500 m).

The results of combining elevation and distance data are shown in table 4.8. It is clear that a significant number of city wards are at risk of flooding. These include: Map'o-gu, Kangdong-gu, Songdong-gu, Yongdungp'o-gu, Yongsan-gu, Songp'a-gu, Kangnam-gu, and Dongjak-gu. Only one ward (Chung-gu) is considered to be safe from flooding by virtue of its high elevation and distance from the river. This ward contains much of the central business district, including Seoul City Hall and many government buildings, as well as hotels, foreign embassies, and the headquarters of major firms.

Ideally, it should be possible to identify the numbers of people and buildings that occupy different elevation and distance zones. However, technical limitations of the Geographic Information System that was used in the analysis precluded this possibility. To determine the accuracy of the preceding analysis it was checked against the record of flood damage for three recent years (1984, 1987, 1990) (table 4.9). This shows that there is a high degree of agreement between wards that are assessed as having a high risk of flooding and actual losses.

It was also observed that water rose to a height of more than 11 m at the Han River Bridge on three occasions between 1960 and 1991. This indicates that the potential for flooding in Seoul is large. Flood events with a frequency of once in 400 years could rise to 13.85 m above sea level at this bridge.

Table 4.6 Elevation zones in Seoul wards

Code No.	Ward	Total area Ha	%	Under 10 m Ha	%	10–15 m Ha	%	Over 15 m Ha	%
1	Tobong-gu	4,580.75	100	4.85	0.11	60.71	1.33	4,515.18	98.56
2	Nowon-gu	3,558.98	100	523.00	14.70	445.76	12.52	2,590.22	72.78
3	Map'o-gu	2,450.94	100	1,667.74	68.32	287.47	11.78	485.74	19.90
4	Yangchon-gu	1,757.45	100	905.63	51.44	163.75	9.30	691.07	39.26
5	Kangdong-gu	2,611.61	100	899.06	34.43	273.61	10.48	1,438.93	55.09
6	Songdong-gu	3,365.56	100	2,105.73	62.57	331.96	9.86	927.88	27.57
7	Unpyong-gu	3,152.11	100	292.58	9.28	199.98	6.34	2,659.54	84.38
8	Songbuk-gu	2,458.33	100	120.12	4.89	123.04	5.01	2,215.18	90.10
9	Chongno-gu	2,317.64	100	96.36	4.16	115.84	5.00	2,105.45	90.84
10	Chungnang-gu	1,849.87	100	528.82	28.59	124.47	6.73	1,196.58	64.68
11	Sodaemun-gu	1,925.32	100	100.94	5.24	147.35	7.65	1,677.03	87.11
12	Tongdaemun-gu	1,502.33	100	1,049.27	69.84	174.38	11.61	278.68	18.55
13	Kangso-gu	4,333.93	100	3,126.81	72.14	300.15	6.93	906.98	20.93
14	Chung-gu	875.19	100	–		1.18	0.13	874.01	99.87
15	Yongdungp'o-gu	2,437.54	100	2,052.52	84.21	173.86	7.13	211.15	8.66
16	Yongsan-gu	2,160.43	100	1,180.36	54.63	211.19	9.78	768.89	35.59
17	Songp'a-gu	3,412.39	100	2,020.54	59.21	205.97	6.04	1,185.87	34.75
18	Kangnam-gu	3,962.04	100	935.72	23.62	470.15	11.87	2,556.17	64.51
19	Socho-gu	4,884.51	100	676.41	13.85	173.10	3.54	4,035.00	82.61
20	Dongjak-gu	1,605.11	100	217.08	13.52	84.34	5.25	1,303.70	81.23
21	Kuro-gu	3,478.80	100	1,636.48	47.04	321.80	9.25	1,520.52	43.71
22	Kwanak-gu	3,024.97	100	10.31	0.34	49.16	1.63	2,965.51	98.03
	Total	61,705.84	100	20,150.33	32.66	4,439.22	7.19	37,109.28	60.15

Source: Calculated by the author.

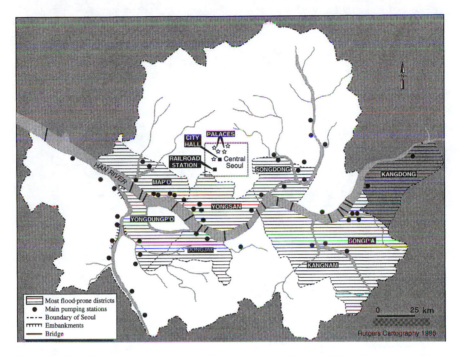

Fig. 4.5 Han River valley in Seoul (Source: Seoul Metropolitan Government and various others)

Statistical risk assessment

Clearly there is a significant potential for serious flooding in Seoul. Floods are likely to become worse as the amount of green space declines and undeveloped land is replaced by impermeable paved surfaces that shed runoff quickly and speed up the arrival of flood waves in populated areas downstream.

To what extent does conversion of green space contribute to increasing future floods? The procedure employed to answer this question is shown in figure 4.6. Several simple regression analyses were performed for Seoul using historical data on green spaces and flood damage. It is important to note that there are insufficient data to produce a reliable statistical analysis of the relationship between green space and flood damage for Seoul at this time. What follows is illustrative of the procedure that might be used with more confidence if better information becomes available. Those who might wish to adopt this method elsewhere must be aware that it requires long-term data, which may be difficult to acquire in places where human and material resources are limited.

Table 4.7 Distance from Han River in Seoul wards

Code no.	Ward	Total area Ha	%	Within 500 m Ha	%	Within 1 km Ha	%	Beyond 1 km Ha	%	Han River Channel itself Ha	%
1	Tobong-gu	4,580.76	100	–		–		4,580.76	100	–	
2	Nowon-gu	3,558.97	100	–		–		3,558.97	100	–	
3	Map'o-gu	2,450.94	100	485.97	19.83	464.36	18.95	1,093.71	44.62	406.90	16.60
4	Yangchon-gu	1,757.45	100	0.60	0.03	71.11	4.05	1,685.75	95.92	–	
5	Kangdong-gu	2,611.61	100	397.44	15.22	323.77	12.40	1,147.63	43.94	742.77	28.44
6	Songdong-gu	3,365.57	100	593.17	17.63	554.08	16.46	1,876.26	55.75	342.06	10.16
7	Unpyong-gu	3,152.10	100	–		–		3,152.10	100		
8	Songbuk-gu	2,458.33	100	–		–		2,458.33	100		
9	Chongno-gu	2,317.64	100	–		–		2,317.64	100		
10	Chungnang-gu	1,849.88	100	–		–		1,849.88	100		
11	Sodaemun-gu	1,925.32	100	–		–		1,925.32	100		
12	Tongdaemun-gu	1,502.33	100	–		–		1,502.33	100		
13	Kangso-gu	4,333.94	100	463.39	10.69	429.84	9.93	3,135.39	72.34	305.31	7.04
14	Chung-gu	875.19	100	–		4.74	0.54	870.45	99.46	–	
15	Yongdungp'o-gu	2,437.53	100	307.19	12.60	255.88	10.50	868.09	35.61	1,006.37	41.29
16	Yongsan-gu	2,160.43	100	392.96	18.19	341.84	15.82	1,068.26	49.45	357.38	16.54
17	Songp'a-gu	3,412.39	100	298.33	8.74	305.89	8.96	2,532.22	74.25	275.94	8.09
18	Kangnam-gu	3,962.04	100	277.63	7.01	268.99	6.79	3,172.17	80.06	243.24	6.14
19	Socho-gu	4,884.51	100	203.91	4.17	192.84	3.95	4,343.50	88.93	144.26	2.95
20	Dongjak-gu	1,605.12	100	263.84	16.44	286.84	17.87	975.35	60.76	79.09	4.93
21	Kuro-gu	3,478.80	100	–		–		3,478.80	100	–	
22	Kwanak-gu	3,024.97	100	–		–		3,024.97	100	–	
	Total	61,705.84	100	3,684.43	5.97	3,500.19	5.24	50,617.90	82.46	3,903.32	6.33

Source: Calculated by the author.

Table 4.8 Distribution of flood hazard zones by ward

Code No.	Ward	Areas within 500 m of Han River	Under 10 m above sea level	Flooding experienced (inundated area)
1	Tobong-gu		0	
2	Nowon-gu		0	
3	Map'o-gu	0	0	0
4	Yangchon-gu	0	0	
5	Kangdong-gu	0	0	0
6	Songdong-gu	0	0	0
7	Unpyong-gu		0	0
8	Songbuk-gu		0	
9	Chongno-gu		0	
10	Chungnang-gu		0	0
11	Sodaemun-gu		0	0
12	Tongdaemun-gu		0	0
13	Kansgo-gu	0	0	0
14	Chung-gu			
15	Yongdungp'o-gu	0	0	0
16	Yongsan-gu	0	0	0
17	Songp'a-gu	0	0	0
18	Kangnam-gu	0	0	0
19	Socho-gu	0	0	0
20	Dongjak-gu	0	0	0
21	Kuro-gu		0	0
22	Kwanak-gu		0	

Simple regression analysis

A series of simple regression analyses was carried out for the city of Seoul involving area of green space (open vegetated land) as the independent variable and a series of measures of flood loss as the dependent variables. The results are shown in table 4.10. These calculations reveal that several measures of loss are strongly explained by the amount of green space. Indeed, the presence of green space explains 31 per cent of the total variation in deaths and a comparable amount of the building damage. This finding is consistent with the hypothesis that green space helps reduce the risk of floods.

Regression equations:

$$\text{Log (Death and} \ldots \ldots) = 4.28198 - 0.000125644 \times \text{Green Space}$$

$$(R^2 = .31068. \text{ SIGNIF F} = .1936)$$

$$\text{Log (Building Damage)} = 5.28678 - 0.000161053 \times \text{Green Space}$$

$$(R^2 = .27955. \text{ SIGNIF F} = .0772)$$

Table 4.9 Flood damage in exposed areas, 1984, 1987, and 1990

Damaged area	1984			1987			1990		
	Inundated areas (ha)	Inundated dwellings (dwelling units)	Cause of damage	Inundated areas (ha)	Inundated dwellings (dwelling units)	Cause of damage	Inundated areas (ha)	Inundated dwellings (dwelling units)	Cause of damage
Goduck							132.0	557	No levee, etc.
Chunho							28.0	3,190	Excess water in pumping station
Sungnae-Pungnap	449.0	15,053	No levee				181.0	15,080	"
Chamsil	17.3	204	Excess water in pumping station				44.6	4,185	No levee, etc.
Banpo	10.2	300	"	76.0	600	Excess water in pumping station	2.5	239	Excess water in pumping station
Daebang							17.0	1,322	Rise in the water level of the Han River
Yung-dungp'o	25.8	1,112	"	230.0	6,781	"	62.6	2,618	Excess water in pumping station
Guro	39.3	835	Rise in the level of outside water	180.5	13,135	"	124.0	6,596	Excess water in pumping station, etc.

Station	Area	Volume	Cause	Area	Volume	Cause
Gayang	15.8	933	Rise in the water level of the Han River	2.0	174	Rise in the water level of the Han River
Magock Gonghang	229.0	11,942	Rise in the level of outside water	5.0	157	"
				1.0	61	Inadequate pumping of inside water
Nanji	18.0	1,065	Excess water in pumping station	50.0	950	Rise in the water level of the Han River
Eunpyung	35.4	2,194	"	8.4	510	Rise in the level of outside water
Mangwon	58.7	1,269	Rise in the water level of the Han River	6.0	610	Excess water in pumping station
Bongwon	53.0	4,170	Long-term rainfall and excess water in pumping station	51.0	4,862	"
Yongsan	5.9	232	Excess water in pumping station	13.0	1,005	"
Oksu	1.0	168	Excess water in pumping station	2.0	239	"

Table 4.9 (cont.)

Damaged area	1984			1987			1990		
	Inundated areas (ha)	Inundated dwellings (dwelling units)	Cause of damage	Inundated areas (ha)	Inundated dwellings (dwelling units)	Cause of damage	Inundated areas (ha)	Inundated dwellings (dwelling units)	Cause of damage
Sungdong	288.0	3,290	Excess water in pumping station				160.2	10,232	Excess water in pumping station
Myunmok	96.4	5,516	"	76.1	5,025	Excess water in pumping station	50.9	3,191	"
Jayang	25.4	523	Rise in the level of outside water				4.0		"

Source: Seoul Metropolitan Government.

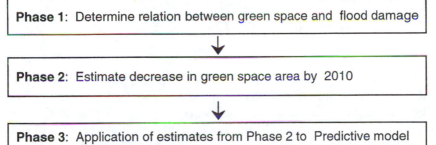

Fig. 4.6 Procedures employed to determine the relation between flood damage and land use

Table 4.10 Results of simple regression analysis of green space and flood losses, 1980–1991

Dependent variable	Constant	Independent variable (green space)	R^2	SIG F
Total Damage	16145.98543 (0.3022)	−0.55955 (0.3608)	.084	.3608
Log (Total Damage)	5.32885 (0.0566)	−0.000130613 (0.000097714)	.15159	.2109
Life Damage	245.0611 (0.4005)	−7.34254 (0.3608)	.04276	.5191
Log (Life Damage)	3.85621 (0.0934)	−0.000100735 (0.2439)	.165	.2439
Death & Life Loss	129.19862 (0.3975)	−0.00401596 (0.5029)	.04608	.5029
Log (Death & Life Loss)	4.28108 (0.1001)	−0.000125644 (0.1936)	.31068	.1936
Public Facilities Damage	174.91123 (0.9090)	−0.000107922 (0.9986)	.0000	.9986
Log (Public Facilities Damage)	5.28678 (0.0291)	−0.000161053 (0.0772)	.27955	.0772
Other Building Damage	5860.98634 (0.1967)	−0.20861 (0.2413)	.13429	.2413
Log (Other Building Damage)	5.03357 (0.1016)	−0.000127812 (0.2672)	.15094	.2672

Source: Calculated by the author.

The amount of green space is known for every year during the period 1980–1991. A regression equation was developed to allow annual estimates of the amount of green space remaining to be made up to the year

2010. It is of the form:

$$\text{Green Space} = 48634.342 - 309.88532 \times \text{year}$$

$$(R^2 = .9417. \text{ SIGNIF F} = .0000)$$

The results of projecting this equation into the future are shown in table 4.11. This provides estimates of the increased flood losses that are likely to occur as green space disappears. A visual display of the relationship between future land-use changes and flood risks can be obtained by superimposing table 4.12 on the land-use plan contained in the General Development Plan of Seoul Towards the Year 2000. At present these maps and tables on flooding should be regarded as inexact guides to the future. They can be refined as more and better data on damage and land use become available.

Responses to flooding

City leaders in Seoul face a dilemma over what to do about flooding. Much of the potentially developable land is exposed to significant risks of flooding, but in a large, densely populated, fast-growing city it is difficult to leave that land in open space uses. Already the costs of rehabilitating flood-damaged property and land are substantial (table 4.12); over a recent eight-year period they exceeded US$50 million. There is no adequate insurance programme for floods, so most (70 per cent) of the costs are borne by individual victims.

At present, the city of Seoul relies on a combination of forecasting and protective engineering works to help reduce flood losses. Flood-risk zoning is not practised and public officials appear to believe that there are no standard planning methods for incorporating this kind of zoning into urban planning strategies. However, this study shows that much can be done toward that end even with the limited information that is now available.

Flood forecasting

The Action Plan for Preventing Flood Damage (1994) recognizes that meteorological information about water levels and discharge volumes should be provided to the public through the mass media (Seoul Municipal Government, 1994, p. 9). However, because it does not take long for excess water that accumulates in upland catchments to create a flood wave in the populated lowlands of the Han valley, it has proven difficult to institute a system for disseminating effective flood warnings.

Table 4.11 The estimated impact of green space on flood risks, 1980–2010

Year	Green space (ha)		No. injured		No. of deaths		No. of claims for damage		No. of claims for building damage		No. of claims for public facility damage	
	Actual	Estimated	Actual	Estimated	Actual	Estimated	Actual	Estimated	Actual	Estimated	Actual	Estimated
1980	31,115	23,843.52	2	28.47	0	19.59	24	163.90	2	27.97	13	96.82
1981	29,547	23,533.63	7	30.59	7	21.10	188	179.91	4	31.38	0	106.07
1982	26,921	23,223.75	1	32.87	1	23.08	15	197.48	3	35.20	9	116.20
1983	26,477	22,913.86	11	35.32	5	25.24	188	216.77	7	39.49	163	127.32
1984	26,177	22,603.98	95	37.95	43	27.61	20,274	237.94	3,084	44.29	3,612	139.48
1985	25,585	22,294.09	14	40.78	3	30.20	173	261.19	20	49.69	92	152.79
1986	24,404	21,984.20	3	43.82	0	33.03	287	286.70	7	55.74	252	167.38
1987	23,725	21,674.32	106	47.08	39	36.13	15,595	314.70	376	62.53	4,734	183.36
1988	22,888	21,364.43	2	50.59	0	39.52	16	345.44	15	70.14	0	200.86
1989	22,504	21,054.55	0	54.36	0	43.22	119	379.19	94	78.68	6	220.04
1990	21,821	20,744.66	76	58.41	44	47.28	22,528	416.22	580	88.27	8,067	241.10
1991	21,692	20,434.78	0	62.76	0	51.71	993	456.88	32	99.01	158	264.09
1992		20,124.89		67.44		56.56		501.51		111.07		289.31
1993		19,815.01		72.47		61.86		550.49		124.60		316.94
1994		19,505.12		77.87		67.67		604.27		139.77		347.20
1995		19,195.24		83.67		74.01		663.29		156.79		380.35
1996		18,885.35		89.91		80.96		728.08		175.89		416.67
1997		18,575.47		96.61		88.55		799.19		197.31		456.46
1998		18,265.58		103.81		96.85		877.26		221.34		500.05
1999		17,955.70		111.54		105.94		962.95		248.29		547.79
2000		17,645.81		119.86		115.88		1,057.01		278.53		600.10
2001		17,335.92		128.79		126.74		1,160.26		312.45		657.42
2002		17,026.04		138.39		138.63		1,273.58		350.50		720.17
2003		16,716.15		148.70		151.63		1,397.99		393.18		788.94
2004		16,406.27		159.78		165.86		1,534.54		441.06		864.28

Table 4.11 (cont.)

Year	Green space (ha) Actual	Estimated	No. injured Actual	Estimated	No. of deaths Actual	Estimated	No. of claims for damage Actual	Estimated	No. of claims for building damage Actual	Estimated	No. of claims for public facility damage Actual	Estimated
2005		16,096.38		171.69		181.41		1,684.43		494.78		946.80
2006		15,786.57		184.48		198.43		1,848.96		555.03		1,037.21
2007		15,476.61		198.23		217.04		2,029.57		622.62		1,136.25
2008		15,166.73		213.00		237.39		2,227.81		698.45		1,244.74
2009		14,856.84		228.88		259.66		2,445.42		783.50		1,363.60
2010		14,546.96		245.94		284.02		2,684.28		878.92		1,493.81

Source: Calculated by the author.

Table 4.12 Costs of rehabilitating flood damage (million won)

	Paid by victims and loans[a]	Payments by government	Totals
1984	15,445	273	15,718
1985	0	5	5
1986	0	5	5
1987	6,663	370	7,033
1988[b]	–	–	–
1989	1	19	20
1990	7,836	11,694	19,530
1991	67	30	97
Totals	30,012 (71%)	12,396 (29%)	42,408 (100%)

Source: Seoul Metropolitan Government.
a. Losses of less than 100,000 won register as 0.
b. Data not available.

Protective engineering works

Pumping stations and river improvements constitute the main counter-measures against floods that are practised in Seoul. According to the Seoul Municipal Government's Action Plan for Preventing Flood Damage (1994, p. 3), the main efforts are directed toward strengthening and maintaining flood disaster prevention facilities such as pumping stations, levees, and reclamation works. Although there are many reservoirs in the surrounding hills, nearly all are designed for water supply purposes – not to assist flood control. Many people in Seoul believe that additional flood-control reservoirs are necessary if flooding is to be curtailed.

Land-use controls

Little attention has been paid to risk zoning as a means of reducing flood damage in Seoul. For example, risk zoning and other hazard land-use controls are not included in the Second Comprehensive National Physical Development Plan (1982–1991) (Government of the Republic of Korea, 1982). This document is intended to ensure orderly improvement and balanced growth of the capital region by affecting the distribution of population and industry. Under the plan's provisions, Seoul is divided into five areas characterized by different physical conditions and different degrees of prospective development: an improvement promotion area; a restricted development area; a development promotion area; a resource conservation area; and an area reserved for development (Korean Planners' Association, 1980, p. 34).

Seoul also possesses a General Development Plan that was prepared by the Ministry of Construction in 1990. It is intended to guide the city's emergence as a fully fledged world city. Under the provisions of this plan, the Seoul of the future will be: (1) an international city with a mature spatial structure that is built around information and intelligence-oriented industries; (2) a city that meets citizens' needs for improved living; (3) a city that offers a diversity of opportunities for citizen participation; (4) a city that discharges metropolitan functions; and (5) a city of upgraded culture, recreation, and social welfare. Although the criteria for commercial, industrial, residential, and open space land uses are spelled out in some detail, natural hazards – including flooding – are not included among them. Likewise, flood-risk zoning or related land-management measures are not mentioned in the local-level Action Plan for Preventing Flood Damage prepared by the Seoul Municipal Government (Seoul Municipal Government, 1994, pp. 1–2).

Unless they are substantially amended, existing development plans will probably place substantially more people and buildings at risk of floods. For example, according to the General Development Plan of Seoul, two of the most frequently inundated wards – Map'o-gu and Kangso-gu – are targeted to receive significantly increased populations. In this document, areas near the Han River are generally designated for low-density residential development. It is clear that the criteria that are used to identify potential development districts in Seoul do not include flood risks. As a result, undeveloped areas of Nowon-gu, Map'o-gu, Kangnam-gu, Kangso-gu, and Songp'a-gu that are at risk of floods are designated as suitable for future development. Nor are other frequently experienced natural hazards, such as wind and landslides, taken into account. Clearly, controls on new buildings and gradual relocation of residents currently at risk are worthwhile measures that should be adopted by city leaders.

Citizen participation might well be employed to good effect in the design of flood-sensitive city development plans. The 1981 Urban Planning Act of Korea requested municipal governments to provide citizens with opportunities for participation by means of compulsory public hearings and public displays of proposals. However, this mechanism still does not operate effectively for a number of reasons. These include: lack of communication between officials and the public; citizen indifference; and opposition by planning professionals. Research on flood-risk perception by local populations could provide important inputs to the participation process and should be encouraged.

Conclusion

The recent rapid growth of Seoul has been accompanied by the development of a serious flood problem in areas adjacent to the River Han. If present trends continue, large numbers of people and buildings will be at risk in the twenty-first century. Thus far, the city has relied almost totally on structural engineering works to provide protection against floods. However, the limits of these measures are already evident, as indicated by increased pumping failures and subsequent floods. Non-structural flood-reduction measures have been neglected in Seoul.

This study is the first to identify relationships between flood damage and land use in Seoul. Its shortcomings are many, but – even in the present preliminary form – it has produced useful information that can be used to guide future land-use plans and development programmes.

REFERENCES

Clifford, Mark. 1989. "Through the roof: South Korea's soaring real estate prices spark protests." *Far Eastern Economic Review*, 8 January, pp. 102–103.

Government of the Republic of Korea. 1982. *Second Comprehensive National Physical Development Plan (1982–1991)*. Seoul.

Kim, K. G. 1990. *Land use changes in the urban fringes: The case of the Seoul Capital Green Belt, Republic of Korea*. UNESCO Research Report, Seoul.

Koo, Chamun. 1989. "New housing construction and the poor: A study of chains of moves in Seoul (Korea)." Ph.D. dissertation, University of Southern California.

Korean Planners' Association. 1980. *The year 2000: Urban growth and perspectives for Seoul*. Seoul.

Lee, Kyu B. 1988. "Urban housing, land development, and urban public enterprise." *Municipal Affairs* 23(10): 31–40.

Moon, H., Y. S. Lee, and T. H. Yoon. 1991. "Seasonal variation of heavy metal contamination of topsoils in the Taejun Industrial complex (II)." *Environmental Technology* 12(5): 413–419.

Parsley, E. 1993. "Walk, don't run: US evacuation plan could be stalled by traffic." *Far Eastern Economic Review* 1 April, p. 24.

Republic of Korea and Asian Development Bank. 1982. *Han River Basin Environmental Study, Final Report*, vol. 1. Seoul.

Seoul Metropolitan Government. 1989 *Seoul Statistical Yearbook*.

———— 1992. *Seoul Metropolitan Administration*. Seoul.

Seoul Municipal Government. 1994. *Action Plan for Preventing Flood Damage*. Seoul.

Shin, E. B., S. K. Lee, and D. W. Song. 1989. "Acidity of rainwater in the Seoul area." In L. Brasser and W. C. Mulder (eds.), *Man and his ecosystem. Pro-*

ceedings of the 8th World Clean Air Congress, The Hague, 11–15 September 1989, vol. 2. New York: Elsevier Science Publishers, pp. 263–267.

Shin, Hang-Sik, Ja Kong Koo, and Jong-Oh Kim. 1989. "Design of economic solid waste management system in Seoul, Korea." *Resource Management and Technology* 17(1): 9–14.

Sohn, D., M. Heo, and C. Kang. 1989. "Particle size distribution of heavy metals in the urban air of Seoul, Korea." In L. Brasser and W. C. Mulder (eds.), *Man and his ecosystem. Proceedings of the 8th World Clean Air Congress*, The Hague, 11–15 September 1989, vol. 3. New York: Elsevier Science Publishers, pp. 633–638.

UNEP/WHO (United Nations Environment Programme and World Health Organization). 1992. *Urban air pollution in mega-cities of the world*. Oxford: Basil Blackwell.

Williams, Jack F. 1993. "Cities of East Asia." In Stanley D. Brunn and Jack F. Williams, *Cities of the world: World regional development*, 2nd edn. New York: HarperCollins, pp. 431–477.

Won, Jaimu. 1990. "Multi-criteria evaluation approaches to urban transportation projects." *Urban Studies* 27 (February): 119–138.

Working Group for the Middle-Range Development Plan of Seoul. 1979. *Comprehensive plan for Seoul: Toward the year 2000*. Seoul.

5

Environmental hazards in Dhaka

Saleemul Huq

Editor's introduction

Since its founding as an independent state in 1971, Bangladesh has been affected by numerous public problems that seem to defy effective solutions.[1] Prominent among these are natural hazards and disasters. Tropical cyclones, riverine floods, riverbank erosion, and tornadoes have inflicted a continuing sequence of heavy losses (Brammer, 1987, 1990a, 1990b; Choudhury, 1984; Chowdhury, 1993; Hossain, 1993; Islam, 1974; Jabbar, 1990; Khalil, 1992, 1993; McDonald, 1991; Matsuda, 1993; Montgomery, 1985; Paul, 1993; Rahman and Bennish, 1993; Zaman, 1993). Scientists are divided about the mix of natural and societal factors that is involved in these disasters, and sharp disputes about appropriate hazard-management policy have sometimes soured relations between the government of Bangladesh and the international aid community (Anon., 1988, 1993; Bingham, 1992; Custers, 1992; Haque, 1993; Ives and Messerli, 1989; Pearce, 1993, 1994; Rasid and Paul, 1987; Sklar, 1993; Tickell, 1993).

Dhaka, the capital of Bangladesh, faces many profound challenges. Not the least of these are a very rapidly growing population and a position amid some of the world's most flood-prone lands.[2] Dhaka is already bursting at the seams trying to house, feed, employ, and care for a vast pool of extremely poor inhabitants. Within two decades it is likely to be counted among the world's largest cities (Miah and Webber, 1990; Meier and Quium, 1991).

119

Flooding is perhaps Dhaka's best-known hazard, though by no means the only one. In recent years most of Dhaka has been inundated by prolonged record-setting floods, some of which were clearly visible on space satellite photographs (Rasid, 1993a). During 1988 approximately 900,000 of the city's residents found shelter in refugee camps during the height of one such flood episode (Christian Science Monitor, *13 September 1988). A strategy for flood control has been developed but its prospects for success are uncertain. Not only are there formidable technical, economic, and political obstacles to be overcome, but the approach is narrowly conceived and not well coordinated with the city's other (limited) efforts to improve environmental quality and raise levels of living. Very few of the public agencies that are responsible for guiding the future development of Dhaka have taken account of environmental hazards and their effects. Non-governmental organizations are attempting to fill some of the gaps in the umbrella of public agency responses, and some noteworthy successes have been achieved. However, it will require a major shift in the commitments of urban and national leaders and a vast increase in resources before the environmental hazards of Dhaka are brought within a range that would be regarded as acceptable in mega-cities of the developed world.*

Urbanization

Bangladesh remains a predominantly rural country, although the level of urbanization is increasing. The urban population is around 20 per cent and growing at a rate of 7 per cent per year. Forty-five per cent of urban Bangladeshis are concentrated in five cities, with the largest number (25 per cent) in Dhaka. It is estimated that in 1991 over 6 million people lived within the 400 km^2 of the Dhaka Statistical Metropolitan Area. If present trends continue, by the year 2000 this region's population will be around 8 million, rising to over 12 million by 2025 (table 5.1). The city's population growth is fuelled primarily by migration from rural areas and many of the migrants are located in 1,125 slums and squatter settlements. It is estimated that about 1 million people now live in such places, often at densities of more than 650 people per acre. These are the people most vulnerable to floods, epidemics, crime, and other environmental hazards.

Dhaka is sited on the north bank of the Buriganga River, approximately 160 km upstream from where the Ganges river system empties into the Bay of Bengal (fig. 5.1). Historically the city's fortunes have waxed and waned in response to geopolitical events that often originated far outside its boundaries (Ahmed, 1986). It was established as the capital

Table 5.1 Population, area, and density in Dhaka City, 1951–2025

Year	Population	Area (km^2)	Density (inh./km^2)
1951	335,926	72.52	4,632
1961	550,143	90.65	6,069
1974	1,607,495	323.75	4,965
1981	3,440,147	401.45	8,569
1985	4,200,000	401.45	10,462
1989	5,500,000	401.45	13,700
1991	6,100,000	401.45	15,195
2001[a]	8,400,000	401.45	20,924
2025[a]	12,500,000	401.45	31,137

Sources: Based on various Bangladesh statistical sources.
a. Projected.

of Bengal in A.D. 1610 by Governor Islam Khan during the period when Mughal kings ruled the Indian subcontinent. Later, when Britain took control in the eighteenth century, the state capital was shifted to Calcutta and Dhaka's pre-eminence faded. Thereafter Dhaka enjoyed a brief resurgence when Bengal was partitioned in 1905 and it again became a capital – this time of East Bengal. For a few years the city expanded and there were new investments in infrastructure. However, when the two parts of Bengal were recombined (1911), this promising development was cut short. Following the partition of India in 1947, Dhaka again became a capital city (of East Pakistan) and there was a further major spurt in urban growth. An even larger increase took place in 1971, when the independent country of Bangladesh was created with Dhaka as national capital (table 5.1). At that time there were between 1 and 2 million people living in the city.

The locations of residential, commercial, and industrial land uses have changed in step with the city's expansion. In pre-Mughal times, Dhaka was a small market town that possessed 52 bazaars, each of which provided a focus for its own residential neighbourhood. When the Mughals came to power, Chawkbazar emerged as the central business district and high-class residential areas began to expand along the bank of the Buriganga. Later, during the British period, Chawkbazar gradually surrendered its retail functions and by 1930 it had become a wholesale trade centre. At the same time, the riverfront also lost its desirable residential character and became both a low-class residential area and the city's main commercial centre. In the early part of the twentieth century, large and small industrial areas emerged at Hatkhola (glass), Shadhana Aush-

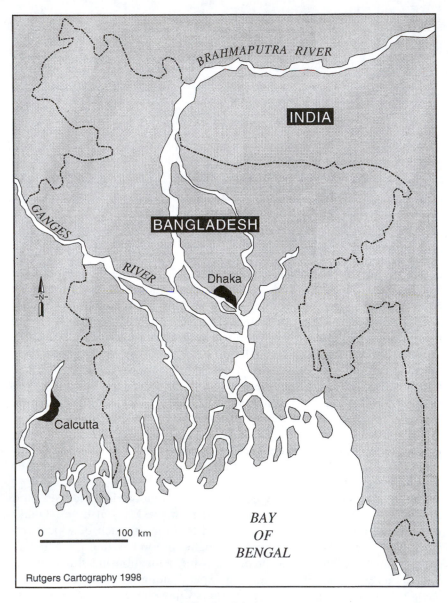

Fig. 5.1 Dhaka

adhalaya (pharmaceuticals), and elsewhere (shell cutting, brass, precious metals, and weaving).

After the creation of Pakistan in 1947, Dhaka's land-use pattern underwent rapid change. New high-class residential areas were created to the north of the city at Dhanmondi, Lalmatia, Gulshan, and Banani. Former high-class areas (e.g. Gandaria, Purana Paltan, Wari, and Shengunbagicha) were taken over by middle-class residents. The central business district expanded and new industrial areas were added at Postagola and Demra. By the time Bangladesh became independent, the old part of Dhaka was increasingly congested and unable to expand. New planned and unplanned developments spread north to the vicinity of the international airport and in other directions where high ground (above flood level) was available. New residential and commercial areas were also built on earth-filled lower-lying sites. Lower-class areas (bustees) are now found along the railroad line between Gandaria and Mahakhali and in a variety of other places. Although much of the industrial development is concentrated in formal industrial zones, the garment industry occupies widely scattered premises throughout the city.

Future growth of Dhaka

The dominant direction of expansion for Dhaka has long been towards the north over a series of old (Pleistocene) river terraces. It is widely believed that future growth will continue this trend, but there is little agreement about the details. The Dhaka Metropolitan Area Integrated Development Project has identified three possible growth scenarios. The first of these assumes the existence of a comprehensive flood protection plan for areas that are at risk of inundation from five nearby rivers. If constructed, this might permit expansion of settlement into areas that lie immediately east and north of the present city. The second scenario emphasizes the development of a number of locations that are now on the periphery of the present built-up zone. The third scenario is an extension of the second and assumes that additional land will be developed north of the city. One estimate of the city's layout in the year 2025 is portrayed in figure 5.2.

In-migration from rural areas accounts for the lion's share (74 per cent) of Dhaka's recent growth. Natural increase (18 per cent) and annexation of surrounding territory (8 per cent) have had less effect on the size of the city's population. Natural hazards are one of four factors that explain Dhaka's magnetic attraction for rural dwellers (Choguill, 1987). Floods, cyclones, and shifting river beds displace huge rural populations and frequently are stimuli for beginning the trek to the city. In addition, Dhaka is widely perceived as a place that will provide superior economic opportunities and access to cultural enrichment. A long-standing neglect of small towns by public policy makers also contributes to the pressure on Dhaka. Finally, the agrarian structure of Bangladesh – and its larger rural economy – are still adjusted to overseas colonial era markets. For

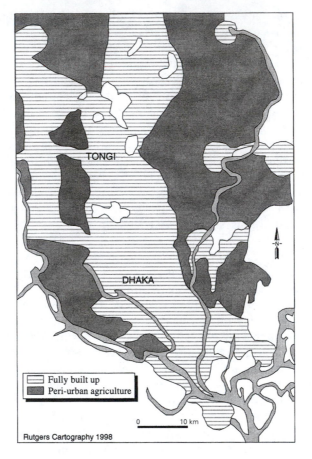

Fig. 5.2 Dhaka in 2025 (Source: Dhaka Metropolitan Government, Dhaka Metropolitan Area Integrated Development Project)

example, despite the poor international demand for jute, a disproportionate amount of the country's resources go into jute production for export. Only limited effort goes into building up Bangladesh's ability to feed itself and to supply other domestic demands.

Dhaka is affected by pressures of the contemporary global economic system. The bulk of present exports consist of manufactured clothing, which is subject to quota restrictions in the United States and Europe. As a result, many industrial entrepreneurs are shifting production into other fields, including shoes, toys, pharmaceuticals, and petrochemicals. Electronic circuits and computer software are also being produced. Demand for new factories, workshops, backyard manufacturing units, and make-shift offices is blossoming, thereby pushing people out of houses in the old part of Dhaka and into new residential suburbs and peri-urban slums. Rich agricultural land in the Savar and Tongi corridors is likely to be converted to urban uses in the near future. The remaining vacant land within the Dhaka Metropolitan Area will probably become filled by industrial premises, houses, offices, hotels, and entertainment premises.

Land ownership and income are directly correlated in Dhaka. Upper- and upper-middle-income households constitute 30 per cent of the city's residents but control over 80 per cent of the land and housing assets (Islam and Khan, 1987). The highest-income population makes up around 2 per cent of the households but occupies 20 per cent of all the residential land. At the opposite end of the socio-economic spectrum, 70 per cent of the city's families are classed as middle- and low-income groups; nearly 30 per cent of them fall below the hard-core poverty line. They live in slums and squatter settlements, often being repeatedly pushed to ever more distant peripheral sites (Shakur and Madden, 1991). These people are marginalized both by market forces and by public inaction (Islam and Khan, 1987).

In summary, metropolitan Dhaka will probably continue to expand by filling the remaining undeveloped areas within the municipal boundaries. More and more people who live on the city's outskirts will begin com-muting to the main centrally located facilities. Since urban services are now stretched to breaking point, the prospect is that the quality of life in Dhaka will only deteriorate unless large capital investments can be attracted to the city.

Hazards and disasters

Historical data on hazards and disasters in Dhaka are almost totally lacking. This makes it difficult to project likely patterns of risks and potential losses in the expanding city of the future. None the less, it is possible to make some pertinent generalizations about future hazards. For the purpose of this analysis these can be divided into two categories of events: routine hazards and surprises.

Routine hazards

Flooding

Flooding is both a routine hazard in Dhaka and an extreme one. For example, in 1995 small-scale to moderate floods occurred in and around

Dhaka during the months of April, July, August, and November. These were mostly a result of strong outflows from the Brahmaputra river system, which caused water in other rivers to back up near the city. Outbreaks of diarrhoea were prevalent in Dhaka during mid-August because floods carried untreated sewage into drinking water. Much routine flooding occurs as a result of drainage backups and storms, but the residents of Dhaka are affected by a number of other routine hazards including: environmental pollution, traffic congestion, and a suite of hazards associated with inadequate housing, slums, and squatter settlements (Rahman, 1993). These are essentially everyday problems that threaten human safety, health, and well-being although, sometimes, their intensity and extent may assume surprising proportions.

Drainage backups are a general and worsening problem throughout the city. Dhaka was built on a series of dissected river terraces above adjacent floodplains. In earlier centuries the city was crossed by major drainage canals (khals) such as the Dholai khal and the Begunbari khal, which carried away runoff. Increased urbanization has resulted in the low-lying areas becoming filled with residential, industrial, and other urban land uses. Drainage is impeded and backups of water are endemic in a large part of the city. Many neighbourhoods are impassable or inaccessible after normal rains. When downpours occur, such as during annual monsoons, the problems are compounded. Widespread and lengthy disruptions of roads, telecommunications, electricity supplies, and water supplies are common. Many low-class and middle-class residences go under water at these times. As the city expands, the pressure on remaining open spaces becomes more intense and these problems grow worse.

Cyclones and other large storms

Cyclones and other large storms affect Dhaka frequently, disrupting the city and causing extensive losses to people and property. Owing to a lack of records it is not possible to give more than notional indications of total storm impacts. It is clear that the residences of low- and middle-income families are particularly susceptible to damage. They most often possess roofs made of metal sheeting or of bamboo matting, which are easily blown off by high winds. Hardship for the occupants is a common result. The Building and Housing Research Directorate and some non-governmental organizations are seeking to develop low-cost housing alternatives that will offer better protection against storms. This is a much-needed project. In general, upper-class residents of Dhaka do not face similar problems because the quality of housing construction is much superior (Syed, 1993).

Waste disposal

Waste disposal is a chronic problem of major proportions in Dhaka that has an impact on many other hazards, including flooding (Hoq and Lechner, 1994; Bhide, 1990). Waste problems are primarily caused by unplanned land-use changes and poor management of sewage and garbage. The rising potential for industrial effluents to affect residential populations is a good example. Originally, most industrial areas were located on the outskirts of the city, but they have now been surrounded by encroaching residential settlements that were not subject to zoning controls or other land-use regulations. The Hazaribagh and Tejgaon industrial areas are particularly vulnerable because there large numbers of tanneries and other effluent-generating industries rub shoulders with many housing developments. Similar processes are also occurring in areas like Demra, Fatullah, Narayanganj, Tongi, Jaydevpur, and Savar.

Environmental pollution

Water pollution and land pollution have already reached alarming proportions. Watercourses are often choked with debris (e.g. animal viscera, vegetable scraps, plastic bags) and they act as breeding grounds for mosquitoes (Hasan and Mulamoottil, 1994). About one-third of Dhaka's households are connected to sewer systems. Another one-third employ septic tanks or pit latrines. The rest either have no sewage disposal facilities or use surface latrines (United Nations Department of International Economic and Social Affairs, 1987). It is not unusual to see overflows of sewage in the streets and drains during the rainy season. Areas of polluted stagnant water have frequently been trapped within the urban area behind the embankments that protect against river flooding. The garbage disposal system is also woefully unable to cope with existing demands. The entire city possesses 186 garbage trucks, and disposal sites are becoming difficult to find in the vicinity of the built-up area (Rahman, 1993).

Air pollution is a growing problem traceable to vehicle emissions and open burning of kerosene, firewood, and other biomass as well as to brick kilns and the incineration of rubber tyres in poor neighbourhoods. Many ornamental trees that had flanked the principal streets were cut down during the past two decades both to provide fuel and to reduce the incidence of bird droppings (Hasan and Mulamoottil, 1994). It is estimated that there are around 160,000 motor vehicles in the city, of which approximately 50,000 are driven by highly polluting two-stroke engines. Aerosol lead is a particular problem because most motor vehicle engines are poorly maintained and use leaded gasoline, which is the only kind

Table 5.2 Vehicular modes on Dhaka's roads

Type of vehicle	% of total
Powered	
Motor car	18.74
Bus	2.13
Truck	1.26
Motor cycle	5.92
Auto-rickshaw	7.00
Non-powered	
Rickshaw	59.38
Bicycle	3.81
Other	1.76

that can be produced by the country's single oil refinery. Recently, airborne lead levels in Dhaka have exceeded those of Bombay and Mexico City (Reuters, 1997).

Traffic congestion

Traffic clogs the roads of Dhaka and the vehicular accident rate is high. The development of new streets and roads more or less kept pace with expanding population until Bangladesh became independent, but a wide gap has since opened up. Nearly 65 per cent of the traffic is composed of non-powered vehicles (i.e. rickshaws, bicycles), which are not physically separated from cars, buses, and other powered vehicles (table 5.2). Traffic jams are common at peak hours and the circulation system is often thrown into disarray by political demonstrations and natural disasters. With road traffic projected to double during the decade from 1990 to 2000, the prospects for improvement are poor. As in many cities of the developing world, noise is a pervasive urban problem that has not been addressed by public authorities. Vehicle exhausts, horns, and loud speakers commonly operate at high decibel levels and the growth of new suburbs close to the airport is raising the salience of aircraft noise as a serious problem.

Fire

Fire is another everyday hazard in Dhaka. In most years the city experiences several major fires that consume hundreds of houses and small businesses at a time. For example, in 1992, three large fires destroyed 640 houses. The following year two fires burned down over 200 houses. A

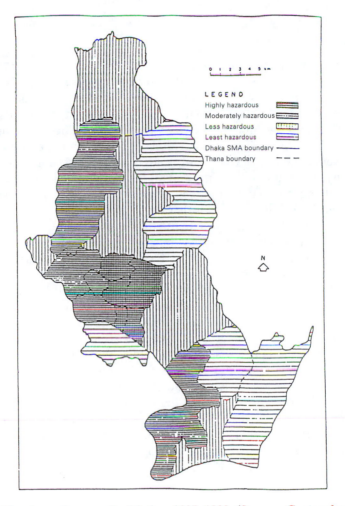

Fig. 5.3 Fire hazard zones in Dhaka: 1987–1988 (Source: Center for Urban Studies, University of Dhaka)

recent study found that there is a direct correlation between population density and fire frequency. Slum communities and the densely developed sections of old Dhaka are most at risk. The big fires of 1992 and 1993 affected slum districts and most of them began in the premises of small plastics and rubber companies. A combination of hazardous materials, congested living and working spaces, and narrow streets is involved in most fires (fig. 5.3).

Surprises

Floods

In some years the monsoon rains are abnormally intense and prolonged, thereby giving rise to surprise floods. This last occurred in 1987 and 1988. In 1987 an estimated once-in-50-year flood inundated the southern two-thirds of Dhaka for several months (fig. 5.4). Almost two-thirds (66 per cent) of all slum dwellings in the city were affected. The following year, a 100-year flood took place and covered 77 per cent of the city (fig. 5.5). Both events disrupted road, rail, and air transportation and severed electricity, telecommunications, and water supplies. As Dhaka expands into low-lying areas that now act as emergency retention basins, it is likely that there will be more of these surprises.

After the big floods of the late 1980s the government of Bangladesh formulated a number of Flood Action Plan studies, one of which focuses specifically on Dhaka. A Dhaka City Flood Control and Drainage Project has begun in haste as part of an overall Flood Action Plan (FAP) that was funded by 15 donor countries and coordinated by the World Bank. The national FAP has occasioned much controversy both inside Bangladesh and in the international development assistance community. It relies on large-scale, sophisticated, and expensive engineering technology – reminiscent of the Dutch system of structural measures for flood control. There is only limited attention to protecting coastal areas that are at even greater risk of cyclone-driven floods than inland areas are of riverine flooding, and the equally serious hazard of drought is largely ignored. Some critics contend that the FAP is intended to divert water away from relatively well-off cities such as Dhaka and into poverty-stricken rural areas that already bear a disproportionate share of flood losses (Robinson, 1993, pp. 170–171; Blaikie et al., 1994, pp. 138–144). It has also been argued that the floods of 1987 and 1988 should be regarded as highly unusual events that do not warrant the launching of a massive "tech-fix" project whose side-effects may be as upsetting as the problems it is intended to resolve.

The Dhaka City Flood Control and Drainage Project is one element of the larger national strategy. It is a huge engineering and drainage scheme that covers a 265 km^2 tract around Dhaka bounded by several rivers. Plans call for the construction of a continuous embankment along the edges of the tract and pumping stations within it to remove any water that accumulates there. The western side of this giant ring-dyke (30 km long) was completed in 1990 and sluice gates were added in 1993/94. A second phase (29 km), on the eastern side of Dhaka, is still incomplete. Part of the project, the Dhaka–Narayanganj–Demra Polder south of Dhaka, surrounds a largely agricultural area on the periphery of the city and is

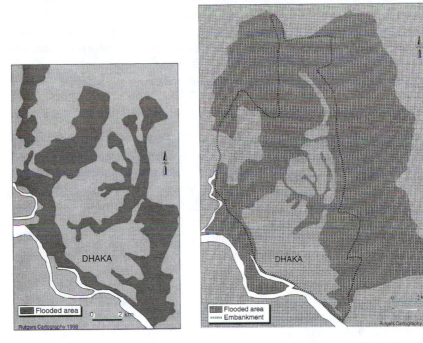

Fig. 5.4 Dhaka maximum flooded area, 1987 (Source: Dhaka Metropolitan Government)

Fig. 5.5 Dhaka maximum flooded area, 1988 (Source: Dhaka Metropolitan Government)

intended to improve flood control, increase agricultural yields, and enhance fishing. However, one of the main effects of this project has been to encourage urban invasion of the floodplain in an area that is not served by municipal services, including water supply and sewers (Rasid, 1993b; Rasid and Mallik, 1993, 1996). In addition, the construction and maintenance costs and the operational requirements of the entire scheme are very heavy for such a poor country.

Cyclones and tornadoes

From time to time, cyclones and tornadoes have affected Dhaka. When this occurs, cyclone damage is usually widespread, whereas tornado damage is confined to small areas. However, the city is acquiring more high-rise structures that may be at risk in the future.

Earthquakes

Although there have been no major earthquakes in recent decades, Dhaka is also known to be susceptible to serious earthquake hazards (Reuters, 1995).

Crime

Finally, it should be noted that the incidence of urban crime has been increasing very rapidly in Dhaka. In an attempt to curb the trend towards violent crime, in 1992 the government of Bangladesh introduced stringent laws directed against illegal ownership of firearms.

Management of hazards and disasters

Environmental planning and management are improving in Dhaka, but the city has a long way to go before it possesses the tools and institutions necessary for effective urban management let alone the capacity to address the human dimensions of environmental hazards. For example, the city's master plan is seriously outdated. It was prepared in 1959 before Bangladesh became independent, when the metropolitan population was around 2 million, yet it remains the only legal document for land-use regulation (United Nations Centre for Human Settlements, 1996, p. 256). Recently, efforts have been made to separate planning and building regulations into two comprehensive codes and to readjust complicated landholding boundaries to permit easier development of fragmented and irregular plots (United Nations Centre for Human Settlements, 1996, pp. 299, 303).

There have been a number of initiatives to reduce the environmental hazards of Dhaka at various times during the past 150 years. For example, in August 1840 a Municipal Committee was established to address problems of sanitation, economic decline, and corruption. The Committee undertook to create embankments along the rivers, to reclaim low-lying flood-prone areas, and to improve navigation by modifying river channels, but its efforts were undercut by legal and financial difficulties. At the same time, many poor residents moved out of Dhaka to settle on adjacent flood-prone territories because the municipal authorities raised city taxes (Ahmed, 1986).

Today, the Dhaka Municipal Corporation has overall responsibility for government services but its authority is limited (e.g. no zoning powers). There is very little coordination among departments and agencies whose missions overlap, such as the Water Development Authority (flood control), the Urban Development Directorate (housing), and the Public Health Engineering Department (mosquito control, sanitation) (Hasan and Mulamoottil, 1994). Most of these units are hard-pressed to meet their obligations and they often are glaringly deficient. Perhaps as a result, non-governmental organizations have begun to assume leadership of some disaster-management functions during recent years.

The Dhaka City Planning Authority (Rajuk) is charged with responsibility for planning the future expansion of the metropolitan area. It has authority to regulate land use, including high-rise buildings and filled land – two of the potential hazard trouble spots of the future. Dhaka Municipal Corporation has responsibility for garbage disposal and street lighting among other matters. Dhaka Water and Sewer Authority is currently unable to meet the city's needs for potable water; projections indicate that the gap between demand and supply is likely to grow wider. This agency is also involved in efforts to improve drainage for already developed parts of Dhaka. However, undeveloped districts are filling up with new settlements at a rate that will likely outpace the effects of these efforts. The Fire Brigade and the Civil Defence Authority play roles in emergencies, but they are all too often unable to intervene effectively.

In recent years, non-governmental organizations have begun to address some of the deficiencies of the formal hazard-management agencies and to offer alternatives. Residents of the Dhaka–Narayanganj–Demra Polder were particularly adept in forming neighbourhood groups to assist with flood-fighting and recovery tasks during the floods of 1987 and 1988.

Conclusions

The hazard dilemmas of Dhaka are multifaceted and interrelated. At their heart is the fact that this is a large, rapidly growing city on a congested site that lacks safe areas for expansion, in a very poor country that has set a high priority on fast economic development. As a result, there are many competing priorities for resources that might be devoted to hazard reduction and a narrow range of feasible hazard-reduction alternatives. The impacts of existing risks are increasing and settlement is encroaching into flood-prone areas, while the vulnerability of poor in-migrants grows and the efficacy of public responses remains low.

Clearly there is a need for more information about the genesis and status of environmental hazards in Dhaka. The social composition of this mega-city makes it likely that there are very substantial numbers of vulnerable people already resident therein but it is also possible that the general conditions of human existence are better in Dhaka than throughout the rural districts outside it. None the less, it would be unfortunate if arguments about the need to improve safety and security for rural Bangladeshis were to obscure the reality that hazards are growing worse in Dhaka and other urban centres. This is likely to happen as the city's population grows and marginal lands are occupied by housing and industry. Whether the hazards will increase at a faster rate than those of rural areas remains to be seen but the trends are not favourable.

As a first step towards improving the prospects for a reduction in hazards in Dhaka, a quantum increase in basic research on the human ecology of hazard is necessary. Baseline information about the distributions of populations at risk, their capabilities to adjust to hazard, and the range of available means for coping with hazards is required. Case-studies of neighbourhoods that are particularly at risk and groups whose vulnerability is increasing are especially warranted. Moreover, a careful analysis of the similarities and differences in hazard-proneness and coping strategies between rural and urban centres in Bangladesh would pay rich dividends.

Acknowledgements

The editor would like to acknowledge the assistance of Sudha Maheshwari and C. Emdad Haque in updating and modifying the original manuscript to incorporate recent information.

Notes

1. Two contrasting images of Bangladesh have been fashioned over the past two decades. One is the view, variously attributed to John Kenneth Galbraith, Henry Kissinger, and Alexis Johnson, that Bangladesh is "an international basket case"; in other words, a state that is unable to care for itself and must rely on the charity of others. The second is an image of "Golden Bengal" – a land of future plenty whose potential has not yet been realized (*New York Times*, 23 June 1985).
2. Dhaka is sited on the southern fringes of an area that is seldom flooded but immediately across the river from lands that are normally below 6 feet of water during the wet season (*National Geographic Magazine* 183(6), 1993, p. 125).

REFERENCES

Ahmed, Sharif Uddin. 1986. *Dacca: A study in urban history and development*. London: Curzon Press.
Anon. 1988. "Bangladesh floods: Drowned by politics." *The Economist*, 17 September, pp. 38–40.
———— 1993. "Floods and droughts could increase in South Asia." *Bulletin of the American Meteorological Society* 74(7): 1400.
Bhide, A. D. 1990. *Solid waste management in Dhaka, Khulna, Natore*. World Health Organization Assignment Report, BAN CWS 001. Geneva.

Bingham, Annette. 1992. "Muddy waters." *Geographical Magazine* 64(8): 21–25.

Blaikie, Piers, Terry Cannon, Ian Davis, and Ben Wisner. 1994. *At risk: Natural hazards, people's vulnerability, and disasters.* London and New York: Routledge.

Brammer, H. 1987. "Drought in Bangladesh: Lessons for planners and administrators." *Disasters* 11(1): 21–29.

—————— 1990a. "Floods in Bangladesh. 1. Geographical background to the 1987 and 1988 floods." *Geographical Journal* 156(1): 12–22.

—————— 1990b. "Floods in Bangladesh. 2. Flood mitigation and environmental aspects." *Geographical Journal* 156(2): 158–165.

Choguill, Charles L. 1987. *New communities for urban squatters: Lessons from the plan that failed in Dhaka, Bangladesh.* New York: Plenum Press.

Choudhury, G. R. 1984. "Management of sedimentation in Bangladesh." *Water International* 9(4): 155–157.

Chowdhury, A. M. R. 1993. "The Bangladesh cyclone of 1991: Why so many people died." *Disasters* 17(4): 291–304.

Custers, Peter. 1992. "Banking on a flood-free future? Flood mismanagement in Bangladesh." *Ecologist* 22(5): 241–247.

Haque, C. Emdad. 1993. "Human responses to riverine hazards in Bangladesh: A proposal for sustainable floodplain development." *World Development* 21(1): 93–107.

Hasan, Samiul and George Mulamoottil. 1994. "Environmental problems of Dhaka city: A study of mismanagement." *Cities* 11(3): 195.

Hoq, M. and H. Lechner. 1994. *Aspects of solid waste management: Bangladesh context.* Dhaka: German Cultural Institute.

Hossain, M. 1993. "Economic effects of riverbank erosion: Some evidence from Bangladesh." *Disasters* 17(1): 25–32.

Islam, M. A. 1974. "Tropical cyclone, coastal Bangladesh." In G. F. White (ed.), *Natural hazards: Local, national, global.* New York: Oxford University Press. pp. 19–24.

Islam, Nazul and Amanut Ullah Khan. 1987. *Regional research on the absorptive capacities of metropolitan areas in Asia.* Dhaka: United Nations Centre for Regional Development.

Ives, Jack D. and B. Messerli. 1989. *The Himalayan dilemma: Reconciling development and conservation.* London: Routledge.

Jabbar, M. A. 1990. "Floods and livestock in Bangladesh." *Disasters* 14(4): 358–365.

Khalil, G. Md. 1992. "Cyclones and storm surges in Bangladesh: Some mitigative measures." *Natural Hazards* 6(1): 11–24.

—————— 1993. "The catastrophic cyclone of April 1991: Its impact on the economy of Bangladesh." *Natural Hazards* 8(3): 263–281.

McDonald, Hamish. 1991. "Learning from disaster." *Far Eastern Economic Review*, 30 May, pp. 23–24.

Matsuda, I. 1993. "Loss of human lives induced by the cyclone of 29–30 April, 1991 in Bangladesh." *GeoJournal* 31(4): 319–325.

Meier, Richard L. and A. S. M. Abdul Quium. 1991. "A sustainable state for urban life in poor societies: Bangladesh." *Futures* 23(2): 128–145.

Miah, Md. Abdul Quader and Karl E. Webber. 1990. "Feasible slum upgrading for Dhaka." *Habitat International* 14(1): 145–160.

Montgomery, R. 1985. "The Bangladesh floods of 1984 in historical context." *Disasters* 9(3): 163–172.

Paul, Bimal Kanti. 1993. "Flood damage to rice crop in Bangladesh." *Geographical Review* 83(2): 150–159.

Pearce, Fred. 1993. "West sinks Bangladesh flood plan." *New Scientist*, 21 August, p. 4.

——— 1994. "Experts condemn Bangladesh flood plan." *New Scientist*, 6 August, p. 8.

Rahman, M. H. 1993. "Waste management in Greater Dhaka City." *International Journal of Environmental Education* 12(2): 129–136.

Rahman, M. Omar and Michael Bennish. 1993. "Health related response to natural disasters: The case of the Bangladesh cyclone of 1991." *Social Science & Medicine* 36(7): 902–914.

Rasid, H. 1993a. "Areal extent of the 1988 flood in Bangladesh: How much did the satellite imagery show?" *Natural Hazards* 8(2): 189–200.

——— 1993b. "Preventing flooding or regulating flood levels? Case studies on perception of flood alleviation in Bangladesh." *Natural Hazards* 8(1): 39–57.

Rasid, H. and Azim Mallik. 1993. "Poldering vs compartmentalization: The choice of flood control techniques in Bangladesh." *Environmental Management* 17(1): 59–71.

——— 1996. "Living on the edge of stagnant water: An assessment of environmental impacts of construction-phase drainage congestion along Dhaka City Flood Control Embankment, Bangladesh." *Environmental Management* 12(1): 89–98.

Rasid, H. and Bimal K. Paul. 1987. "Flood problems in Bangladesh: Is there an indigenous solution?" *Environmental Management* 11(2): 155–173.

Reuters. 1995. *Reinsurance*, 30 April.

——— 1997. News Service report, 6 April.

Robinson, Andrew. 1993. *Earthshock: Hurricanes, volcanoes, earthquakes, tornadoes and other forces of nature.* London: Thames & Hudson.

Shakur, Tasleem and Moss Madden. 1991. "Resettlement camps in Dhaka: A socio-economic profile of squatter settlements in Dhaka." *Habitat International* 15(4): 65–83.

Sklar, Leonard. 1993. "Drowning in aid: The World Bank's Bangladesh Flood Action Plan." *Multinational Monitor* 14(4): 8–13.

Syed, M. H. B. 1993. "Land policies for housing in Dhaka." *International Institute for Housing Science and Its Applications* 17(3): 173.

Tickell, Oliver. 1993. "Avoiding the Mississippi experience." *Geographical Magazine* 65(9): 6–7.

United Nations Centre for Human Settlements (Habitat). 1996. *An urbanizing world: Global report on human settlements 1996.* Oxford: Oxford University Press.

United Nations Department of International Economic and Social Affairs. 1987. *Population growth and policies in mega-cities: Dhaka*. Population Policy Paper No. 8. New York.

Zaman, M. Q. 1993. "Rivers of life: Living with floods in Bangladesh." *Asian Survey* 33(10): 985–996.

6

Natural and anthropogenic hazards in the Sydney sprawl: Is the city sustainable?

John Handmer

Editor's introduction

Sydney is the largest in a chain of well-watered coastal cities that surrounds the dry interior of Australia. Remarkably free of severe natural hazards in previous centuries, it is now beginning to experience increasing problems with wildfires and floods as new, underserviced suburban housing spreads out and pushes inland toward the Blue Mountains. It is also affected by a wide range of worsening technological and social hazards that challenge municipal myths and international stereotypes about the high quality of urban living in this mega-city. As Sydney struggles to find an appropriate formula for sustainable urban development in the twenty-first century, it is beginning to confront issues that have heretofore received too little attention from urban specialists. Chief among these is the role of environmental hazard in the concept of sustainability. Without an adequate accounting of environmental risks and hazards, the search for sustainability runs the risk of pursuing an unobtainable utopia.

Sydney in the early 1960s ... was no more than a harbour surrounded by suburbs – its origins unsavoury, its temper coarse, its organisation slipshod ... [but, there is now] a new vision of Sydney ... resplendent, festive and powerful ... the most hyperbolic, the youngest in heart, the shiniest [of the cities left by the British Empire].

(Morris, 1993, pp. 4–5)

Introduction

Sydney is the largest and most important city in Oceania. With 4 million people and international institutions of politics, commerce, finance, communications, and the arts, it is arguably the best example of a "world" metropolis in the southern hemisphere. The city is being propelled into an increasingly important role in the global economy by several perceived advantages. These include modern information technology and a time zone that encourages synchronous economic relations with Japan and the emerging markets of East Asia (Lamont, 1994). At the national scale, Sydney is Australia's pre-eminent city and the major destination for immigrants. At the metropolitan scale, it is an inexorably expanding low-density sprawl of bungalows (i.e. single-storey ranch-style houses) that cover an area exceeded only by New York, Los Angeles, and London.

Although Sydney is subject to a wide range of environmental threats, it is not especially susceptible to natural hazards. None the less, many Australians conceive of their nation as one that regularly does battle with a harsh natural environment, characterized by devastating floods, bushfires, and drought (Hughes, 1986, pp. 13, 110, 559). This attitude is reflected in general expectations that natural hazards occur – even in the country's major cities – and that society will set aside resources to cope with them. Though less deeply embedded in the country's self-image, technological hazards, pollution hazards, and social hazards are also important attributes of Sydney. In addition, there are substantial and growing lifestyle inequalities among city residents, but they have not yet produced the vast ghetto areas and gross neighbourhood disparities of many other mega-cities, especially those in less developed countries.

At issue in this book is whether the massive urban agglomerations that are fast becoming the favourite habitat of humanity are also becoming more vulnerable to disaster. And, if this is so, with what implications for sustainability (WCED, 1987; Harrison, 1992)? Increases in the number of environmental disasters are often seen as evidence of weak or declining sustainability. But such damaging events may be offset by increasingly sophisticated responses and loss-redistribution mechanisms that operate, at least in richer countries, through global networks of reinsurance and commerce. Which trend dominates in Sydney and what is the connection between hazard management and urban sustainability? In order to answer these questions it is first necessary to clarify some broader issues.

Hazard in the modern mega-city

Are mega-cities necessarily more hazardous?

Major cities seem to be intrinsically hazardous because they combine large concentrations of people with vast – almost unknowable – ranges of risk-laden commercial, industrial, and transportation activities (Alexander, 1993). In addition, many cities have grown with scant regard to aspects of the biophysical environment such as flooding, unstable slopes, and topography that exacerbate air, ground, or water pollution. Uneven effectiveness of institutions for managing cities and urban hazards complicates the picture (Blaikie et al., 1994). For example, in many places poor urban construction practices and inadequate building standards persist in the face of substantial risks because of ignorance or uncertainty about hazards. Elsewhere, sound standards exist on paper but cities lack the institutions necessary for effective administration and enforcement. Alternatively, appropriate legal, bureaucratic, professional, and educational institutions may exist but fail to function. Finally, even the best efforts of hazard managers may be frustrated by the fact that rules and regulations – the cutting tools of programmatic action – usually fail to affect the informal sector of city life. Taken together, these factors are likely to produce a city that is increasingly susceptible to hazards.

Vulnerability in the context of sustainability

Susceptibility to hazard does not guarantee that losses will occur. Affected communities must also be vulnerable. Vulnerability is a complex concept that connotes both susceptibility to immediate loss and the inability to draw on physical, human, psychological, or institutional resources that might be applied to prevent or offset such losses. In the intellectual world of hazards research and the operational world of disaster management, the concept of vulnerability is frequently paired with its opposite – the concept of resilience (i.e. the ability of a stressed system to resist change and return to a pre-disaster state). In the past, city-wide disaster-related changes have usually been few and have rarely amounted to more than "fine-tuning" of the total urban system. However, very substantial changes have occurred in city neighbourhoods and in the lives of individuals, families, and other groups that were affected by such hazards as earthquakes (Yong et al., 1988), toxic chemical releases (Shrivastava, 1992), and dam failures (Erikson, 1976). Vulnerability to disaster is increased by factors such as poverty, war, and the denial of basic human rights (Red Cross/Red Crescent, 1994). These chronic problems preoccupy many urban residents, leaving them unprepared for periodic disasters. In this connection, slow-onset hazards that are almost imperceptible at the outset (e.g. famines, epidemics, environmental degradation) may pose the most difficult problems.

The relationship between urban vulnerability and urban sustainability is complex and little understood. As defined by the Bruntland Commission (WCED, 1987), sustainability is a prescription that rests on two moral principles – intergenerational equity and intragenerational equity – plus the assumption that environmental integrity should be maintained (WCED, 1987). The concept of sustainability is itself subject to much debate, with the result that standard policy-implementation tools and institutions have yet to come to grips with its wide-ranging implications (Dovers, 1995). Perhaps more important, the idea of sustainability poses special problems for disaster managers, whose primary function is to ensure the safety and security of people who are subject to forces that are capable of undermining existing settlements and social arrangements.[1] Resilience in the face of sudden-onset hazards (i.e. quickly getting back to normal after a destabilizing event) may not be a good indicator of a city's ability to achieve sustainability in the long run (Handmer and Dovers, in press). The reverse is more nearly correct; it may be necessary to make fundamental changes in the way a city operates to ensure that it is in a better position to cope with future disruptions.

Different types of support network operate at different scales, so issues of scale are important in determining urban sustainability. For example, some of the resources that are used to assist disaster recovery originate at national and international levels of the private sector (e.g. insurance and government aid); others employ local kinship networks to channel private donations and NGO remittances from equally far-flung sources. When a disaster is highlighted by the mass media, especially the media of developed countries, there can be near-global mobilization of resources. This is especially so where major cities are concerned, because these are generally linked to national power structures through densely patterned systems of elected leaders and non-governmental organizations.

If a wealthy community suffers damage it will usually be well insured and is likely to have adequate surplus resources for reconstruction and recovery. If these means do not suffice, such a place will almost certainly be very well connected to effective political and bureaucratic systems that can deliver demanded assistance. Paradoxically, however, frequent one-way transfers of resources to prop up wealthy urban areas may be indicators of *un*sustainability!

Sydney as a mega-city

Spatial and temporal setting

Modern Australia was founded on 26 January 1788 at "Sydney Cove," with the arrival of the First Fleet. This convoy of 11 ships brought some 1,000 men and women convicts with their goalers on an eight-month voyage from England. The objective was to establish an urban penal colony in the vicinity of Port Jackson. In the eyes of the native people, this was an invasion that dispossessed them of their land; arguably they still constitute Sydney's most underprivileged group.

Port Jackson was chosen for settlement in preference to the flat swampy area of Botany Bay, just to the south, which is now the site of the international airport and a major industrial complex. Key rationales at the time were the need for secure anchorages, good building sites, and a nearby source of fresh water. The port is an extensive, well-sheltered deepwater harbour with a number of arms, separated by prominent and well-vegetated sandstone ridges. Three-quarters of the city's popuation lives within 5 miles of tidewater (Rose, 1993, p. 205). North of the harbour and along the coastal suburbs the sandstone plateaus are steeply scarped. Many of the more rugged slopes have provided fine building sites for the wealthy; they also provide parks and pockets of bushland – with attendant bushfire risks. The coastal area is also well supplied with excellent beaches. Not surprisingly, the beaches, harbour, and national parks within and on the outskirts of the city are a major recreation resource. Together with the harbour and its unmistakable images of bridge and opera house (photo 6.1), the ocean beaches and adjoining expensive suburbs are globally recognized symbols of Sydney that are used to attract international commerce and to market the city.

The climate too has always been a part of Sydney's image. In the words of an officer who accompanied the First Fleet:

The climate is undoubtedly very desirable to live in. In summer the heats are usually moderated by the sea breeze ... and in winter the degree of cold is so slight as to occasion no inconvenience, once or twice we have had hoar frosts and hail, but no appearance of snow... Those dreadful putrid fevers by which new countries are so often ravaged, are unknown to us. (Tench, 1790)

However, Sydney extends well beyond the harbour – some 50 km or more from the central business district (CBD) (fig. 6.1). Much of the northern part of the city is not greatly dissimilar to the area that surrounds the harbour. But the sprawling suburbs and newer dormitory areas on the

Photo 6.1 Sydney Harbour Bridge and Opera House (reproduced with the permission of the National Library of Australia)

undulating clay plain to the west show none of the attractiveness and offer few of the services found in the central and northern districts. The west is drained by major rivers and streams, the Hawkesbury/Nepean to the north and Georges to the south. These often deliver damaging and disruptive (but not devastating) floods. In addition, they are gradually being poisoned with wastes generated by the growing city (Borthwick and Beek, 1993; Warner, 1991).

Open space is another historic attribute that was bequeathed to Australia's biggest city. "Sydney is indeed fortunate that the army and navy appropriated so much foreshore land in the 19th century and that . . . they have resisted the temptation to build on it" (Spearritt and DeMarco, 1988, p. 122). Today, the city retains much natural bushland around the harbour and in close proximity to the city centre. Other types of national park are located near the central area and on the northern and southern boundaries of the metropolitan district. These fulfil recreational and aesthetic purposes and provide habitat for indigenous species. They also increase the potential for bushfires, and offer refuge for the feral fauna and flora that are making inroads on native species and reducing natural biodiversity.

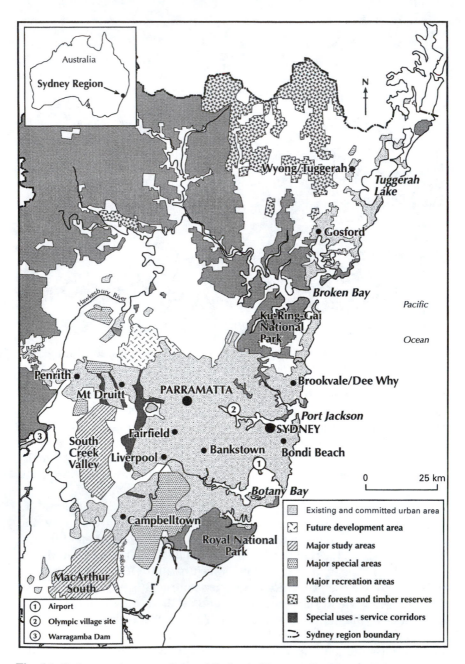

Fig. 6.1 Sydney and surrounding public lands (Source: Government of New South Wales and others)

Consolidation and expansion

Early Sydney was essentially an unplanned settlement (Hughes, 1986, p. 296), but Lachlan Macquarie – a British governor – was responsible for drawing up a system of housing codes, street widths, and land plats that guided the growth of today's inner city, beginning in the 1820s. By the 1830s, increased trade and commerce fuelled continuous steady expansion (fig. 6.2). The "free settlers" and "emancipists" (freed convicts) formed a society that offered far more opportunities and social mobility than England's – although native people were not included on equal terms. Local government was well established by the mid-1800s, and it came to be quite different from the British model. In Sydney, many of the traditional functions of British city government were passed to or shared with separate statutory authorities; these include roads and bridges, transport, water supply, sewerage and drainage, and aspects of flood mitigation. Such bodies are still important features of Australian government and they go some way towards providing area-wide coordination of public services in the absence of municipal government on the metropolitan scale. They can usually be relied on to execute their narrowly defined missions; but they are also slow to respond to new agendas of environmental concern and public participation. Indeed, lack of political accountability has sometimes led to power struggles between the separate public authorities and the elected governments (Day, 1991; Lowe, 1984; Smith and Handmer, 1991).

In Australia, the chief role of local government has been to promote economic growth. Social reformist concerns about slums, which led Ebenezer Howard and others to pioneer the profession of urban planning as a means of human betterment, were not part of the vision of Australian leaders during the 1890s. During the 1920s and 1930s, land-use planning gradually gained legislative acceptance in Australia, but the Great Depression and the Second World War cut short opportunities for the beneficial exercise of public power through effective planning.

Post-war expansion

Rapid population growth dominated public agendas in Australia and Sydney during the post-war period. This was fuelled by federal government immigration policies that reflected the popular slogan "populate or perish." Paranoia about a possible inflow of migrants from Asia, entrenched in the "white Australia policy" and other racially inspired programmes, disappeared only slowly. As late as 1961 even the major Australian current affairs weekly, *The Bulletin*, sported the motto "Australia for the white man" (Horne, 1989). None the less, by the early

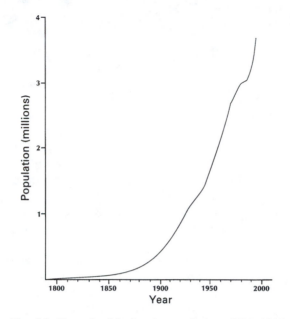

Fig. 6.2 Growth of Sydney's population, 1790–1995

1970s these views had been replaced by a firm public commitment to "multiculturalism."

Sydney was, and still is, a favourite destination for immigrants. Currently population growth rates are around 0.7 per cent per annum, and new arrivals are accommodated by westward expansion of the city (fig. 6.3). For example, the area of Cabramatta is popularly known as "Vietnamatta" (Mellor and Ricketson, 1991). Strategic planning might have improved the quality of living in these western districts, but it was never attempted; growth was simply accommodated rather than managed. Partly as a result, western and southern sectors of Sydney rank low on a wide range of socio-economic and environmental indicators; the poor facilities of the western suburbs are legendary.

Sydney today

Any examination of contemporary Sydney must take account of the city's strong population growth and remarkable areal expansion over the past few decades. Today, in the words of a local government representative:

Only Los Angeles can match [Sydney's] area. No major city in the world matches its low population density. Overseas visitors are surprised by a metropolitan area stretching 140 kilometers from new release areas near Wyong south to Appin. (Latham, 1992, p. 72)

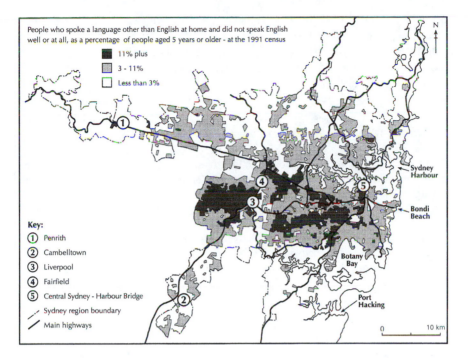

Fig. 6.3 Competency in English (Source: Farrell, 1991, from 1991 National Census)

The city is a very low-density sprawl of nearly 4 million people, covering an area exceeded by only a few mega-cities. Most of Sydney's population (about two-thirds) live in detached houses on their own land blocks. The remainder occupy terraces, townhouses, apartments, or boarding houses (Spearritt and DeMarco, 1988). Property lots are now half the size of the early "quarter-acre blocks." Apart from a few commuter areas in the Blue Mountains and parts of north Sydney, development has generally avoided difficult topography. The city is bounded to the north and south by national parks that function as limiting green belts, but many people commute from well beyond these.

Settlement has encroached into flood-prone zones and fire-prone areas while the physical environment has become seriously degraded. Combined with increasing use of private transport and a lack of interest in environmental issues (until the 1970s), the result has been steadily increasing risk, exposure, and vulnerability to hazard. Decentralization has been fuelled by escalating central-city rents, but attempts to redevelop central districts have also generated local protests that raised awareness of heritage and environment issues. For example, the "Green

Bans" of the early 1970s saw unlikely alliances between the communist Builders Labourers Federation and a variety of groups with heritage and community interests (Roddewig, 1978). They were effective at blocking the redevelopment of areas such as the "Rocks" – the site of Sydney's original settlement.

Whether Sydney will continue to expand as a low-density city is not clear. Certainly, there are countervailing trends at work. In the words of a contemporary student of Australian urban history:

Economic scarcity and the threat of environmental catastrophe have made the suburban sprawl seem as profligate and dangerous as it once seemed safe and boring. If the tide has turned against the suburban way of life, however, it is not just because we can no longer afford it, but because we have also begun to question the social aspirations and political arrangements that so long supported it. Declining levels and changing sources of immigration, lower fertility, and smaller government have produced a new urban agenda in which urban consolidation comes to seem not only virtuous but attractive. (Davison, 1995, p. 69)

Contemporary urban issues

Three overlapping issues have dominated recent debates about urban policy in Sydney: the acceptability of public sector planning as a means of guiding land development; links between low settlement densities and a declining quality of life; and the dominant role of motor vehicle transportation.

Planning

Post-war suburban development in Sydney was driven by the Australian dream of a detached home on a substantial block of land, connected to the rest of the world via long ribbons of tarmac. A private car rapidly became an essential part of the dream. The seemingly never-ending sprawl has, according to Leonie Sandercock (1979), turned land speculation into the national hobby: "One of the most sensational themes in the history of Australian cities has been the story of land speculation and the corrupt behaviour of politicians and public officials ... associated with that speculation." Town planning schemes are accused of providing a "punters' guide" to speculation because they identify areas to be developed far in the future. Indeed, speculation is one factor that has undermined the planning process because land is purchased in anticipation of rezoning, thereby driving up prices ahead of actual development. Among other results, purchasers may exert pressure for modification of plans to enhance their profits, and price inflation consumes funds that would have been available for public authorities to provide services (Stilwell, 1993). Until recently, profits from land speculation were untaxed.

Writing in 1966, a leading Australian planner, Peter Harrison, commented that, despite a "hopeful period" after the Second World War, which resulted in "a good deal of intelligent optimism about reconstruction," particularly in Sydney, planning had little influence over post-war urban growth. Instead, the pattern was one of private speculative development serviced by public authorities (quoted in Ashton, 1993, p. 90). There are two main types of urban plan in Australia: statutory plans and strategic plans. Today, as in the past, statutory plans are the main development control instruments; they are little more than zoning schemes that set allowable uses. Australian planning has long seemed to be concerned about small-scale issues and to be essentially restrictive; it typically places controls on development applications made by individuals. This situation is changing only slowly despite important new reforms such as the Environmental Planning and Assessment Act 1979. This law has attempted to broaden the focus of planning beyond imposing controls on buildings and infrastructure to include developments of all types and to address issues of environment and social policy (Pearson, 1994; Stein, 1986). Strategic plans are much more complex than local statutory plans. They are policy documents that express visions of the future; they are intended to be proactive and performance oriented; and they typically encourage thinking about different means of implementation. There has been a long history of attempts to implement effective proactive strategic planning. But, when the plans were in conflict with the wishes of major developers and politicians, the plans typically gave way.

In the post-war period, there have been three major attempts at strategic planning for Sydney, with a fourth currently under way. All have fallen well short of their aims. For example, while generally accommodating further sprawl, the 1988 Metropolitan Strategy suggested that urban consolidation was to play an increasingly important role. Nevertheless, urban expansion remains the main thrust, and the identification of areas for further development fuels land speculation as before. Positive innovations include the fact that this plan paid attention to natural hazards and other constraints posed by the natural environment. Revisions currently under way incorporate considerations of "sustainability" and there is emphasis on environmental and social equity issues. But even if the plan advocates active intervention in the development process and even if it receives broad endorsement by the public and the bureaucracy, history indicates that its chances of success are small. For, in Sydney, planners are politically weak and developers are well connected (Ledgar, 1976).

Urban density and environment

Concern that high population densities in mega-cities tend to exacerbate natural hazards may be misplaced in Sydney, as well as in other large cities of Australia and the United States. For example, in the outer sub-urbs of many Australian (and American) cities, densities are as low as 10 people per hectare. Such cities typically have very high private transport usage. By way of comparison, European cities have around 40–70 people per hectare and Tokyo over 100 (fig. 6.4).

Several processes are working to reduce Sydney's population density. One of these is gradual decrease of the central-city population. This has come about partly because of attempts to relieve inner-city crowding by constructing new homes elsewhere. Changes in lifestyles, central-city gentrification, and the falling size of the Australian family have also played a part. So too have changes in business practices. Even though a substantial proportion (30 per cent) of Sydney's CBD office space lies empty (*Time*, 7 February 1994), high rents, traffic congestion, limited parking, commuting time, and other issues have encouraged the partial suburbanization of work and commerce. The CBD's share of the metro-politan workforce fell from 23 per cent in 1971 to 12.6 per cent in 1986 (Stilwell, 1993) – although the change in absolute numbers was much smaller. None the less, the inner city remains a substantial population node. The 1991 census shows that inner-city areas – especially south of the harbour and in the eastern suburbs – have densities well over double those in the outer areas. Moreover, pockets of inner-city deprivation remain: for example, the Kings Cross area has long functioned as the city's red light district and is an important focus for the narcotics trade.

The counterpart to declining densities in the central city is the growth of very low-density, underserviced suburbs. During the post-war period of rapid house construction, much of western Sydney was developed with minimal urban facilities (Neutze, 1977). Environmental quality received little attention: many areas were left unsewered for decades and flood hazards were ignored by leaders who pleaded ignorance about them despite a substantial anecdotal history. As car use has grown, so has air pollution – with critical episodes occasionally exacerbated by smoke from bushfires. Even though the bulk of the air pollution is generated around the inner city, it tends to drift to the western suburbs. These are also the areas that have found themselves battling against toxic waste tips (Smith, 1990), as well as enduring high unemployment and crime.

By the late 1970s it became clear that population movements and resulting demographic patterns in Sydney had attendant social costs. The New South Wales state planning authority argued that the expansion of outer urban areas should be curbed because inner districts possessed

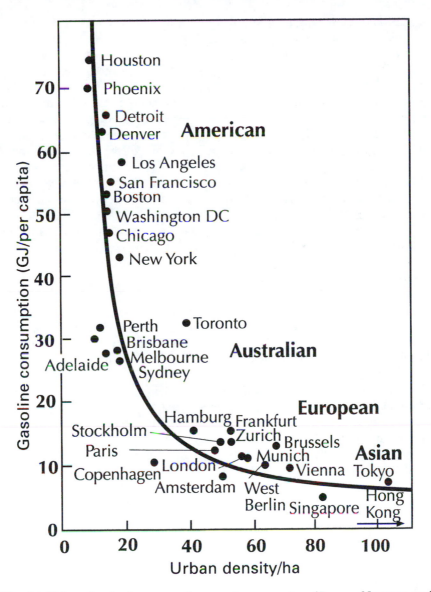

Fig. 6.4 Urban density by per capita petrol consumption (Source: Newman and Kenworthy, 1989)

well-developed infrastructures that could easily accommodate additional growth at low cost. This policy would also have appealing environmental benefits because commuting and pollution would be reduced (New South Wales Department of Housing, 1991). The state government's response to this argument was at first to permit the construction of additional houses on existing single-family land parcels (so-called "dual occupancy"). Then it required local government to facilitate "urban consolidation" (Pearson, 1994; Ryan, 1991). Both programmes are very unpopular and have provoked an intense struggle between state and local governments (Pearson, 1994; Ryan, 1991). Sometimes the outcomes have been perverse, with new areas in the *outer* suburbs being developed for dual occupancy. One commentator on Australian urban affairs believes that the new policies are merely "a desperate effort, under commonwealth (federal) pressure, to save money" (Hugh Stretton, quoted in Collins, 1993, p. 32). Others are more strident and accuse the government of "taking away the only benefit that working class people have accrued from the wealth created in Australia, namely house and land, urban consolidation represents a relegation of lower socioeconomic groups to the status of landless poor" (Troy, cited in Collins, 1993, p. 34).

Transport

Problems of urban density are closely connected with the rising importance of automobile transportation. By 1960, car ownership in Australia was near universal and about half of all trips in Sydney were made by car. Public transport seems appropriate for Sydney in light of its functional layout. It is not a radial city focused on a single centre; it is polycentric. However, public transport has been severely affected by municipal budget cuts and other measures; it is particularly poor in western suburbs (Black and Rimmer, 1992), although it survives in many parts of the city. According to the 1991 national census, a quarter of employed people commute via public transport. Newman and Kenworthy (1989, reported in Stilwell, 1993, p. 57) estimate that only about 8 per cent of all travel is taken on public transport. Sydney seems destined to continue shifting away from public transportation. It is undergoing a transformation from a city that was served by excellent light and heavy railroads, to one that is increasingly dominated by the automobile, with all the associated pollution and congestion problems.

Environmental hazards in Sydney

Sydney experiences a broad spectrum of rapid- and slow-onset hazards associated with natural, technological, and social agents (tables 6.1 and 6.2). Often the distinctions between these different types of hazard are not clear cut; nevertheless they are useful ideal types. The discussion that follows emphasizes hazards propagated by natural agents in the city.

Table 6.1 Some hazards in Sydney

Sudden onset	Slow onset
Terrorism	Environmental degradation
Fires	Crime and social pathology
Floods	Epidemics (poverty and disease)
Earthquakes	Climate change
Severe storms	Homelessness
Dam break	Drought
Toxic spills, especially nuclear	Noise (e.g. airport)
Severe air pollution episodes	
Transportation accidents	
Heatwaves	

Table 6.2 Some significant episodes

Date	Event	Impact
1900	Bubonic plague	103 dead. City functions severely affected by quarantine measures
1919	Influenza pandemic	Over 12,000 dead, 1.3 million cases. Australia wide
November 1961	ANA Viscount crashed into Botany Bay	15 dead
18 January 1977	Granville train crash	83 dead
13 February 1978	Terrorist bombing at Hilton Hotel CHOGM Meeting	3 dead. Meeting moved, army involved. Ananda Marga blamed
August 1986	Flooding in SW Sydney	
March 1990	Hailstorm	A$319 million insurance losses
January 1994	Bushfires	4 dead. Approximately 200 buildings burned

"Continuous" low-level activity
Ongoing low-level terrorist activity – not directed against the state
Chronic air and water pollution
Contaminated soil
Crime

Rapid-onset hazards

Sydney is not affected by many conventional rapid-onset hazards. It does not suffer from winter storms and freezing temperatures or from hurricanes. Extreme heat occurs, but is rare. Neither does Sydney lie in a zone of marked seismic activity; slope instability and subsidence are also not significant problems. Coastal erosion and potential sealevel rise pose difficulties for some areas, especially the northern beaches, but do not affect much of the metropolis.

Nevertheless, some climatic hazards are significant: these include severe storms, floods, droughts, and bushfires. Rapid-onset technological hazards are also a serious problem. Britton judges that Sydney contains most of Australia's chemical industry, "particularly areas surrounding Botany Bay ... and western areas like Rhodes, the Homebush Bay complex and Seven Hills" (Britton, 1991).[2] In general, risk has remained more or less stable, with some exceptions such as bushfires and extreme droughts. Exposure, on the other hand, has increased with population growth and settlement expansion. Although potential damage is increasing steadily because of greater wealth and greater exposure, losses are reduced by timely warnings and emergency response; the impact of damage is managed through loss redistribution. Severe storms are the largest agent of insured losses (Insurance Council of Australia, 1994). Arguably, water-supply infrastructure for drought avoidance constitutes the greatest investment of hazard-management resources. The flood risk, which is considered in more detail below, also includes the potential for dam failure.

Among the non-indigeneous population, awareness of natural hazards dates back to the earliest years of European settlement. Despite warnings from Aborigines, the new residents of outlying townships were affected by floods from the beginning. Floods of the early 1800s prompted Governor Lachlan Macquarie to issue a far-sighted proclamation. This urged settlers to "one and all adopt the firm resolution of forthwith erecting their habitations on the high lands cheered with the animating hope and fair prospect of ... securing themselves from [future flooding]" (1819). Although the Governor had extensive executive power, there were no legislative, bureaucratic, and technical bodies to implement and enforce his directive of non-structural floodplain management. Bushfires as well were experienced from the earliest days. The 1803/1804 summer saw a number of destructive fires, which led to the first government order on bushfire protection.

With the very important exception of a water supply that was designed to carry the city through the most severe drought (Rees, 1982), there has been little if any concession to – or accommodation of – natural hazards in Sydney. The city spread as if land was infinite, with little regard for well-known hazards such as floods and bushfires or for considerations of sound environmental management. Conversely, public health concerns about the biological hazards of disease and epidemics loomed large in Sydney, as elsewhere, until the second half of the twentieth century. The global influenza pandemic of 1919 took a heavy toll in Australia, bringing

12,400 deaths and 1.3 million cases of infection to a population of 5.5 million (Curson, personal communication, 1989). Epidemics of smallpox (1881–1882), scarlet fever (1875–1876), and plague were particularly worrisome in the late 1800s. According to Curson (1985, p. 7), "the [plague epidemic of 1900] caused widespread social and economic dislocation, and was undoubtedly the greatest social upheaval of 19th century Sydney." Many people fled to the Blue Mountains north-west of the city, and vaccines were in short supply. "Curfews were imposed upon the infected zones of the city and people's right of movement severely restricted" (Curson and McCracken, 1989, p. 8). More than 1,750 people were forcibly quarantined. Shipping between Australia and New Zealand and elsewhere was seriously disrupted by these and other precautions. There were longer-term impacts as well. An extensive slum area affected by the disease was demolished as unsanitary. Nevertheless,

[T]he real enemies which produced the highest morbidity and mortality throughout the nineteenth century were a series of less visible, more insidious diseases such as gastroenteritis, dysentery, diarrhoea, bronchitis, tuberculosis, venereal disease and a variety of eye and skin infections. (Curson, 1985, p. 9)

These problems receded with improvements in public health and the installation of a piped water supply and sewerage system (Coward, 1988). However, because the sewers discharged into the ocean, this "solution" has itself gradually become a complex environmental problem.

Contemporary public health concerns include certain types of cancer, AIDS, chlamydia and other sexually transmitted diseases, and myalgic encephalitis. Some observers are also worried about the emergence of a new influenza strain similar to that of the 1919 pandemic (Frank Fenner, personal communication, 1991). AIDS and cancer receive most funding, probably because they attract media attention and are the focus of well-organized lobbying. Skin cancer prevention is also well publicized, especially because of Sydney's outdoor lifestyle and concern over the deteriorating ozone layer. Ultra violet (UV) levels are now part of the daily weather forecast, and, in combination with publicity over the ozone "hole," give the impression of a substantially increased risk. Some potentially important diseases, including the Pandora's box presented by asbestosis, receive relatively little public attention.

Concern with drought is manifest in the form of efforts to ensure that Sydney maintains a highly reliable supply of water. This is reflected in the construction of numerous dams (Rees, 1982), with more planned (Bell, 1993). As evidenced in the escalating and profligate per capita use of water by Sydney residents, the security of this supply appears to be eroding. Exposure, as measured by numbers affected, is also increasing rapidly. Consequently, it might be concluded that the risk of drought is increasing. But this may overstate the problem's severity. The very high per capita water-use rates offer a cushion against which consumption could be reduced relatively painlessly. About half the water (well over half in a hot summer) is used on the spacious gardens that surround most houses. These gardens are themselves a manifestation of public preferences for Northern European greenery in an area of unreliable rainfall with high evaporation rates – a reflection of the city's origins. Universal adoption of native gardens would result in a dramatic reduction in water consumption, although such gardens might pose a larger fire risk.

The security of Sydney's water supply is closely tied to the Warragamba Dam. Failure of the Warragamba Dam could reduce Sydney's water supply by 65 per cent. Combined with drought, this situation would be catastrophic. Although remote, this possibility received more scrutiny following the release of new Australian Bureau of Meteorology estimates of maximum probable rainfall in the mid-1980s. These suggested that extreme floods would be much larger than previously thought. As a result of this finding, many dams were judged to be seriously "underdesigned." Instead of surviving the maximum probable flood, Warragamba Dam was judged to have an annual chance of failure due to flooding of $1:500$ (Taylor and McDonald, 1988). The dam has since been upgraded, but further changes may be needed. An anonymous official has suggested that the state Cabinet is wrestling over whether to spend money on a new dam or on sewage treatment – the latter desperately needed from an environmental perspective.

Like the prospect of dam failure, the chance of a major earthquake seemed remote until one hit the city of Newcastle, just to the north of Sydney, on 28 December 1989. At 5.6 on the Richter Scale, it was not exceptional in Australia: seismologists estimated that an earthquake of similar intensity occurs on average somewhere in Australia or the surrounding continental shelf every year or so. Insured damage amounted to about A$1,000 million (Insurance Council of Australia, 1994). Much stronger earthquakes, up to Richter 7.0 (Rynn, 1991), have been recorded but only remote areas have been seriously affected. In the past, known earthquake hazards had been largely ignored in building regulations and emergency planning, but the Newcastle event changed this situation. Building regulations have been revised, micro-zonation maps are being prepared for major cities, and studies of exposure to earthquakes are under way. One study for Sydney, sponsored by the insurance and reinsurance industry, illustrates that little of the metropolitan area is situated on ground highly susceptible to earthquake-shaking (Blong, 1994). Much of the city is on low-hazard material defined as "shallow soils on competent bedrock" (Blong, 1994, p. 8) – material that transmits but does not amplify earthquake energy. Probable maximum loss (PML) estimates were made for residential structures only; the most conservative approach produced PML losses of just under A$15 billion for an earthquake of Richter magnitude 6.6. (By comparison, the January 1995 Kobe earthquake measured 7.2.) This figure omits some categories of loss (e.g. contents of residences; indirect economic losses) and some outlying districts; most importantly it excludes all commercial, industrial, and government buildings and infrastructure. Total economic losses of a large earthquake would therefore be much higher.

Widespread flooding was experienced throughout New South Wales in the early 1950s and 1970s and intermittently throughout the 1980s (Handmer, 1985). There is some evidence that heavy rainfall in eastern New South Wales has increased since the mid-1940s and with it the consequent risk of flooding (Bell and Erskine, 1981). There is nothing new or unexpected about severe flooding in Sydney, but the hazard has received little attention from urban planners. The first recorded floods on the Hawkesbury River to the north-west of the city occurred in 1799, and there have been many re-occurrences. For example, "in March 1806 the farms along the Hawkesbury River, on which the food supply of Sydney depended, were devastated by flood. The water covered 36,000 acres and destroyed all the standing crops, along with the farmers' tools, livestock and seed reserves" (Hughes, 1986, p. 125). After the Second World War, the outskirts of Sydney began to invade these flood-prone areas as well as similar risky reaches of the Georges River to the south-west. Both rivers have the potential to produce extreme flooding many metres deeper than any so far experienced; associated losses of life and property would be very high. Smith et al. (1990) estimate that a maximum probable flood on the Georges River would result in about A$1,230 million of direct and indirect damage, mostly to industry and commerce. Given the depth of flooding, a properly functioning warning system is essential to reduce potential death tolls. At present, warnings of 6 and 12 hours are available for the Georges River. However, the present warning system did not function well during the floods of 1986 or 1988 (Handmer, 1988). Fortunately these were relatively minor events; despite that fact, they caused substantial disruption over much of Sydney in combination with widespread severe weather. The flood-warning system has since been steadily improved and upgraded.

Inner, eastern, and northern areas of Sydney are subject to flash-flooding from urban storm-water drainage overflow and urban creeks. The main disruption is to transport. Older parts of the city are worst affected for two reasons: natural drainage systems have long since vanished underground; continued development has overtaxed the artificial surface drainage that replaced them.

Bushfires are by far the most spectacular hazard in Sydney. These routinely affect areas on the outskirts of Sydney, and occasionally occur around Lane Cove National Park in the inner city. They have increased in number and impact as low-density suburbs spread into bush-filled areas that are particularly fire prone because of their broken topography. About 6 per cent of bushfires result from lightning (see Chapman, 1994, p. 30), but most are caused by human action – all too often through negligence or arson (Lembit, 1995, p. 139). Accordingly, fire risk is increasing. In some places, such as the hills around Canberra, fires have extirpated native vegetation and encouraged invasion by exotic weeds. Much endemic Australian fauna and flora is adjusted to fire and was traditionally used by Aboriginal Australians as a rangeland management tool to assist hunting (Singh et al., 1981). In general, very frequent fires and very hot fires are regarded as most damaging, but the precise ecological effects of fire in Australia are still strongly debated and this has implications for fire management.

Fires around Sydney are democratic agents of destruction, sparing neither the cottages of low-income residents nor the homes of millionaires. But this is changing as expensive suburbs expand into areas of steep, rugged terrain, with houses sparsely spread through the bush. Depending in part on aspect and drainage, such areas are often particularly susceptible to bushfire, and pose a near impossible task for firefighters. In a fierce fire, burning branches or embers from eucalyptus trees are often carried many kilometres by wind, thereby igniting new fires in unpredictable places well ahead of the main fire front. In these circumstances, mobility and rapid response are critical aspects of fire management. Public policy generally concentrates on saving life and property, because a fully developed fire accompanied by strong winds is very difficult to extinguish.

Slow-onset hazards

It is recognized that the classification of "slow-onset" hazards is somewhat arbitrary and that many hazards are hybrids. What follows is suggestive rather than comprehensive and draws on selected examples to illustrate broader issues and problems. Sydney's slow-onset hazards are on the whole not spectacular but – unlike most rapid-onset hazards – many are the subject of continual attention by the mass media, community leaders, politicians, and intellectuals.

Environmental degradation

Looking at Sydney Harbour on a sparkling clear day, it can be hard to believe that the city suffers from chronic air and water pollution, and that virtually untreated sewage all too often floats onto the beaches (Beder, 1991). Some of the environmental insults can be traced to a legacy of previously ignored or mismanaged problems. These range from decisions not to invest in sewage treatment facilities and public transport, to lack of concern over industrial and commercial practices that poisoned land. Perhaps "the more insidious disasters [were] bequeathed ... by generations who freely used arsenic in their tanneries and sheep dips, dieldrin in their woolen mills and cyanide in their mines" (Kissane, 1990).

Contaminated land and groundwater are widespread; they include sites of former factories, waste tips, warehouses, transport and public works depots, farm sheds and yards, and sites where toxic materials were retailed (Britton, 1991; Tiller, 1992). The latter can include a variety of exotic problems such as radiation contamination of land at a former watch factory site or arsenic contamination of a site where railway sleepers (ties) treated with creosote were stored. Leakage of petrol (gasoline) from storage tanks and from endless "spillage" is pervasive among the thousands of abandoned and existing petrol stations that dot the city. Although major petrol producers and distributors are increasingly considering preventive measures, cleaning up the environment is difficult and expensive (Johnsen, 1992; Rowe and Seidler, 1993). The bulk of environmental law focuses on reactive measures and penalties for contamination; prevention receives minimal acknowledgement (Johnsen, 1992).

Petrol is also a major contributor to another of Sydney's intractable environmental problems: photochemical smog. Indeed, petrol consumption is an approximate surrogate for a range of urban environmental problems, including air pollution, traffic congestion, parking issues, and public transport provision. Traffic congestion and its associated pollution have been major issues in Sydney for decades. Newman and Kenworthy (1989) show that per capita petrol consumption is closely correlated with population density. Cultural factors (and the price of petrol) are also important. Although Sydney's consumption is substantially less than consumption in major Texan cities of comparable density, it is still double the consumption of moderate-density cities in Western Europe such as Copenhagen (fig. 6.4).

The bulk of pollution is generated in the inner-city area around Sydney Harbour and Botany Bay where the concentration of traffic is greatest. A combination of topography, prevailing winds, and sunny weather conspires to exacerbate the problem – principally by moving increasingly polluted air to the western region. On a typical sunny (high-pollution) day, the process is as follows. During the night, relatively clean cold air drains from high ground south of Sydney into the Hawkesbury basin, pushing polluted air into the Parramatta River Valley. This flows eastwards, reaching the coast at about 10 a.m., picking up pollution from the morning rush hour as it crosses the central-city area. Later in the day, this polluted air may return with the sea breeze, reaching the far western suburbs of Sydney by mid-afternoon. As it travels across the city, sunlight reacts with the mixture to form additional ozone; by the time the air arrives in the western suburbs, ozone is at peak concentrations.

In the early 1970s it was recognized that the city had a severe photochemical smog problem. A range of government regulations for cars, as well as improved vehicle fuel efficiency and factory closures, led to dramatic improvement in official air quality. Unfortunately, Sydney's air-quality stations were predominantly located in the inner city, so continuing declines in the air quality of western districts did not show up until later (Wright, 1991/92). If present development trends in western and south-western Sydney continue, areas such as Penrith and Mount Druitt could be experiencing ozone levels as high as 20 ppm by the year 2011, with the added danger of sustained high levels, rather than occasional peaks (Wright, 1991/92). In summary, both air pollution risk and exposure are increasing inexorably. Vulnerability trends are more complex; for example, there appears to be an increasing proportion of asthmatics in the population, but a decreasing proportion of smokers. Unfortunately for the western districts, matters are unlikely to improve in the near future. The west is the area of major urban growth, and it is there that most of the planned new land releases to accommodate further low-density sprawl will be located.

Social hazards

Mega-cities are often viewed as areas of deprivation and social hazard. In contrast, this is not a widely held image of Sydney; there is nowhere in the city that a visitor should avoid – at least during daylight. Nevertheless, within Australia, media coverage reinforces the image of Sydney as a crime capital (Murphy, 1994). According to an Institute of Criminology study, there is a widely held perception that the city is much less safe than in the past. This view is most strongly held by the elderly – who are the least likely to become victims (Murphy, 1994). It also receives endorsement from politicians keen to promote themselves as bastions of "law and order" (Brown, 1978). Yet the statistics barely support these perceptions (New South Wales Bureau of Crime Statistics and Research, 1993). Sydney contains some 62 per cent of the state's population, and has about 66 per cent of the recorded crime, but the pattern of crime is variable. Drug-related offences are much more common in rural areas of New South Wales – about twice as common per capita than in Sydney. Sex offences and property damage are also more common outside Sydney.

The city's western suburbs show up as more violent than other areas of the city, especially the "North Shore." Also, about half the state's total

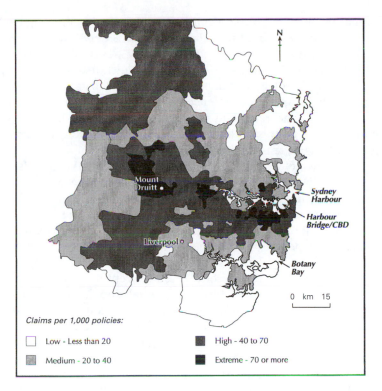

Fig. 6.5 Burglaries in Sydney by postcode (Note: this covers insurance claims only. Source: NRMA, 1989)

narcotics trafficking offences were recorded in the western suburbs of Fairfield and Liverpool, but this may be largely a reflection of intense activity there by narcotics police (New South Wales Bureau of Crime Statistics and Research, 1993). Burglaries follow a broadly similar pattern (fig. 6.5). Responses to crime tend to be very reactive, but this may be changing. The police themselves are emphasizing the development of crime-prevention partnerships with different sections of the community (Murphy, 1994). For many years now, communities all over Sydney have developed "neighbourhood watch" programmes with local police, partly sponsored by the insurance industry. There is also an internationally publicized initiative to rid the city of crime in preparation for the forthcoming Olympic Games: "Sydney's streets will be Australia's safest by the 2000 Olympics" (New South Wales Police Commissioner Tony Lauer, in Murphy, 1994).

Examination of official statistics for trends indicative of declining social cohesiveness produces inconclusive results. Fraud, arson, and robbery with firearms appear to be declining and assaults appear to have increased substantially, but the latter trend is considered to be primarily a reflection of increased crime reporting (New South Wales Bureau of Crime Statistics and Research, 1993). Available statistics on social pathologies are also beset by other problems. To a substantial extent they probably exclude low-level violence, intimidation, extortion, and drug dealing. Much of the racism in Australia takes this form. Less visible criminal activity like this is thought to be quite prevalent in some communities, although the issue is a matter of debate.

From the Second World War until the early 1970s Australia's official unemployment rate was around 1 per cent. But in the wake of falling demand and declining prices for the country's major exports, the Australian economy has been restructured and unemployment has shot up dramatically to about 10 per cent. Along with rising unemployment have come increasing poverty and homelessness. Other social factors are at work as well. A society that was once preoccupied with debating just how egalitarian it was now rarely raises the issue except to comment on spatial disparities. For Sydney these are substantial and increasing (Stilwell, 1993). In a now-familiar pattern, the western and south-western districts of Sydney suffer from higher rates of crime, pollution, poverty, unemployment, and English-language incompetence compared with the area north of the harbour – the "North Shore." Areas of severe deprivation – the "third world city within the first world" (Perlman in MacGregor and Takhar, 1995) – are small but significant and include many Aboriginal Australians. There are also substantial numbers of homeless, as well as flourishing illegal sectors dealing in drugs and prostitution, exploitive informal service and light manufacturing sectors operating beyond the reach of the bureaucracy, and many people living legal but marginal existences.

Responses to social issues have tended to be palliative; on the government's agenda such matters generally rank well below economic reforms and free trade initiatives. Exceptions may occur when major political pressure is mounted, generally through coalitions of interests, and just before elections. The overall trend of social hazards is towards increasing exposure and vulnerability.

Hazard management and response

Important macro institutional issues, such as poverty, are generally ignored in hazard management, as are activities and policies undertaken for non-hazard reasons – even though they may be central to questions of exposure and vulnerability. In principle, hazard management may involve a wide range of interests and orientations: it can tackle issues of risk, exposure, or vulnerability; it can focus on the obvious or try to grapple with the insidious; and initiatives can come from any level or sector of society, including groups outside the formal process of government. Unfortunately, the richness of potential that is inherent in this diversity is typically ignored by professional hazards managers and scholars of the field. They tend to restrict attention to formal organizations with narrowly defined planning or management missions.

Formal hazard-planning and management take a variety of forms, which can be summarized in three general categories: (1) emergency preparedness and response; (2) mechanisms to deal with residual risk; and (3) preventive planning. The first two primarily address issues of vulnerability; the third focuses on management of exposure as well as vulnerability. Hazard-related land-use planning and preventive measures are often caught in a fundamental dilemma between the desire to promote innovation and risk-taking in the cause of economic growth and the need for caution in the face of uncertainty. This is part of a broader tension between economic development interest groups and planning interest groups. Both groups are supported by considerable bodies of legislation and policy (Fowler, 1991). Implementation of sound hazard and environmental policies is often subject to strong resistance from elements of the commercial and political worlds.

A selection of hazards, identified in the previous section as being important in Sydney, is reviewed below to explore management issues and approaches. Sections on bushfires, floods, and environmental pollution illustrate issues of response, comprehensive planning, and complex hazards. The management of residual risk is primarily accomplished by means of loss redistribution policies and programmes that are not hazard specific; they are treated separately.

Rapid-onset hazards

Hazards that manifest themselves as discrete events are generally within the institutional coping capability of modern cities. They are also fodder for media and politicians: disasters in Sydney, especially ones involving a natural agent, are high-profile political events. The city operates professional career emergency services: police, fire brigade, and ambulance – all connected to the sort of facilities expected in a major city of the industrialized world. In addition, very large volunteer organizations form a key component of local community preparedness. According to Britton (1990), some 400,000 Australians, or about 2.5 per cent of the total population, are permanent volunteers with emergency organizations. In addition, many people act as "casual volunteers" during emergencies, without formal membership of any organization. The largest proportion are volunteers in bushfire brigades.

Emergency services are almost exclusively devoted to immediate actions in response to clearly defined events. Although this stance may be singularly appropriate when confronting bushfires or floods, it may have hampered comprehensive planning and preparedness for long-term hazard reduction. It may also suffer from other limitations. For example, most of Sydney lacks approved disaster-management plans for chemical and technological emergencies (Britton, 1991). However, the governmental role in hazard management is changing. Under the State Emergency and Rescue Management Act 1989, attention to hazards goes well beyond response: local governments must address "prevention, preparedness, response and recovery" for all identified natural and techno-logical hazards in their areas. In other words, they must consider risk and exposure, in addition to vulnerability management. Implementation of this act is highly variable, uneven, and slow. Enormous energy has gone into prevention programmes for hazards such as drought and minor flooding, whereas other hazards have received only selective and inter-mittent attention.

Bushfires: Spectacular events and total responses

Response organizations excel when confronted with the most spectacular and newsworthy of events: bushfires. Most bushfire- or wildfire-fighting is done by volunteer brigades, equipped and trained with government funds. Australia-wide there are tens of thousands of trained volunteer fire-fighters. On the fringes of major urban areas such as Sydney, there is often an uneasy relationship between the career brigades and volunteers, and there are many unresolved issues concerning the fire-fighting system. The operation of this system can be illustrated with reference to the widely publicized Sydney fires of January 1994. These fires dominated world media for days (F. Campbell, 1994): CNN and the BBC led with the story for over 24 hours, and Swiss Television erroneously announced that Sydney had been evacuated (*Time*, 24 January 1994, p. 19). How-ever, these were by no means the worst urban bushfires of recent years: the "Ash Wednesday" fires of 1983 in Victoria inflicted more than 10 times as much destruction (2,500 houses versus 200) and nearly 20 times the fatalities (76 versus 4).

Drawing primarily on voluntary brigades, the New South Wales Bush-fire Commissioner was able to mobilize a massive force to combat the fires. Between 27 December and 17 January there were 800 separate fires in the state of New South Wales (Koperburg, 1994). By the end of this period some 800,000 ha had been burned, including vast areas of national park. Acting on advice from the Bureau of Meteorology that the hot, dry, and windy weather would continue and the fire situation would worsen, the Commissioner requested extra resources.

There are, of course, official channels for requesting resources, in particular from other jurisdictions. The Commissioner used these, but primarily he mobilized a personal network around Australia, and – with the cooperation of the mass media – requested assistance from the private sector. The result was an influx of fire-fighting units from all over the country. For example, 55 units drove over 1,000 km from South Australia, and 116 fire-fighters flew the 3,300 km from Perth to take part in combating Sydney's fires. In total, over 20,000 fire-fighters were involved, including military units and career fire services, as well as hundreds from the National Parks and Wildlife and Forestry agencies, and countless volunteers in various support roles such as catering and emergency accommodation. The bulk were employed in the Sydney region (*Bush Fire Bulletin*, 1994, p. 47).

Helped by rain, the fire ended. Fire-fighters were hailed as heroes and paraded through Sydney. The Commissioner had publicly predicted the loss of 2,000 or 3,000 houses, whereas only 200 buildings and 4 lives were lost. To many, this was the Commissioner's achievement. However, the fire highlighted communications and coordination problems (Cunningham, 1994) and raised questions about whether a formal planned response would have been more effective. The fires out, attention turned to why they had happened. The causes were deemed to be a combination of fuel loads and weather patterns and limited cases of arson. Intense media attention gave the impression that the burned area had been "destroyed" and created pressure for simple political responses. Singled out was the apparent failure to reduce the risk of bushfires by burning off excess fuels – so-called "hazard-reduction burning" or "controlled burns." Environmentalists and national park managers became convenient targets. "Environmental protection laws and policies were seen as the cause of abundant flammable native Australian vegetation throughout suburbia" (Harding, 1994). But it is not clear that widespread hazard-reduction burning is of great benefit to the city; specific local action may be required to lower the fire risk (Cunningham, 1994).

Reasons for concern about hazard-reduction burning are complicated and involve links to air pollution, the ecological role of fire, and the potential for ignition of uncontrolled fires. For example, it is clear that bushfires are major contributors to Sydney's air pollution; the big fires of 1994 drove up pollution readings to their highest levels since the last major bushfires 13 years earlier (Bishop, 1994). Unfortunately, controlled burning also affects air pollution; on 1 June 1994, it was reported that the "smog" over Sydney was due to controlled burning in national parks. Legally, controlled burning is exempt from the provisions of the Clean

Air Act 1961. The New South Wales Health Department investigated links between asthma and air pollution, after high particulate levels caused by controlled burns in May 1991, and in the wake of media speculation (Churches and Corbett, 1991). Results were inconclusive, but the editor of the Health Department's *Bulletin* warned of smoke's negative health impacts (*Public Health Bulletin*, 1991).

Fauna and flora suffered in the 1994 fire, although not as much as first feared. Although fire is an integral part of most Australian ecosystems, the appropriate frequency and intensity of controlled burns is fiercely contested (Benson, 1995; Robinson, 1995; Williams et al., 1994). Moreover, compared with "natural" fire regimes, controlled burns are generally of a lower intensity and greater frequency and occur at a different season (Whelan, 1995). Finally, over a quarter of all bushfires result from the escape of "controlled burns," although many of these burns did not have the required permits (Chapman, 1994). Other complications in the controlled-burning debate include differing perceptions of hazard among interest groups. To the bushfire manager, long grass may indicate a dangerous fuel level; to the stock owner, it may be a valuable grazing resource.

The roles of settlement patterns, housing construction, and the general expansion of suburbia into high-hazard areas received little attention following the January 1994 fire. Guidelines on these subjects are issued by the state planning department and other agencies, but local governments have generally omitted considerations of bushfire hazard from the development-control process even for very high-risk areas such as northwest-facing sites on top of escarpments (Cunningham, 1994). One outcome of the 1994 fires may be increased application of fire-related building and planning codes.

The bushfire hazard has evolved from a straightforward single issue into a complex and highly contested nexus of problems involving environment, health, safety, settlement patterns, and other issues. Despite this context, fire-hazard management addresses a far narrower range of issues. Hazard-reduction burning tops the list, spurring calls for more resources (Phelan, 1994) and the reduction of legal obstacles. Other important management alternatives involve emergency response, loss redistribution, and the provision of public information. In short, fire management tends to be driven by the specifics of different fire events whereas a broader and more subtle strategy might be more appropriate.

Some of the more complex issues that should be addressed may be taken up over the long term by local bushfire-management committees; these are newly required by 1994 amendments to the Bushfires Act 1949. They are to include representatives of all groups that have a significant role in bushfire management, including the conservation movement

(Koperburg, 1994). At the state level, bushfire-related matters are co-ordinated by a committee, which now has representatives from the conservation movement as well as private landowners.

Flooding: Effective institutional response to hazard

Unlike management of bushfires, management of flood hazard is the responsibility of a powerful engineering bureaucracy with a strong interest in risk reduction rather than emergency management. Since Sydney's founding there have been political pronouncements about reducing flood losses by careful land-use management, but it is only since the late 1970s that this rhetoric has been incorporated into coherent policy. And, like bushfire management, the context of flood-hazard management has evolved into a complex, multi-objective process.

A number of factors combined to ensure that engineering responses dominated public policy on flooding from the 1950s onwards. These include: a bureaucracy in which engineering agencies are strongly represented; weak planning bodies; favourable funding arrangements for structural works; an absence of environmental concern; and supportive politicians. Levees were a favourite mechanism. By the 1970s, legislation and policies on environmental management and planning had strengthened and the time was ripe for a reappraisal of the state's approach to flood-hazard management. That reappraisal occurred in the wake of destructive floods and prompted the development of a state policy that aimed to reduce exposure to flooding. The new policy promoted "removal of urban development from flood prone areas wherever this is practicable and appropriate" (New South Wales Department of Environment and Planning, 1982). Under this scheme, floodplains were defined by the 1:100 flood. The policy was highly prescriptive, with an emphasis on deterrence: councils were "strongly advised" to adhere to the standards prescribed by the policy. In particular, they were reminded of their potential legal liability for flood damage suffered by developments they have approved in flood-prone areas.

The new policy was unpopular with land and development interests, and many local governments agitated for change. Protest was galvanized by a letter sent to each of the 3,500 property owners in one community advising them about their flood-prone status. Over 1,000 people, mainly concerned about the effect of the policy on property values, attended a lively public meeting. State government leaders were faced with the prospect of a political controversy just weeks before an election (see fig. 6.6). As a result, the state Premier announced a change of policy, and promised to abolish both the 1:100 flood standard and the flood-mapping programme. After being returned to power, the government approved a new policy on 11 December 1984 – which remains in force today.

IS YOUR HOME ON WRAN'S HIT -LIST?

The Wran Labor Government is going to reduce the value of your home!

If you live in the area shaded on this map, your home could be declared "flood prone" under the present Labor Government.

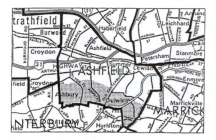

Once the Labor Government has mapped your area – and they could be mapping it now – the value of your property could be reduced by up to 50%.

If your street is declared "flood prone"

YOU WILL NOT BE ABLE TO SELL YOUR HOME
YOU WILL NOT BE ABLE TO EXTEND YOUR HOME
YOU WILL NOT BE ABLE TO BORROW MONEY ON THE SECURITY OF YOUR HOME
YOU WILL NOT BE ABLE TO INSURE YOUR HOME AGAINST FLOODING

HAS YOUR HOME BEEN FLOODED IN THE LAST 100 YEARS?

As a result of Labor Government policy introduced last year, your home may be declared "flood prone" if the Government thinks it may be subject to flooding once in every hundred years.

In areas affected by the "one-in-a-hundred-year" flood policy, development will be severely restricted and property values will plummet with serious impact on insurance and finance.

A LIBERAL GOVERNMENT WILL:

■ Abolish the concept of the one-in-a-hundred-year flood.

■ Enable you to sell your home at full value.

   A Liberal Government will limit the definition of flood prone lands to actual flood channels (derived on the basis of a flood once every 20 years), and progressively acquire such land at full market values.

■ Allow local councils to approve extensions to homes in these areas.

■ Begin effective flood mitigation works to reduce the flood risk to all properties.

A vote for Nick Greiner on March 24 will stop Labor's Flood Prone Lands Policy washing the value of your home down the drain.

LET'S CLEAN UP N.S.W.

Authorised by S.G. Litchfield, The Liberal Party of Australia, N.S.W. Division, 234 George Street, Sydney 2000. Printed by Dashing Printing, 6 Bridge Street, Sydney.

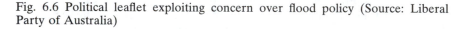

Fig. 6.6 Political leaflet exploiting concern over flood policy (Source: Liberal Party of Australia)

Compared with the 1977–1984 policy, which emphasized floodplain avoidance and uniform application of policy criteria, the new policy "recognizes, firstly, that flood liable land is a valuable resource which should not be unnecessarily sterilized by preventing development, and secondly, that by carefully considering the circumstances of each proposal, developments can be allowed which might otherwise have been unnecessarily refused or unreasonably restricted" (New South Wales–FAC, 1985, p. 3). Now decisions about the development of flood-prone land are made on the basis of several criteria, which include flood damage, safety, and economic, social, and environmental factors, not on the sole basis of flood frequency. The switch reflects a strong move towards a multi-objective approach, wherein each application is treated on its "merits." The new policy emphasizes compliance with procedure rather than prescription of outcomes. It is inherently more flexible, and recognizes that flood-hazard management involves conflict and trade-offs, and must have the cooperation of local government.

The legal situation was clarified by legislation giving local government immunity from liability provided they act in "good faith" and follow the state policy as set out in the *Floodplain Development Manual* – a set of comprehensive guidelines. Explicit provision is made for involving various stakeholders, including the "public," in the development of flood-plain management plans. Councils are encouraged to establish floodplain management committees through which local community groups and individuals can effectively communicate their views. By means of the environmental impact assessment process, the public are also involved in the review of flood-control proposals.

Because land use is largely under the control of local government, implementation of a state policy depends on obtaining local cooperation. This is achieved through inducements and other features that are designed to build up the local capacity to implement the policy (May and Handmer, 1992). In addition, implementation of the policy requires increased use of negotiation and conflict-resolution skills, as well as a commitment to opening the decision-making process to broad public involvement. As the policy is being implemented essentially by the same engineering bureaucracy as the previous highly prescriptive policy, it seems likely that some changes in organizational style and culture will eventually occur. The "merits" approach is now being extended to other areas of environmental management, such as estuaries and coasts. One appealing aspect of the approach is its flexibility, and therefore longevity, in the face of changing political priorities. Although the full range of flood-mitigation strategies is employed, including purchase of some chronically flood-prone properties, the most recent policy generally permits development but requires that ground floors be elevated above the level of the 100-year flood. Although it offers protection against lesser floods, this approach will lead towards gradually increasing exposure to floods greater than 1:100 up to the maximum probable. As a result there will be a gradual escalation of the potential for catastrophic losses during extreme floods.

The reduction of vulnerability is achieved by means of warning and emergency response systems, and by various compensation arrangements (see the section on residual risk, below). The same approaches are used to curb exposure. Flood-warning systems are coordinated by the state's Flood Warning Consultative Committee. A great deal of energy is going into the design of total warning systems, the dissemination of warning messages, and tailoring adjustments to the needs of those who are at risk (Elliot et al., 1995). New and upgraded systems are often jointly funded by local authorities and by state and federal agencies. One reason for the success and momentum of warning systems is that roles are clearly allocated: the federal Bureau of Meteorology has a legislative duty to provide warnings, and the State Emergency Service has a legislative duty to coordinate flood emergency planning and management.

More recently, the same government organizations have turned their energies to the chronic flash-flooding of urban streams and storm-water drains in Sydney. The approach has included the establishment in 1989 of an overarching body, the Upper Parramatta River Trust, to overcome problems that arise because not all councils cooperate. The Trust is a construction authority with full fund-raising powers. It also attempts to coordinate development controls among its four constituent local governments, which serve a population of some 200,000. At the state level, authorities have been pursuing a strategic approach to the problem. Areas with serious recurring problems of exposure or risk reduction have been targeted for remedial action, and a feasibility study has been completed into the possible provision of a radar-based flash-flood warning system for the greater Sydney region. This would be linked with the relatively new Sydney severe weather service at the Bureau of Meteorology.

Slow-onset hazards: Environmental degradation

It can be argued that a suitable institutional framework for the management of environmental degradation exists in Sydney. There are many important pieces of legislation. Some are quite specific, such as those that are concerned with air, water, noise, and hazardous chemicals. Others, which are more relevant to issues of hazard, affect the broad context of environmental management. Four of these are of particular importance. The Environmental Planning and Assessment Act 1979 established the state's modern planning system, and broadened its original focus on land-use zoning to include most other aspects of natural and social environments. The Local Government Act 1993 set out the roles and responsibilities of local government, including their primary role in land-use management. The Environmental Offences and Penalties Act 1989 widened the definition of "environment" and prescribed severe penalties for harming it. Finally, the State Emergency and Rescue Management Act 1989, discussed above, also applies to planning for environmental emergencies.

This framework is suited to well-defined, rapidly manifest problems. But there are serious doubts about its ability to address macro-scale issues that lack obvious cause-and-effect relationships. In many cases of environmental degradation, the problems are insidious, they are poorly defined, they occur over time-periods that exceed customary planning horizons, and the identities of culpable parties are not obvious. Ideally, in such circumstances the focus should be on prevention. The issue of photochemical smog in Sydney illustrates some of the difficulties.

Photochemical smog

Automobile pollution receives considerable media attention. Much of this focuses on the fact that a substantial proportion of Sydney's cars are old, thereby pre-dating tight emission regulations and the introduction of lead-free petrol. This picture ignores the half-hearted nature of emission controls. For example, heavy vehicles are exempt from most regulations but contribute about one-third of the total emissions of nitrogen oxides (Wright, 1991/92). Emission standards for cars lag many years behind those in California, a region that is widely used as a comparative yard-stick. Attempts at reducing consumption or shifting to cleaner fuel confront political agendas that aim to keep gasoline prices as low as possible. They are much lower than prices in Europe, although higher than in the United States.

Air pollution policy is also hampered by scientific uncertainty about the most effective approach. Although the general dynamics of Sydney's smog are understood, complex relationships between nitrogen oxides and ozone, and their joint effect on the distribution of smog over the Sydney region, are unclear. For example, reduction of nitrogen oxides in the western areas may lower ozone levels there but increase them in eastern Sydney (Wright, 1991/92). Lowering the levels of nitrogen oxides from car exhausts is difficult and expensive – but technically feasible.

There has been recent investment in public transport, especially harbour ferries and the city's rail network, which is critical for those commuting from over 100 km away. It is also planned to extend the railway to the international airport. But much of the city is not serviced by rail, and Sydney's tram (light rail) lines were ripped up in the 1950s. Today, buses compete for space with other traffic. Pedestrianization and city-wide strategies for traffic calming are spreading, sometimes impelled by local citizen actions that block streets and otherwise try to keep residential areas free of arterial traffic. On the other hand, new tollways have just been completed, and new connections in the highway network are still being implemented, including an underwater tunnel to relieve chronic congestion near the Harbour Bridge.

It is difficult to envisage an effective solution to Sydney's smog problem during the next several decades, so avoidance of the western areas where ozone concentrates may be the only real option. That alternative challenges conventional thinking about the city's form and density, and it might be counterproductive because most Sydney-dwellers work far from their homes by choice as much as from necessity (Moriarty and Beed, 1988). "Telecommuting" may become important in the future but has yet to have any impact on the problem.

Managing the residual risk: Loss redistribution

Loss is a measure of the residual effects of hazard after ameliorative adjustments. Management of all hazards, no matter how proficient, cannot eliminate the chance of loss. The better managed the hazard, the less visible are the residual losses; but they still exist and may be of sufficient magnitude to require further societal responses. In Sydney, the burden of loss redistribution is shared between public, non-governmental, and private sectors. All of these operate compensation programmes for clearly defined hazard events. Victims can obtain relief from government welfare agencies, charitable groups, or insurance companies. These programmes are coordinated by the New South Wales State Disaster Welfare Committee, within the framework of the state Disaster Welfare Plan (Jarvis, 1989). The system is relatively well insulated from political pressure and media bias. But the spotlight of media attention can sometimes affect the level of support that is provided; recent examples include the 1990 Nyngan flood and 1994 Sydney fires. Disasters in Sydney can readily become political events, with politicians rushing to the site and promising generous assistance.

Household insurance policies in Sydney automatically include fire, storm, and earthquake in the list of covered perils – but not flooding (or nuclear accidents). Coverage is believed to be near universal in the city, but over half of all policy holders may be seriously underinsured (Les Lester, personal communication). The insurance industry itself is re-insured offshore and has been subject to repeated severe blows over the past decade, primarily from natural disasters. Occasionally, the viability of the whole industry has been questioned, but there is little evidence that the sustainability of a major city has been in jeopardy. In summary, the present sophisticated loss-sharing system reduces vulnerability to natural hazards by greatly enhancing recovery; in the short term, and for clearly defined events, resilience is strong. Of course, an event of extreme magnitude, or a series of extreme events, would likely exceed this capacity.

Loss redistribution and welfare also occupy the premier position among adjustments to "social" hazards such as homelessness, unemployment, and crime. Here the system is sadly inadequate. Governmental agencies are faced with the impossible task of dealing with the distributional effects of national government economic policies that are themselves not geared to reducing social inequities. Because both major political parties are committed to economic restructuring and continued shrinkage of the public sector, it falls to non-governmental organizations, churches, community leaders, prominent academics, small political parties, and independent members of Parliament to articulate the concerns of marginalized groups.

Although inadequate for social hazards, the whole concept of compensation for residual risk is at its weakest when faced with environmental problems. Unfortunately, it is quite likely that restitution of severely degraded or contaminated land is simply not possible. The costs of hazards can be transferred to the future, to be borne by another government or another generation – rather than spread among Sydney's present residents. Allowing significant social problems to go unchecked violates the sustainable development principle of intragenerational equity.

Opportunities for intervention to enhance hazard management and urban sustainability

Who might intervene?

Knowledge is a precondition for intervention in any public issue. The fact that much has already been learned about Sydney's experience with hazards suggests that reasoned intervention to change hazardousness and unsustainability is therefore worthwhile. But intervention by whom? Some of the major influences on sound urban hazards management and urban sustainable development are to be found in institutions and activities that lie well beyond the control of cities. Economic fortunes, war or civil unrest, international agreements, the degradation of a critical resource base, major changes in the policies of national government or international business, can all have profound influences on the vulnerability of a major city. Many of the policies of the Australian government have direct effects on Sydney, sometimes exceeding those of local governments. For example, national economic policies and national welfare or other redistributive arrangements are important in determining the amount of unemployment and the extent of poverty. These in turn affect social hazards and capacities to respond to natural disasters. The discussion that follows focuses on the role of states and localities and the various institutions that influence these levels of government, for it is among those actors that the most important scope for intervention in the nexus of urban hazard occurs.

In this context, the terms "top–down" and "bottom–up" are misleading oversimplifications of managerial alternatives. Action directed at environmental hazards has come from creative tension between different arms of the same government, and from a variety of vertical alliances or coalitions. For example, citizens, local governments, and some state and national politicians have come together to work out strategies for noise pollution control at Sydney airport (*The Australian*, 9 December 1994, p. 1). Much action, especially on the environmental front, results from activity by environmental non-government groups (ENGOs), such as Greenpeace and the Sydney Total Environment Centre. In Australia, public membership and support of such groups are high (Papadakis, 1993).

Setting the context and facilitating action: The state

Subject to the national and international constraints mentioned above, state government has a unique role in establishing legal and policy frameworks, promoting the implementation of policy, and engaging in strategic planning. New South Wales has gone beyond this to establish committees that plan and coordinate all aspects of hazard management at state, regional, and local levels, including warning systems, disaster welfare, and recovery. By means of its strategic planning framework and appropriate inducements, the state government has also been able to facilitate – and to some extent direct – local activity in hazard and environmental management. It has also clarified the legal basis of hazards management so that public commitments to emergency planning and management have a footing in common law. But for some issues, such as land use, local government has primary responsibility and it has been difficult for the state to play a leadership role.

State government is also in a position to undertake action unilaterally in some areas. These include establishing and enforcing standards and providing information. State or national standards and regulations deal with many hazard- and environment-related matters, including building codes, water quality, tolerable levels of pollution, emissions and contaminants, vehicle and aviation safety, and so on. However, these standards are frequently developed in conjunction with major stakeholders. Inevitably, this approach raises questions about the proper level of standards and about enforcement.

State government is also an active provider of public information about natural hazards, although many local government and business organizations that are recipients would rather hide this kind of material. Environmental agencies (and the media) are particularly active suppliers of information about anthropogenically induced climate change, water pollution, and other hazards. Information about technological issues is less readily available. Biological hazards are well publicized, though, with the exception of AIDS, the information is often less than is warranted by the threats.

The predominance of information about natural and environmental hazards occurs in part because related public agencies have accepted that this is part of their mission. It is also a politically easy option. After major disasters, such as the 1994 Sydney fires, public inquiries are usual. Lesser events receive less political attention, but are nevertheless frequently examined for potential lessons. Occasionally, inquiries are conducted in the absence of precipitating events, and may deal with broad issues such as poverty, or emergency planning and management (Australia, Senate Standing Committee, 1994). The reports of all these inquiries are valuable sources of information and advice, although implementation of recommendations is usually partial at best.

State government involves a mixture of political actors and administrative agencies. The skills, experience, and actions of the technological agencies are often highly important, but they are insufficient without sound political leadership and strategic thinking; these latter qualities are currently weak in Sydney. Despite legislation that formalizes state responsibilities for incorporating environmental considerations into the planning system (e.g. the Environmental Planning and Assessment Act 1979 and a raft of supporting legislation), degradation of most elements of the metropolitan biosphere continues. Among other reasons is the difficulty of shifting the orientation from events and projects to comprehensive and integrated long-term programmes. Here, resistance is not simply driven by the concerns of specific agencies or constituencies; the whole state budgetary process works to reinforce sectoral planning and existing agency boundaries.

Opportunities for state government to make significant improvements in the management of Sydney's hazards are of four main kinds: continued refinement of the legislative and policy framework; provision of information and other stimuli to public debate; building the capacity and commitment of local government and state institutions to undertake the tasks that must be addressed; and the development of strategic planning for Sydney. History suggests that the last will be the most difficult to achieve. Even where sound state policy exists, a major challenge has always been to find ways of encouraging recalcitrant councils to implement it.

Mediating between bottom and top

Between state governments and local residents there are a host of powerful mediating groups. These include: local government (councils), the media, major NGOs, business organizations, unions, and even the courts. These groups may coordinate or facilitate local action, implement state policy, promote their own agendas, or quite literally mediate between central authority, local communities, and the various interest groups. In a participative democracy with a politicized planning system, mediation and coalition-building are essential to make the system work. Coalitions or alliances increase the salience of an issue, sometimes to the point of setting the political agenda, influencing government policy, and affecting its implementation.

For many aspects of hazard and environmental management, local governments are key actors. They are often in a difficult position – caught between the state, which sets policy, and citizens, to whom they are accountable. In the local planning process, councils must mediate among many competing interests, including stakeholders of local and non-local business, environmental groups, other community actors, and state governments. It is in this process that a real opportunity lies to reduce future exposure to hazard, especially by means of appropriate land uses. However, in Sydney there is great resistance to zoning regulations that severely limit economic uses of land, and compromises are often struck. In any case, local government decisions can generally be overruled by the state.

In Australia there is a strong national principle that response to disaster must come first from local communities. The volunteer groups are the most obvious manifestation of this principle. The state of New South Wales has made hazard-planning and management mandatory for local governments and most councils cooperate with state agencies, but – in typical Australian style – there are few obvious punishments for failure to meet state expectations. Not all hazards receive the same level of attention or involve the same mediating groups. Charitable organizations typically deal with socio-economic problems such as homelessness, poverty, and substance abuse. They also act as advocates for victims. ENGOs coordinate action about environmental issues. In some cases the environmental groups are locally based, in others they involve international or national organizations such as Greenpeace or The Wilderness Society. The latter types of group have generally opted to address industrial hazards, especially waste disposal issues. However, such groups have not been very active on so-called "brown" (i.e. urban) issues in Sydney; they have preferred to concentrate on "green" (i.e. rural and open space) issues. In other cases, alliances have formed between elements of the media, ENGOs, and affected communities. Day (1991) documents some interesting cases in the Sydney region.

Greenpeace had a major influence on Sydney's successful bid to stage the Olympics in A.D. 2000. The bid rested in part on a strong environmental theme, expressed particularly forcefully in the design of the Olympic village, which will be built at Homebush Bay near the head of the estuary that includes Sydney Harbour (Bell, 1993). Greenpeace's village proposal is based on the application of comprehensive environmental criteria covering "water, energy, and biodiversity, as well as transport and land-use issues." As such it attempts to address many of the slow-onset hazards that are dealt with inadequately at present: energy consumption; water consumption; transport; and other aspects of the natural and social environment. If, as seems likely, the village is built according to these criteria, following the Olympics it will serve as a model of how a sustainable settlement can be compatible with modern living. The Greenpeace proposal attempts to extend some of the concepts to the Sydney region (Bell, 1993). It demonstrates that major initiatives can come from the non-government sector, and, to the extent that Greenpeace represents its members, from the community.

Networking: Community groups

Sydney contains many resident action groups (RAGs). These are more or less formal bodies of citizens that campaign on behalf of specific limited objectives on topics such as urban development, traffic, environmental quality, and flooding (Costello and Dunn, 1994). Local governments usually are the first to feel the weight of RAG efforts and have often sought to institutionalize their protests by bringing them into the conventional political process. This tends to reduce the potential for radical protests in favour of trade-offs, bargaining, mediation, and coalition-building. Paradoxically, the success of RAGs has begun to weaken their influence in Sydney because state government – which is less susceptible to pressure than local government – has moved to reduce the planning powers of local governments.

Some local groups have succeeded in changing state, and even federal, policy. Changes to flood policy for south-west Sydney have already been mentioned. A battle to prevent the establishment of a chemical industrial zone on Kurnell Peninsular, where James Cook made the first English landing in Australia, resulted in important changes to industrial location policy (Smith, 1990). Most recently, the issue of aircraft noise from an expanded Sydney airport has motivated people to protest. Community groups and local councils, with the support of locally based state and federal politicians, mounted a very high-profile campaign and threatened to blockade the airport. They demanded expeditious construction of a new airport on the outskirts of the city. This action helped to force policy changes affecting noise reduction, enforcement of existing noise-control procedures, new noise-insulation measures, and a commitment to speed up the new airport (*The Australian*, 9 December 1994, p. 1).

Through the formation of umbrella groups and networking, RAGs greatly increase their power and help ensure that experience gained by the separate groups is not lost. After preventing the establishment of perceived noxious facilities in Kurnell, the Kurnell Action Committee went on to monitor the companies involved and to warn other RAGs of their activities. The Community Action Network also argues for termination of the most noxious aspects of industry (Costello and Dunn, 1994). Such groups are exerting strong pressure for reductions in risk and exposure. It is they rather than the government who are trying to grapple with the causes of hazard rather than settling for a compensation-driven policy.

Work by Costello and Dunn (1994) shows that, in Sydney, community action may greatly expand participation in governance, particularly by women. In this and other ways RAGs are a powerful vehicle for the expression of community aspirations; and may be able to force changes that have positive hazard and environmental outcomes. At issue now is how to build empowering coalitions, networks, and alliances that include the marginalized inhabitants of Sydney. The apparent effectiveness of RAGs in Sydney poses a dilemma for state government. On the one hand it is desirable to encourage public participation in decision-making, but on the other hand no government wishes to see its policies rejected. In the short run, it seems unlikely that Australian states will give much support to RAGs. However, the federally sponsored National Landcare Program provides an interesting model (A. Campbell, 1994). Here, federal money is made directly available to groups of landholders for environmental activities – bypassing the states.

Hazard management and urban sustainability

Sydney is a typical example of a modern sprawling industrial city: it is characterized by low population densities, car dependence, growing societal inequalities, and environmental degradation. Despite a long history of attempts at strategic planning, the results have been mediocre. With a few exceptions, hazards – natural or otherwise – have not been taken into consideration by urban managers. There are in effect two Sydneys. The CBD is the site of the "international" city – an ugly unplanned mess of semi-vacant buildings flanked by a magnificent harbour, a famed bridge, and a celebrated opera house. Its lack of planning epitomizes the prevailing public attitude to development. Paradoxically, the other city – the vast western suburbs that are home to the greatest concentration of Australians – was "planned," but with little consideration for the services that residents might need or want, or the environmental implications of inadequate infrastructure.

Despite the lack of proper planning, the city is remarkably free of serious sudden-onset natural hazards, and it is coming to grips with technological hazards. Bushfires are a possible exception that has the potential to wreak major destruction, but they have yet to inflict severe damage. This fortunate state of affairs is due in part to Sydney's relatively benign geophysical setting and its well-developed emergency services. It is also due to the ability of the city to bounce back from disaster by drawing effectively on its own resources and those of its global networks. The hazards that are well managed share certain characteristics: rapid onset, high visibility, and amenability to immediate response.

Other more slowly evolving hazards are raising serious problems. Although there are legal frameworks for addressing hazards such as environmental degradation and crime, it is doubtful that the institutional capacity or political will exist to tackle them. Poverty, too, appears to pose enormous challenges for the welfare system. Coming to grips with the more intractable issues of profligate water usage, limited sewage treatment, and chronic air pollution appears to be low on the agenda of a government where planning has a tradition of subservience to the demands of major development. Some of these gradual hazards are quite literally invisible. This includes much air pollution and many kinds of land and water contamination. Others are intangible in other ways, for example anxiety induced by fear of crime, ill health from living in stressful or contaminated environments, as well as hidden networks of criminal activity and corruption. Whether invisible or intangible, hazards such as these can be powerful determinants of action and can lead to the gradual abandonment of areas by public authorities and by those with the resources to leave. These less spectacular, less discrete hazards are unambiguously anthropogenic in origin; successful management involves addressing the structure of institutions and society.

The management of most hazards is complex, and in some cases has evolved from dealing directly with a single issue, such as bushfire or sewage, into complex systems of response to multifaceted environmental problems. Most often, responses are straightforward and fall into one or more of three categories: loss redistribution, building regulations, and emergency response. These approaches ignore complexity and treat the hazard or disaster as a discrete entity. This general statement masks some solid attempts at comprehensive management, most notably of the flood hazard, and recognition of the need to negotiate solutions between the increasing range of conflicting interests.

It is difficult to escape the conclusion that the present management approach to natural hazards can probably continue despite occasional events that bring the city to a near standstill; it is sustainable. What is of far greater concern is the city's declining sustainability in the face of inexorable environmental degradation, growing social inequity, and the possibility of rising criminal activity. These extremely complex "macro" policy problems do not respond well to the conventional policy toolkit. The standard approaches are good at dealing with "micro" issues, such as local air pollution or flood mitigation, problems that are spatially and temporally discrete, and not especially complex or uncertain. The major issues of sustainability are not like this: they are "multi-faceted, complex, fraught with uncertainty and ignorances, spatially and temporally diffuse, highly connected to other issues" (Dovers, 1995).

Unfortunately, there is no evidence that Sydney's institutions are capable of handling the more complex, slower-onset, less visible hazards. Solutions to these are not obvious and will, presumably, not be easy. Perhaps the best hope for the future lies in coalitions of diverse sectors and levels of society brought together by a common interest in Sydney's future; arguably, the successful Olympic 2000 bid had this ingredient. But even more important is the need for imagination and vision in the search for solutions (Watson, 1993). In the words of Prime Minister Paul Keating:

What we want is to put a bit of poetry into the souls of our town clerks and shire engineers – to get our planners and architects to come up with ideas with a sense of the miraculous. (*Time*, 7 February 1994, p. 33)

Acknowledgements

Much of the research reported here was undertaken while I was a Fellow at the Centre for Resource and Environmental Studies, Australian National University. My thanks go to members of that organization, especially its director, Professor Henry Nix, and to Christina Jarvis and Sarah Platts for excellent research support. The chapter has benefited from discussions with Ken Mitchell, Dennis Parker, Steve Dovers, and Chas Keys.

Notes

1. This is exemplified by the contents of recent reviews of disaster management in the United States (National Academy of Public Administration, 1993) and Australia (Australia, Senate Standing Committee, 1994). Where mitigation is considered, opportunities for significant change exist. But usually changes are very minor, at least in terms of the pattern of human settlement and activity.
2. During recent years the following technological hazards have been identified in the Sydney metropolitan area: (1) 8,000 tonnes of hazardous waste stored at the ICI site in Botany; (2) 23,000 tonnes of liquefied propane gas stored in the Port Botany area; (3) some 18,000 chemical storage sites; (4) nuclear wastes buried under the Royal Australian Air Force base in Richmond, University of New South Wales, Sydney naval headquarters, and sites near the Australian Nuclear Science and Technology Organization at Lucas Heights (Britton, 1991).

REFERENCES

Alexander, D. 1993. *Natural disasters*. London: UCL Press.
Ashton, P. 1993. *The accidental city: Planning in Sydney since 1788*. Sydney: Hale & Iremonger.

Australia, Senate Standing Committee on Industry, Technology, Transport, Communications and Infrastructure. 1994. *Disaster management*. Canberra: Parliament of Australia.

Beder, S. 1991. "Controversy and closure: Sydney's beaches in crisis." *Social Studies of Science* 21: 223–256.

Bell, F. C. and W. D. Erskine. 1981. "Effects of recent increases in rainfall on flood runoff in the Upper Hunter Valley." *Search* 12: 82–83.

Bell, K. 1993. *Strategy for a sustainable Sydney*. Surrey Hills: Greenpeace Australia.

Benson, J. 1995. "Fire mitigation strategies: Fire based." In: C. Brown and L. Tohver (eds.), *Bushfire! Looking to the future*. Sydney: Envirobook, pp. 98–106.

Bishop, K. 1994. "Choking formula – Control burning plus no wind equals smog." *Sydney Morning Herald*, 1 June.

Black, J. and P. Rimmer. 1992. *Sydney's transport: Contemporary issues, 1988–1992*. Current Issues Background Paper. Australia: New South Wales Parliamentary Library.

Blaikie, P., T. Cannon, I. Davis, and B. Wisner. 1994. *At risk: Natural hazards, people's vulnerability and disasters*. London: Routledge.

Blong, R. 1994. *Earthquake PML: Household buildings, Sydney*. Sydney: Grieg Fester (Australia).

Borthwick, J. and P. Beek. 1993. "Our river of discontent: Big city pressures threaten a classic waterway." *Geo* 15(2): 90–103.

Britton, N. R. 1990. "Disaster volunteers – What do we know about them?" *Newsletter of the International Hazards Panel*, Enfield: Middlesex University.

———— 1991. *Organizational and community response to a technological emergency: Case study of a major incident within a metropolitan Australian city*. Occasional Paper No. 6. Center for Disaster Management, University of New England.

Brown, D. 1978. "Crime and the visiting 'expert'." *Legal Service Bulletin* 3(3): 102–105.

Bush Fire Bulletin. 1994. Special issue, "The January Bush Fires." Granville: Department of Bush Fire Services.

Campbell, A. 1994. *Landcare: Communities shaping the land and the future*. Sydney: Allen & Unwin.

Campbell, F. 1994. "Fireballs, frenzy and hyperbole." *Wildfire* 3: 15–17.

Chapman, D. 1994. *Natural hazards*. Melbourne: Oxford University Press.

Churches, T. and S. Corbett. 1991. "Asthma and air pollution in Sydney." *New South Wales Public Health Bulletin* 2(8): 72–74.

Collins, T. 1993. *Living for the city: Urban Australia, crisis or challenge?* Sydney: Australian Broadcasting Corporation.

Costello, L. N. and K. M. Dunn. 1994. "Resident action groups in Sydney: People power or rat bags?" *Australian Geographer* 25(1): 61–76.

Coward, D. H. 1988. *Out of sight: Sydney's environmental history 1851–1981*. Canberra: Department of Economic History.

Cunningham, C. 1994. Evidence to the Coroner's Inquiry into the 1994 Sydney fires. Department of Geography, University of New England.

Curson, P. H. 1985. *Times of crisis: Epidemics in Sydney, 1788–1900*. Sydney: Sydney University Press.

Curson, P. and K. McCracken. 1989. *Plague in Sydney: The anatomy of an epidemic*. Australia: New South Wales University Press.

Davison, Graeme. 1995. "Australia: The first suburban nation?" *Journal of Urban History* 22(1): 40–74.

Day, D. 1991. "Resolving conflict in the NSW water sector." In J. W. Handmer, A. H. J. Dorcey, and D. I. Smith (eds.), *Negotiating water: Conflict resolution in Australian water management*. Canberra: Centre for Resource and Environmental Studies, Australian National University, pp. 168–189.

Dovers, S. R. 1995. "A framework for scaling and framing policy problems in sustainability." *Ecological Economics* 12: 93–106.

Elliott, J., J. W. Handmer, C. Keys, and J. Salter. 1995. *Flood warnings: An Australian guide*. Mt. Macedon: Australian Emergency Management Institute.

Erikson, K. T. 1976. *Everything in its path: Destruction of community at Buffalo Creek*. New York: Simon & Schuster.

Farrell, D. 1991. *Sydney ... A social atlas*. Canberra: Australian Bureau of Statistics.

Fowler, R. 1991. "Environmental law in Australia." In J. W. Handmer, A. H. J. Dorcey, and D. I. Smith (eds.), *Negotiating water: Conflict resolution in Australian water management*. Canberra: Centre for Resource and Environmental Studies, Australian National University, pp. 73–91.

Handmer, J. W. 1985. "Flood policy reversal in New South Wales, Australia." *Disasters* 9(4): 279–285.

———— 1988. "The performance of the Sydney flood warning system. August 1986." *Disasters* 12(1): 37–49.

Handmer, J. W. and S. Dovers. In press. "A typology of resilience: Rethinking institutions for sustainability." *Industrial and Environmental Crisis Quarterly*.

Harding, R. 1994. "The Sydney bushfires." *People and Place* 2(1): 40–50.

Harrison, P. 1992. *The third revolution: Population, environment and a sustainable world*. New York: Penguin.

Hewitt, K. 1994. *Interpretations of calamity*. Boston, MA: Allen & Unwin.

Horne, D. 1989. *Ideas for a nation*. Sydney: Pan.

Hughes, Robert. 1986. *The fatal shore: The epic of Australia's founding*. New York: Alfred A. Knopf.

Insurance Council of Australia. 1994. "Major disasters since June 1967: Revised to 31 Dec 1993." (Insurance Losses.) Melbourne: Insurance Council of Australia.

Jarvis, K. 1989. "Flood relief in NSW: Aims and changes." In D. I. Smith and J. W. Handmer (eds.), *Flood insurance and relief in Australia*. Canberra: Centre for Resource and Environmental Studies, Australian National University, pp. 113–118.

Johnsen, H. 1992. "The adequacy of the current response to the problem of contaminated sites." *Environmental and Planning Law Journal*, August: 230–243.

Kissane, K. 1990. "Toxic time bomb." *Time Australia*, 18 June, pp. 10–16.

Koperburg, P. 1994. "Who has the last word?" In C. Brown and L. Tohver (eds.), *Bushfire! Looking to the future*. Sydney: Envirobook, pp. 27–32.

Lamont, L. 1994. "Now Sydney's the hub of global village." *Sydney Morning Herald*, 10 December, p. 1.

Latham, M. 1992. "Urban policy and the environment in Western Sydney." *Australian Quarterly* 64(1): 71–81.

Ledgar, F. 1976. "Planning theory in the twentieth century: A story of successive imports." In G. Sedden and M. Davis (eds.), *Man and landscape in Australia: Towards an ecological vision*. Australian Commission for UNESCO. Canberra: AGPS.

Lembit, R. 1995. "Contributing factors to the 1994 fires." In C. Brown and L. Tohver (eds.), *Bushfire! Looking to the future*. Sydney: Envirobook, pp. 137–141.

Lowe, D. 1984. *The price of power*. Melbourne: Macmillan.

MacGregor, S. and S. Takhar. 1995. "Employment, poverty and social integration in mega-cities: Mechanisms of exclusion and inclusion." Institute of British Geographers Conference, 3–6 January, Newcastle, University of Northumbria.

Macquarie, L. 1819. Quoted in T. Williamson, *Housing in flood prone areas*. Melbourne: Department of Housing and Construction, 1975.

May, P. and J. W. Handmer. 1992. "Regulatory policy design: Cooperative versus deterrent mandates." *Australian Journal of Public Administration* 51(1): 43–53.

Mellor, B. and M. Ricketson. 1991. "Suburbanasia!" *Time Australia*, 8 April, pp. 18–26.

Moriarty, P. and C. Beed. 1988. "Transport characteristics and policy implications for four Australian cities." *Urban Policy and Research* 6(4): 171–180.

Morris, J. 1993. *Sydney*. Harmondsworth, Middx: Penguin.

Murphy, D. 1994. "Night of the no-go zones." *The Bulletin*, 3 May, pp. 24–29.

National Academy of Public Administration. 1993. *Coping with catastrophe: Building an emergency management system to meet people's needs in natural and manmade disasters*. Prepared for FEMA. Washington D.C.: National Academy of Public Administration.

Neutze, G. M. 1977. *Urban development in Australia: A descriptive analysis*. Sydney: Allen & Unwin.

New South Wales Bureau of Crime Statistics and Research. 1993. *New South Wales recorded crime statistics*. Sydney: New South Wales Bureau of Crime Statistics and Research.

New South Wales Department of Environment and Planning. 1982. *Zoning of flood prone land*. Circular No. 31, 15 February.

New South Wales Department of Housing. 1991. *Urban consolidation review*. Sydney: Department of Housing.

New South Wales–FAC (New South Wales–Floodplain Advisory Committee). 1985. "Policy summary." Sydney: The Committee.

Newman, P. and J. Kenworthy. 1989. *Cities and automobile dependence: An international sourcebook*. Aldershot: Gower.

NRMA. 1989. *Household burglary in New South Wales*, Sydney: NRMA Insurance Ltd.

Papadakis, E. 1993. *Politics and the environment: The Australian experience*. Sydney: Allen & Unwin.

Pearson, L. 1994. *Local government law in New South Wales*. Sydney: Federation Press.

Phelan, P. 1994. "Fight fires higher up: Australia needs to address its airborne firefighting capacity." *The Bulletin*, 18 January.

Public Health Bulletin. 1991. Editorial note. 2(8): 73.

Red Cross/Red Crescent. 1994. *World disasters report 1994*. Internation Federation of Red Cross/Red Crescent Societies. Dordrecht: Kluwer.

Rees, J. 1982. Profligacy and scarcity: An analysis of water management in Australia. *Geoforum* 13(4): 298–300.

Robinson, C. J. 1995. "Cleaning up the country." *Wildfire* 5(1).

Roddewig, R. J. 1978. *Green bans: The birth of Australian environmental politics*. Sydney: Hale & Iremonger.

Rose, A. James. 1993. "Cities of Oceania." In Stanley D. Brunn and Jack F. Williams, *Cities of the world: World regional development*. New York: Harper-Collins, pp. 195–224.

Rowe, G. C. and S. Seidler, eds. 1993. *Contaminated sites in Australia: Challenges for law and public policy*. Australia: Allen & Unwin.

Ryan, P. 1991. "Urban consolidation." In Z. Lipman (ed.), *Environmental law and local government in New South Wales*. Sydney: Federation Press.

Rynn, J. M. W. 1991. "The potential to reduce losses from earthquake in Australia." In D. I. Smith and J. W. Handmer (eds.), *Australia's role in the International Decade for Natural Disaster Reduction*. Resource and Environmental Studies No. 4. Canberra: CRES, Australian National University.

Sandercock, L. 1979. *The land racket: The real costs of property speculation*. Canberra: Silverfish.

Shrivastava, P. 1992. *Bhopal: Anatomy of a crisis*, 2nd edn. London: Paul Chapman.

Singh, G., A. Kershaw, and R. Clark. 1981. "Quaternary vegetation and fire." In A. Gill, R. Groves, and I. Noble (eds.), *Fire and the Australian biota*. Canberra: Australian Academy of Science, pp. 23–54.

Smith, D. I. and J. W. Handmer. 1991. "Water conflict and resolution: An example from Tasmania." In J. W. Handmer, A. H. J. Dorcey, and D. I. Smith (eds.), *Negotiating water: Conflict resolution in Australian water management*. Canberra: Centre for Resource and Environmental Studies, Australian National University, pp. 47–62.

Smith, D. I., T. Lustig, and J. W. Handmer. 1990. *Losses and lessons from the Sydney floods of August 1986*. Canberra: CRES, Australian National University.

Smith, G. J. 1990. *Toxic cities: and the fight to save the Kurnell Peninsula*. Kensington, NSW: New South Wales University Press.

Spearritt, P. and C. DeMarco. 1988. *Planning Sydney's future*. Sydney: Allen & Unwin with the New South Wales Department of Planning.

Stein, P. Justice. 1986. "Environment and planning: The great leap backwards." *Law Society Journal* 24(5): 62–65.

Stilwell, F. J. B. 1993. *Reshaping Australia: Urban problems and policies*. Leichhardt, NSW: Pluto Press Australia.

Taylor, J. N. and McDonald, L. A. 1988. "Issues arising from revised spillway design floods for dams." *ANCOLD Bulletin* 79: 53–71.

Tench, W. 1790. *Narrative of the Expedition*, pp. 118–130. In: C. M. H. Clark (ed.), *Select documents in Australian history 1788–1850*. Melbourne: Angus & Robertson, 1970, pp. 46–47.

Tiller, K. G. 1992. "Urban soil contamination in Australia." *Australian Journal of Soil Research* 30(6): 937–957.

Warner, R. F. 1991. "Impacts of environmental degradation on rivers, with some examples from the Hawkesbury-Nepean system." *Australian Geographer* 22(1): 1–13.

Watson, S. 1993. "Cities of dreams and fantasy: Social planning in a postmodern era." In R. Freestone (ed.), *Spirited cities: Urban planning, traffic and environmental management in the nineties*. Sydney: Federation Press, pp. 140–149.

WCED (World Commission on Environment and Development). 1987. *Our common future*. Oxford: Oxford University Press (The Bruntland Report).

Whelan, R. J. 1995. "Wildlife populations: Impact of hazard reduction burning." In C. Brown and L. Tohver (eds.), *Bushfire! Looking to the future*. Sydney: Envirobook, pp. 107–118.

Williams, J. E., R. J. Wheland, and A. M. Gill. 1994. "Fire and environmental heterogeneity in southern temperate ecosystems: Implications for management." *Australian Journal of Botany* 42: 125–137.

Wright, B. 1991/92. "Smog moves west as Sydney grows." *Ecos* 70 (Summer): 17–22.

Yong, Chen, et al., eds. 1988. *The great Tangshan earthquake of 1976: An anatomy of disaster*. Oxford: Pergamon Press.

7

Disaster response in London: A case of learning constrained by history and experience

Dennis J. Parker

Editor's introduction

London is the oldest of the contemporary mega-cities surveyed in this book. Its longevity affords a special perspective on urban hazard because events with very low recurrence intervals show up in the record of experience and because there are correspondingly more opportunities for social learning of adjustments to a wider range of environmental extremes. Age can impart a different flavour to the assessment and management of environmental risks. One example is heightened concern about safeguarding unique historical buildings and irreplaceable art objects that are part of a city's cultural heritage. Another is the adoption of a world-weary posture of metropolitan indifference to the vicissitudes of nature, something that is often on display in the editorial pages of serious newspapers in big cities. Yet we know little about the relevance of historic hazards to contemporary society, or about the process by which it is transmitted across the generations, or about the weight that is attached to lessons of long-gone experience. Dennis Parker's analysis of contemporary hazards in London opens windows onto all of these topics.

What experience and history teach is this – that people and governments never have learned anything from history, or acted on principles deduced from it.

(Georg Wilhelm Hegel – 1770–1831)

Introduction

London's history as a large urban centre stretches back at least 10 centuries to the year A.D. 1000 when it was home to an estimated population of 25,000. It is a palimpsest of many urban forms, whose development has sometimes been strongly influenced by major disasters such as the Black Death, the Plague, and the Great Fire (1661) as well as the Second World War blitz (1940–1941). Today, this city possesses an extensive suite of natural, technological, and social hazards but it is not as intensely hazard prone as other mega-cities such as Tokyo and Los Angeles.[1]

In view of London's long acquaintance with hazard, its residents might be expected to have developed many successful ways of adjusting. But environmental hazards have rarely been high on the public agenda; nor have adjustments been uniformly or consistently effective. In this city, the relationship between experience, learning, and response to hazard has been complex. A social matrix of cultural, institutional, and political-economic factors has often intervened to weaken the links between experience of hazard and protective actions (Parker and Handmer, 1992a, p. 262).

Evolution and contemporary urban issues

Location and physical setting

Broadly defined, the contemporary London mega-city includes a large rectangular tract of south-east England roughly demarcated by a line from Colchester in the north-east to Milton Keynes in the north-west and Southampton in the south-west (fig. 7.1).[2] The natural hazards of this region are not very obvious to the casual observer. South-east England is more or less geologically stable. Though not unknown, slope instability is rarely troublesome because the terrain is gently undulating and relative relief is usually less than 100 m. This region enjoys a mild, cool temperate climate whose most common hazards are occasional mists and fogs. Consequently, London's climate has typically been perceived as rather benign by residents and outsiders alike. That image is only partly accurate because variability is southern England's dominant climate characteristic. The region occupies a battleground between ever-changing air-masses that originate in a wide range of latitudes from the poles to the tropics. Weather extremes occur, although they are infrequent.

Older parts of London are located in the broad and originally marshy Thames Valley (fig. 7.2). Much of the city is close to the Thames estuary, and the river is tidal up to Teddington in west London. Many riverside

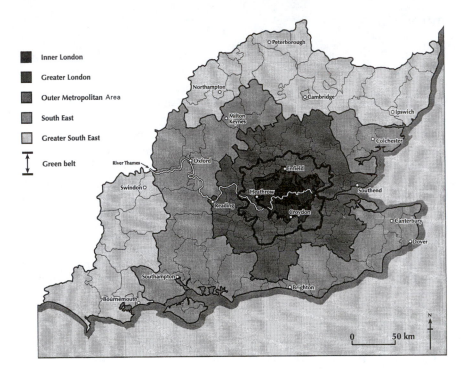

Inner London

Greater London

Outer Metropolitan Area

South East

Greater South East

Green belt

Fig. 7.1 The London mega-city

areas are either just below or just above high-tide level, and tidal flooding is a serious threat. Early neighbourhoods developed on gravel hill-tops and terraces adjacent to the Thames and around the mouths of its tributaries. As the city expanded, it spilled into the lower and middle reaches of tributary valleys that later provided routes for roads, railways, and canals. Catchments became heavily urbanized, greatly modifying their hydrological characteristics. By medieval times London comprised two distinct centres – Westminster and the City – joined by the Strand thoroughfare. Today it retains these twin cores, although during the twentieth century other growth centres have been added, most notably at Heathrow Airport to the west and in the rejuvenated London Docklands to the east.

Although the London urban region now contains around 13 million people, it is no longer the world's largest city. Some projections indicate that it will rank thirteenth in size by the year 2000. The extent of the slippage in international rankings depends entirely upon how London is defined, and its implications are not self-evident because at least some of the central city's population losses were deliberately imposed. Unlike most of its rivals, London has experienced almost 50 years of effective

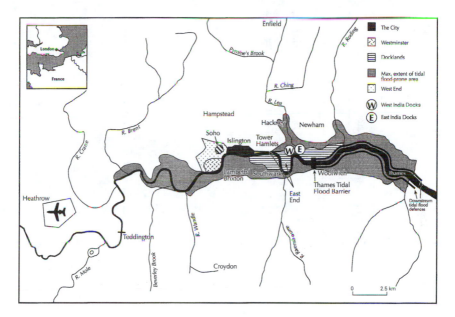

Fig. 7.2 Locational relationships between the historic cores of London and the River Thames tidal floodplain

land-use planning controls, which constrained population growth inside traditional metropolitan boundaries and brought into being a "dispersed metropolis" in the region beyond.

The role of environmental hazards in urban growth: 1100–1945

Environmental hazards have played an important role in London's evolution, and throughout much of the city's existence the vulnerability of individuals has been differentiated by class and wealth. Affluent classes sought to reduce their exposure by moving out of places that were naturally or socially at risk. In the process they gradually created exclusive neighbourhoods, characterized by services and amenities that did not exist in lower-class districts. Compared with later periods, relatively little is known about the hazards of medieval London but they are widely believed to have been numerous. Clout and Wood (1986) portray a society whose residents were intimately acquainted with threats to survival. Densely packed wooden buildings frequently burned down, drinking water was badly polluted, epidemics abounded, and mortality rates were high (table 7.1). In 1212 at least 3,000 died in a fire that destroyed much of the city. Despite these problems, it was not until the latter part of the seventeenth century that a community-wide disaster was recorded

Table 7.1 Selected historic disasters in London or affecting London, prior to 1900

Flood	Windstorm	Seismicity	Epidemic	Air pollution	Other
1091 } 600 p	1091	1692 "Panic" in central London	1349 Black Death, 50% of population died	1873 270–700 k	1666 Great Fire, 13,200 p, 8 k, 200,000 h
1703 } 22 k, £2 million (1703 prices)	1703	1884 Colchester earthquake	1603 Black Death, 30,000 k	1880 700–1,000 k	1867 Ice disaster, 40 k
1774			1625 Black Death, 40,000 k	1882	1878 Princess Alice sinking, 630 k
1809			1636 Black Death, 10,400 k	1891	1898 Albion launching, 39 k
1821			1665 Great Plague, 110,000 k	1892 1,000 k	
1823			1832 Cholera, 7,000 k		
1853			1833 Cholera, 1,500 k		
1875 } 600 p	1875		1849 Cholera, 18,000 k		
1877			1854 Cholera		
1881			1861 Typhoid		
1894					

k = number killed
p = properties damaged
h = number of homeless
£ million = damage estimates

in detail. Then, in successive years there occurred the city's two most celebrated disasters – the Great Plague of 1665 and the Great Fire of 1666. Only the fire is examined here.

The Great Fire occurred after a long hot summer had desiccated timber throughout the city. Wooden buildings provided ample fuel and narrow streets allowed flames to spread easily. Fanned by a strong east wind, sparks ignited riverside cellars and warehouses where inflammable materials were stored. Buildings were destroyed in a 1 km wide zone that stretched for 2 km along the north bank of the Thames. This amounted to about 80 per cent of the walled city area. The fire was halted only by deliberate destruction of buildings to create fire-breaks, aided by a period of calm windless weather (Reddaway, 1940). Although there were few (if any) known deaths, St. Paul's cathedral, 86 churches, and about 13,200 dwellings were destroyed. The disaster was so great that it threatened to undermine London's position as a major trading centre, and there was an accompanying but temporary breakdown of law and order. This was fed by groundless rumours that the city was about to be invaded by Dutch and French forces; the resulting evacuation propelled over 200,000 Londoners into refugee camps.

Destruction caused by the Great Fire provided a major opportunity for the redesign of London. At least seven grand redevelopment schemes were proposed, but none was adopted because merchants feared that a loss of business would occur if they waited for the urban cadastre to be remapped and for formal rebuilding to be completed (Jones et al., 1984). Instead, Parliament enacted a number of smaller schemes. These included: minimum street widths of 14 feet; regulations governing the height of buildings in relation to street width; special fire courts to resolve disputes during the rebuilding; straightening of roads; and tax or compensation programmes for landowners who gained or lost from alterations that occurred in the wake of the fire. Most important of all, the London Rebuilding Act made (fire-proof) stone and brick compulsory construction materials. Thatched roofs were also replaced with tiled ones.

The fire helped to alter the city's social fabric. Many wealthy residents moved to new planned accommodation in what is now the West End. This was partly to escape the fire's devastation, but also to avoid the congestion and perceived social evils of the City and port areas in the East End (fig. 7.2). These migrations tended to segregate rich and poor. None the less, the entire urban area continued to expand, especially as new bridges linked the old city to neighbourhoods on the marshy south bank of the Thames (Clout and Wood, 1986). Other fires also affected much of London. Between 1666 and 1839 at least 57 multiple house fires are known to have occurred. Some parts of the city that were repeatedly

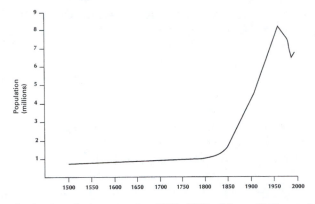

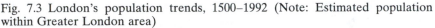

Fig. 7.3 London's population trends, 1500–1992 (Note: Estimated population within Greater London area)

and disproportionately affected include Wapping, Limehouse, Shadwell, and Stepney (Jones et al., 1984).

By the time of the first official population census (1801), London's population had risen to 1,117,000 (fig. 7.3). The city's accelerating growth was tied to Britain's emergence as a colonial trading nation. Shipping of goods through London's port increased greatly in the late eighteenth century, thereby creating demands for new enclosed docks, which were subsequently built in the early 1800s (fig. 7.2). The docks – and the opportunities for casual labour that they provided – formed the basis for an East End society that was generally characterized by insecure employment, poverty, inadequate housing, and disease. Poor and persecuted Jewish, Huguenot, and Irish immigrants who arrived in search of employment added to the disaster vulnerability of East End neighbourhoods because they lacked either the material resources with which to cushion themselves against disaster or the political power with which to call upon leaders for supplemental relief.

At the height of the British Empire, Victorian London (1837–1901) was "the most commanding concentration of people, trade, industry and administration to be found anywhere on the globe," a "supreme assembly of wealth, fashion and political and economic power" (Clout and Wood, 1986). It was also the scene of many hardships; deprivation, poverty, pollution, and disease were critical problems for working-class Londoners. Cholera epidemics regularly killed thousands (table 7.1). An official inquiry in 1842 established links between squalor and disease, but there was little official response until the end of the 1850s, when Dr. John Snow demonstrated a correlation between cholera and polluted drinking water in the Soho district of London. After this date, a series of ultimately successful measures gathered momentum. Abstraction of water from tidal portions of the Thames was banned. Intercepting sewers were constructed to capture effluent that flowed into the river and its tributaries, diverting it to outfalls well downstream to the east of London.

Air pollution was a persistent London hazard from the thirteenth century onwards. Pollution episodes were closely linked to trends in urban and industrial development as well as to variations in climate. Attempts to alleviate the problem met with varying degrees of success and it remained a major contributory cause of death and disease amongst Londoners, especially in the nineteenth century (tables 7.1 and 7.2). Aerial bombardment during the Second World War was the most recent large disaster to affect London. Damage was sustained during two periods: the blitz, which occurred mainly between September 1940 and the summer of 1941, and the flying bomb (V-1 and V-2) attacks, which reached a climax towards the end of the war. The impact on London was profound and comparable to that of a great earthquake. About 30,000 residents were killed and a further 50,000 were seriously injured. Most attacks concentrated on the City of London, where about 40 ha of buildings were almost completely destroyed, and upon the commercial port areas. Working-class residents of tightly packed housing near the East End docks were particularly affected and far more bombs struck there than elsewhere in London (fig. 7.4). The docks never really recovered from this damage and their extinction was sealed by technological changes in shipping that began during the 1960s. Thereafter, the use of bigger ships and containerization shifted port activities downstream away from the immediate vicinity of London.

Urbanization and hazards since the Second World War

Since the end of the Second World War, the hazard potential of London has been strongly – if indirectly – affected by public policies that were aimed at limiting the city's size, controlling the distribution of industry, and revitalizing decayed inner areas.

Limiting growth by means of a green belt

After 1940, the containment of London within a green belt of statutorily protected open-space land became a major public goal. The susceptibility of London's massed population to aerial bombardment during wartime provided a special stimulus to a broad-based clamour for restrictions on the city's growth. In addition, official pronouncements emphasized the declining quality of urban life in a sprawling city without adequate services and the difficulties of preserving rapidly disappearing countryside. London was portrayed as choking on traffic and suffering the "waste" of daily commuting, while many Londoners continued to live in

Table 7.2 Notable hazard events in the London region, after 1900

Tidal flood
- 1928[a] 14,000 ev, 14 k
- 1953 13,000 ev, c. 100 k

Non-tidal flood
- 1926
- 1947 Lea and Thames
- 1968 Mole, Wey, Lea 15,000 p
- 1975 Hampstead, £1 million
- 1977 NW London
- 1981
- 1983
- 1987 Waltham Abbey
- 1992 Silk Stream
- 1992 Lewisham
- 1993 Thornwood
- 1993 Widespread

Drought
- 1921
- 1934
- 1944
- 1976
- 1988–92
- 1993/4

Windstorm
- 1987 10 k, £1.5 billion
- 1990 20 k

Seismicity
- 1931 North Sea epicentre

Air pollution
- 1948 300 k
- 1952 4,000 k
- 1956 480 k
- 1957 340–700 k
- 1986 Chernobyl
- 1994 155 k

Terrorist/terrorist related
- 1973–94 41 k
- 1991 Stock Exchange
- 1991 Downing Street
- 1991 Liverpool Street, 6,700 ev
- 1992 St Mary Axe/Baltic Exchange, £1.5 billion
- 1993 Bishopsgate

Others
- 1908 Woolwich explosion, 16 k
- 1917 Silvertown explosion, 82 k
- 1940–45 Blitz, 30,000 k
- 1975 Moorgate Underground, 31 k
- 1980 Fire/explosion at chemical works, Barking, 10,000 ev
- 1981 Brixton riot
- 1985 Tottenham riot
- 1987 King's Cross Underground fire, 31 k
- 1989 Rail crash, Clapham, 35 k
- 1989 *Marchioness* sinking, 56k
- 1992 Power failure on Underground, 100,000 ev
- 1992 West End riot, £3 million

a. Also a non-tidal flood.
k = number killed
ev = evacuated
p = properties damaged
£1 million/billion = damage estimate

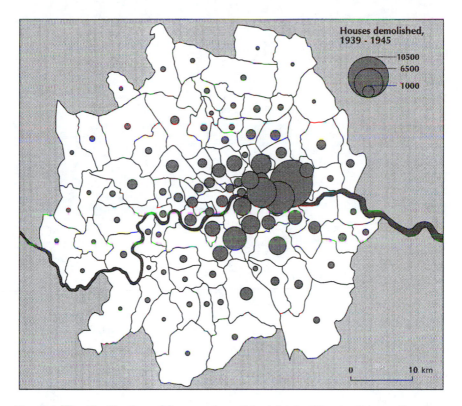

Fig. 7.4 The distribution of houses demolished in the blitz in Greater London, 1939–1945

cramped and aged housing. These arguments became the basis for creating the influential Royal Commission on the Geographical Distribution of the Industrial Population in 1940 (Wise, 1962).

Immediate temporary measures were taken throughout the Second World War to evacuate large numbers of Londoners to outlying towns. This expedient served to break established patterns of urban life and opened the way for post-war voluntary migrations to the outer suburbs (Wise, 1962). Between 1942 and 1944, Sir Patrick Abercrombie's plan for London's post-war development was prepared. This envisaged a mass regrouping of population in a surrounding ring of new or enlarged towns, each with their own workplaces. Inner areas that lost population to these towns would be redeveloped and enhanced. Vulnerability to external attack would be reduced by dispersing population over a wider area.

In 1947 the Town and Country Planning Act in effect nationalized development rights and placed them in the hands of County Planning Authorities, which were empowered to stop development. A "green

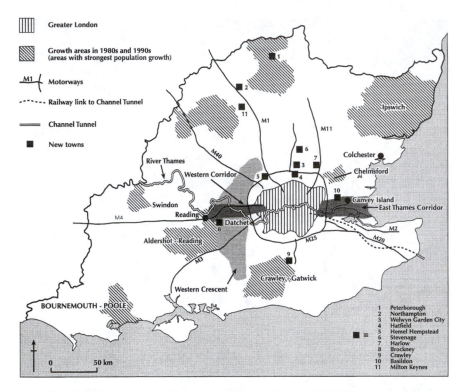

Fig. 7.5 New towns and growth zones in the post-1940 London region

belt" was established around London to halt the city's growth (fig. 7.1). New and enlarged towns were created (fig. 7.5). Many Londoners relocated to towns beyond the green belt, allowing blitzed and blighted areas in London to be progressively redeveloped.

These and other actions taken in support of the Abercrombie plan effectively halted London's physical sprawl but they failed to halt the growth of employment in London. In the 1950s much of the nation's population growth occurred in a ring of communities about 25–55 km from the centre of London, beyond the green belt. Like their wealthy ancestors, and aided by greater mobility, middle-class Londoners found that they could escape urban congestion and live in satellite towns or in the countryside. By the 1960s the highest rates of housing and population growth were 55–110 km from the city centre. A decade later, the fringes of the expanding mega-city were already pushing out of the traditional London region into parts of East Anglia, the south central coast, and the Midlands.

The London green belt's effect on quality of life and exposure to hazards has been complex. The city's centripetal growth, itself hastened by green belt policy, has partly alleviated the build-up of further environmental hazards in the metropolis by reducing concentrations of people and investments at risk. Conversely, and at the same time, there has been an intensification of environmental problems beyond the green belt, including encroachment of development onto exurban floodplains. In other words, many of the outer suburban environmental problems of the 1990s have been exacerbated by the process of London's decentralization. Traffic congestion, loss of open space, poor air quality, and flood hazards are now significant issues in the zone 55–110 km out of central London.

Since 1961 the population of Greater London (i.e. the central city and its inner suburbs) has stopped growing and gone into decline. The population of Greater London declined from 7.8 million in 1966 to 7.0 million in 1991, whereas the population of the Outer Metropolitan Area rose from 5.1 million to 7.4 million over the same period (Hall, 1989, p. 43). Recently there have been signs of population revival in the city's central precincts (Champion, 1987). It was recognized that the overall population of the South-East and the London mega-city would continue to grow, while the population of Greater London was expected to stabilize. In a move that had profound consequences for the London region, the 1970 Strategic Plan for the South East sought to channel investment and population into five major growth areas and seven smaller ones. The intent was to replace a unicentric region focused on London with a polycentric one that includes many other urban nodes. During the 1980s and 1990s population growth was largest in the outer fringes of the mega-city. London now resembles a fragmented urban region similar to that described by Gottmann (1961). Places such as Crawley–Gatwick and the "Western Corridor" to Reading are functionally, if not administratively, a part of London (fig. 7.5), and other UK metropolitan areas are interconnected with London in a progressively more complex web of interdependence.

London and regional balance

London's growth has been affected by Britain's industrial location policy. In the early 1940s it was believed that the London region drained away jobs and industries that might otherwise have supported strong urban centres elsewhere in the country. At the same time, North-East England and other declining industrial regions were losing population and investment. This analysis led to the creation of a system of national control over the location of manufacturing industry. Since then, policies for diverting employment to other regions have waxed and waned. Labour Party governments (which have drawn support disproportionately from

old industrial communities in the North of England) have typically called
for the rejuvenation of lagging peripheral regions, whereas Conservative
Party governments have tended to accept or welcome the dominance of
London. In any event, most industrial location policies appear to have
been ineffectual. For example, Hall (1989) notes that firms that were
given development certificates to relocate outside of London were not
required to invest in the lagging industrial regions. As a result, many
"high-tech" electronics companies merely set up shop in the so-called
"Western Crescent" just beyond London's green belt (fig. 7.5).

A new departure in industrial location policy occurred in the 1990s
when the UK government designated the middle Lea valley of north
London as an area that will receive selective funding for employment and
training initiatives (Government Office for London, 1994). For the first
time, the London area's severe structural problems of industrial decline
and high unemployment were recognized. More recently, the European
Union has designated the lower and middle Lea valley, and adjacent
parts of the East End and Docklands, as areas eligible to receive funds
for purposes of economic regeneration. As argued herein, there is a
relationship between industrial redevelopment policy and the growth of
hazard potential, but rarely, in modern Britain, is the relationship
explored or taken into consideration. This is well illustrated by the case
of the Docklands redevelopment scheme.

Urban regeneration and Docklands redevelopment

Planned decentralization left behind many of the working-class poor in
old and decaying residential and industrial areas of central London.
During the 1960s the problems of these areas attracted growing attention.
The Inner Urban Areas Act of 1978 set up special partnership authorities
for deprived areas such as Hackney, Islington, and Docklands. Participants included borough councils, various central government ministries,
and agencies responsible for employment training. The aim was to foster
economic regeneration and to coordinate policies for transport, housing,
education, community, health, and environmental improvements. During
Margaret Thatcher's Conservative political revolution (1979–1990) the
concept of strategic planning was abandoned and in 1986 the Greater
London Council, the mega-city's governing body since 1965, was abolished. Henceforth the forces of commercial enterprise were free to shape
London's future. Urban enterprise zones were created and urban development corporations were set up to make the necessary public investments that would bring private capital to inner-city areas.

The redevelopment of an important part of London's derelict inner
docks is the most spectacular example of free market urban regeneration
in London (fig. 7.2). The project was begun by the Conservative govern-

ment soon after the election of 1979. A London Docklands Development Corporation (LDDC) was created to slice through bureaucracy and create a separate "city" within London. Using unprecedented powers, the LDDC welcomed private developers to the Docklands enterprise zone. By mid-1986, £279 million of public funds had been invested to attract six times that amount in private capital, much of it for the construction of office space. Some 400 new companies and over 8,000 new jobs appeared in the area, together with sites for thousands of new homes (Hall, 1993). The Docklands initiative has generated fierce controversy. One debate concerns the way in which existing working-class communities were bulldozed and replaced by "Yuppie" neighbourhoods whose housing was unaffordable and whose jobs were unobtainable by the original inhabitants (Coupland, 1992). Another controversy surrounds the style of development, which has been criticized as lacking variety and a human scale (Hall, 1989). The enormous Canary Wharf tower in the heart of Docklands attracted much criticism because of both its visibility and its chequered investment history. Construction began in the mid-1980s during a period of rapid economic growth and prosperity, when there was an acute shortage of office space in London's central business district. By the early 1990s demand for office space had slumped, little of the 4.4 million sq. feet of floor space had been leased, and the original investors went into bankruptcy. Taken over by its banking creditors, the project languished for several years and has only recently been re-acquired by private investors (*New York Times*, 3 October 1995).

Docklands is one reflection of a private property boom that occurred throughout much of London during the second half of the 1980s. This was aided by the deregulation of British financial services in 1986 and has been coupled with an unexpected recovery of population growth in central London (Champion, 1987). Yet there has been no strategic planning for any of this redevelopment that would, for example, consider the impacts of new construction on traffic congestion and pollution or the potential for increased exposure to other environmental hazards such as flooding.

Vulnerability and resilience in a premier global banking and finance centre

London is the premier global centre for international banking and financial services. About 800,000 people are employed in these services throughout the metropolitan area, 300,000 of them in the City (the financial business district). London leads all other places in foreign exchange turnover, the issuance and trading of international bonds, and the trading of overseas equities. Over 43 per cent of the world's currency trading is centred in the mega-city. It is also where the world gold price is fixed daily. In all matters of financial transactions London competes for the supreme world leadership position with New York and Tokyo. Herein the city's position at the hub of a global air transportation system is an undoubted advantage.

London's accession to a high position in the world's economic system has changed the mix and magnitude of environmental hazards to which it is exposed. On the one hand, the financial businesses located in London are to a large degree insulated from fluctuations in an often volatile domestic British economy; risks of labour unrest, government regulatory pressures, and the consequences of business failures are now diluted because investments are spread across many countries. On the other hand, London is a major insurance and reinsurance centre, and insurance investors are now exposed to risks of natural disasters and other crises elsewhere in the world. By itself this is not necessarily a disadvantage if insurance risks are correctly assessed and internalized in the system of premiums that policy holders pay for coverage. But the rising scale of disaster insurance losses during the past decade suggests that these risks are not accurately assessed. For example, the losses of the 1980s were three times higher than those in the 1960s (Downing et al., 1993). Recent research indicates that perceptions of natural risks in London vary widely among insurance firms and that – for some perils – premiums bear little relation to the physical risks (Doornkamp, 1995). Worse may yet be in store. During the 1980s the world's business community was startled by predictions that a large earthquake in Tokyo could precipitate a collapse of the global economic system (Lewis, 1989). This would have particularly grave repercussions for London.

Finally, London's dominant position in global banking and finance may perversely increase the possibility of intentionally created disasters. For example, financial institutions and the transportation system that serves their workers have become targets for groups that seek to attract the attention of the mass media and force changes in public policy. During the 1990s the Irish Republican Army (IRA) carried out a number of major bombings that devastated parts of the City (see below). The IRA is only one of a number of terrorist organizations that are known to have targeted London. One effect of terrorist bombings on the City has been a dramatic increase in local property insurance premiums – a further component of rising disaster losses.

Income inequality and social polarization

Since the 1970s London has experienced a marked growth of income inequality and social polarization. Not only have the better-off grown more wealthy but the worse-off have grown poorer. The lowest-income groups suffered a drop in levels of living during the 1980s that was even more marked than similar trends in New York and other developed mega-cities (Logan et al., 1992, pp. 131–132). Racial and ethnic minorities have been disproportionately affected by these trends because they are heavily concentrated in London, particularly in the inner city. In 1982, Greater London was home to 49 per cent of Britain's West Indians and 34 per cent of its Asians (including 66 per cent of Bangladeshis). In 1987, unemployment amongst Greater London's ethnic minorities was 16 per cent, compared with 9 per cent for the majority population. Because the value of state unemployment benefits and other social support programmes has declined relative to income during recent years, higher unemployment rates almost certainly translate into worsening relative deprivation. Female-headed households have been particularly disadvantaged.

Socio-economic disadvantage is concentrated in three adjacent inner-city boroughs: Islington, Hackney, and Tower Hamlets. These areas have high levels of overcrowding, predominantly rental housing, lack of access to cars, and severe unemployment. They are all experiencing gentrification, which sharpens income gaps between rich and poor. Other boroughs such as Lambeth/Brixton and Newham (which coincides with the East End) also have high levels of deprivation (fig. 7.2). The re-emergence of beggars and large numbers of homeless people, many of them the young unemployed, is the most obvious sign of spreading destitution in London. There are about 3,000 officially designated "street homeless" and a further 30,000 unofficial homeless households that are none the less provided some public support. In addition, about 70,000 single people, childless couples, and others who are functionally homeless are not recognized or served by local authorities (Brownhill and Sharp, 1992).

Housing is a particular problem. Poor tenants become trapped in public housing and unable to afford private sector alternatives. The concentration in London of multinational financial corporations has led to an increase in demand for luxury housing and large-scale investment in "up-market" homes. New clusters of offices and high-income housing occupy sites adjacent to working-class neighbourhoods characterized by high-rise, high-density public housing (Harloe et al., 1992). Income inequality and social polarization are inextricably linked to London's changing market for labour. The metropolitan area has experienced a major shift from jobs in manufacturing and goods-handling to jobs in finance and producer services. This has benefited educationally qualified white workers of both sexes at the expense of men from established working-class communities (Gordon and Sassen, 1992). However, there has also been a growth of low-wage service jobs requiring few qualifications linked to an increase in flexible jobs within a growing informal sector.

Hazards and urbanization: A summary overview

In London over the past 50 years the process of urbanization has been subtly but significantly bound up with the creation and alleviation of environmental hazards. These links were especially clear during the 1940s, when hazard reduction was an explicit element of policies mainly intended to dilute a high-density city. Since then, the locus of hazard potential has moved progressively outwards, with intense growth of hazard potential along certain axes, notably the "Western Corridor." Today, the mega-city as a whole is probably more resilient to environmental risks and hazards, but many inner neighbourhoods and other areas are significantly more vulnerable to loss.

Urban issues – Outlook for the future

It is clear that a number of factors are likely to play important roles in shaping twenty-first-century London. These include: London's role in the global economy; the nature of metropolitan government; the choice of public management philosophies that guide government actions; strategies to resolve a continuing transportation crisis; and the success of measures that may be adopted to cope with burgeoning crises in the city's health-care delivery system, rising crime rates, and declining environmental quality.

London's position as a leading international city

Britain's role in an increasingly interconnected global economy will have a major influence on London's future, including propensities toward increased environmental hazards. The need for maintaining international competitiveness is continuously debated by leaders of government and industry. Although London's share of international banking has declined and Tokyo's has increased, the UK government and the Corporation of London remain committed to ensuring that London retains legal and institutional arrangements that are highly favourable to international financial dealings. These should lead to a continuation of existing pressures for urban expansion.

Britain's role in Europe will also affect London's development. To the extent that the United Kingdom plays a leading role in pan-European economic and political initiatives, London will attract additional investments from mainland countries. A Conservative Party government took Britain into the European Economic Community, but recent Conservative governments have been hesitant to embrace further initiatives, such as the European Monetary Union and the Maastricht Treaty, that would integrate Britain's economy and society more closely with those of its continental neighbours. Both of Britain's leading political parties have repeatedly reaffirmed their commitment to European integration, but strong Eurosceptic groups continue to exist within Britain. Since the accession to power of the Labour Party government in 1997, prospects for a strong commitment in support of European union have probably increased.

London's government

The question of how best to govern the London mega-city has been raised in various guises during the past century. London was the first large city to create an integrated metropolitan government. It did so in 1889 when the London County Council (LCC) assumed responsibility for governing what is now termed Inner London (Bennett, 1991). As the city grew, this body was succeeded by the larger Greater London Council, which continued in place until its abolition in 1986. With the end of metropolitan-wide government, London temporarily lost one opportunity for implementing broad policies for hazard reduction. However, the central (UK) government has had to re-create institutions that oversee at least some of the functions of metropolitan government. During 1993, the work of four departments of state was merged into a single regional office for London supervised by a Cabinet committee and run by a civil servant of deputy secretary rank. Significant numbers of civil servants within this new Government Office for London and the associated Department of the Environment are now involved in metropolitan-wide issues. A debate about whether to resuscitate a formal metropolitan government continues at least in academic circles (Bennett, 1991; Hebbert, 1992) and may re-emerge on the political stage in the future.

Public management philosophy

Margaret Thatcher's governmental reforms initiated a series of fundamental institutional and economic changes that have affected London's vulnerability to disaster. Successive Conservative governments have pursued monetarist economic policies that have focused on controlling inflation and "returning" state-run businesses to the private sector. Hood and Jackson (1992) refer to a "new public management" era, which is characterized by strong emphasis on cost-cutting and financial discipline rather than public investment. In the process, London's public services (transport, sewers, housing stock) have been run down. The consequences of poor maintenance and underinvestment were highlighted during a public inquiry that followed the November 1987 fire in King's Cross underground station, which killed 30 people and injured 50 (Department of Transport, 1988). A trend towards establishing "profit-making" units within public sector agencies such as the National Rivers Authority (NRA), the Home Office, and the Metropolitan Police Service is particularly worrying for hazards researchers because these units have important responsibilities for safety and hazard management. Privatization often causes breakdowns of existing cooperative relations among public agencies. There is also a strong preference for public policies that encourage deregulation, "light" regulation, and self-regulation. Some analysts argue that, as a result, the capacities of regulating and inspecting agencies have failed to keep pace with the growth of vulnerability to the full range of hazards (Horlick-Jones, 1990). In other words, it is argued that new public management initiatives may well have increased London's vulnerability to disasters, especially those of a hybrid type.

London's transport crisis

The most widely experienced public crisis that faces London today – and the one that may continue to head the list in the foreseeable future – is inadequate transportation. There is a high propensity for congestion, shut-down, and large-scale accidents. There are also strong causal linkages between London's growing dependence upon motor vehicles and its worsening air pollution hazard, including the risks of global climate change. Much of Greater London is unsuited to motor vehicles because – unlike Los Angeles and other recently established fast-growing cities – it was constructed in an era of horse-drawn transportation. London's overground and underground rail systems were also developed at a time when it made sense that most lines should converge on the central city. They are not now well suited to the contemporary decentralized mega-city. A large part of the transport system is old and prone to breakdown. Additional problems include: major gaps in the rail and highway networks, poor integration among different kinds of transportation, very high fares, and underinvestment in capital facilities and rolling stock. Much of the entire transportation system operates at or above capacity for relatively long periods. This makes London more vulnerable to disruption from accidents and bad weather. The annual costs to workers and businesses of road traffic congestion in South-East England probably exceed £10 billion. There are about 38,000 road accident victims in London per annum, including over 500 killed (1991 figures). During the 1980s, public inquiries were held into a series of transportation-related disasters, including train accidents and fires in underground stations. Environmental costs are also substantial. Congestion encourages many drivers to leave the highways and take short cuts through residential streets that are ill suited to that use. Although levels of lead in the atmosphere have been reduced, other forms of air pollution and traffic noise have been increasing.

Transportation remains a major public issue, although improvements have been made and more are contemplated. For example, there are proposals to curb the use of private cars, possibly by levying charges for road use. The city can also look forward to continuing calls for greater investment in public transport and the adoption of bicycle-friendly and pedestrian-friendly urban policies. Shifts in central government policy on transport are now detectable. These are apparently driven by the beginnings of recognition that there is a link between road traffic generation and air-quality hazards as well as other forms of environmental damage. During 1994, central government issued policy guidance advising local planning authorities to reduce vehicle trips by means of development controls.

An inner-city health crisis?

There is substantial concern about the future of London's hospitals and ambulance system. These problems are worst in central London, where population has been declining and its composition has been changing. The previous national (Conservative Party) government maintained that there are too many hospitals in the inner city and it restricted hospital operating budgets, postponed capital investments, and sought to close some emergency departments (e.g. St. Bartholomew's). These measures have affected the ability of hospitals to respond to medical emergencies in the central city and may have affected general standards of health in central neighbourhoods. London's emergency services are also hampered by poor labour relations between staff and administrators and by inadequate communications and control systems. For example, ambulance emergency response times are slow: only about 65 per cent of London's ambulances arrive to pick up passengers within the targeted 10 minute response period.

Livability and crime

Many residents express growing disillusionment with the quality of life in London, especially inner areas. This is reflected in a host of problems, including: old infrastructure, visible congestion, poor public transport, noise and pollution, inadequate local authority maintenance of dwellings, and high costs of living. Social polarization and the growth of an urban "underclass" add to this gloomy picture. Street begging, homelessness, and race riots (in 1981 and 1985) are also on the increase. Because of the perceived deterioration in the quality of life, 50 per cent of Londoners have considered moving out of the capital; others have already left, thereby contributing to inner-city decline (Appleyard, 1993). Recorded crime is on the rise in London. Areas of highest risk generally coincide with the most economically and socially deprived parts of inner London, notably the boroughs of Hackney, Newham, Lambeth, and Southwark. In parts of inner London, women are particularly severely affected by the threat of street violence and are in effect under curfew.

One of the central questions that will face London policy makers during the next two decades is whether levels of public disenchantment can be reduced. If disenchantment grows, investments may shift to other regions of Britain. There is now growing interest in the subject of designing defensible spaces for purposes of crime reduction (Coleman, 1985). If implemented, defensible space policies might herald a return to traditional kinds of street design and other urban forms from an earlier London.

Urban sustainability

The concept of urban sustainability implies that city leaders and city residents are committed to long-term planning horizons. London has a long history of pioneering urban design, including the Garden City ideas of Ebenezer Howard, that found practical expression in the city's green belt. By this means urban form was modified, countryside was safeguarded and low-density urban development was encouraged. The main negative features of these policies have been increased hazard potential in outer London neighbourhoods and the energy inefficiencies of large-scale commuting. Today the city is decentralized – a settlement pattern that is not currently favoured by the European Commission, which prefers high-density, mixed-use, compact urban centres (Breheny and Rookwood, 1993).

An eastward shift in the locus of hazard potential

The spatial context of London is likely to change markedly over the next two decades in relation to the rest of Europe and the rest of the United Kingdom. This may well shift the locus of hazard potential in London from the west to the east and south-east. With the opening of a submarine rail tunnel between England and France in 1995, Britain ceased to be an island, and the accessibility of London was dramatically changed. Rail service now brings the centres of London, Paris, and Brussels within two and a half hours of each other. Despite fierce competition from shipping and air services, the Channel Tunnel rail link is likely to become the main gateway for Britons seeking to do business in Western Europe, and communities along the route from London to the Kent coast are likely to receive additional investments from British-based companies that hope to penetrate the continental market and from mainland-based companies that hope to increase trade with the United Kingdom. As a result there will probably be an increase in exposure to environmental hazards on the southern and eastern fringes of London as new developments flow into this region. The magnitude of the shift depends upon how successfully London will be able to compete with other cities as the economic heart of Europe is dragged eastwards towards Germany and the emerging economies of Eastern Europe. It also depends upon precisely where new peri-London developments will be permitted to locate.

Accelerating development of the East Thames Corridor (fig. 7.5) may be one important outcome of these pressures. The lower Thames, with its extensive docks, used to be the economic powerhouse of London. It lost that position during the post-war years, when new industries were increasingly drawn west of the city along the middle Thames Valley (fig. 7.5). Now these districts are experiencing serious problems, such as overloaded transportation systems and encroachment of building into environmentally sensitive areas – including the Thames floodplain. At the same time, depressed communities on the east side of London cry out for regeneration.

There are clear signs that the economic core of London has started to shift eastwards. Some industries have already moved production into eastern locations. During the 1980s, for example, newspaper companies largely abandoned Fleet Street and moved printing and distribution functions east to more spacious quarters. Central government has also begun to encourage the development of the East Thames Corridor, and plans exist to establish a major new "Euro-city" at Ebbsfleet (fig. 7.6). East London has witnessed large new distribution and retail developments encouraged by the completion of the eastern sections of the M25 superhighway. The advent of jet air services at the newly expanded London City Airport has given east London a further boost. The government's vision is of a revitalized growth zone with a series of environmental enhancements that will pull London's centre of economic gravity south-east towards continental Europe (Department of the Environment, 1993a). Unfortunately, many of the prime development sites are in areas prone to tidal flooding, and an eastward movement of investment could well lead to significantly increased flood exposure, similar to that which occurred when London expanded west (fig. 7.6).

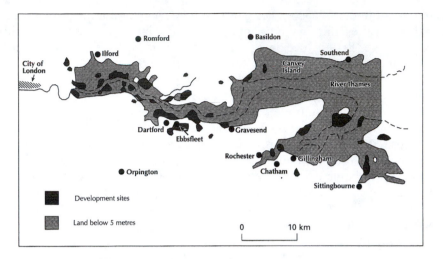

Fig. 7.6 Prime development sites in the East Thames corridor

Table 7.3 Three categories of hazard affecting the London mega-city

Meteorological/geophysical	Technological	Social
Flood	Air pollution	Terrorism
Drought	Hazardous waste	Crime
Windstorm	Transport	Homelessness
Fog	Nuclear accident	Crowds
Subsidence	Fire	Public disorder
Rising sea levels	Industrial hazards	Child poverty
Seismicity		
Snow		
Global warming		

Environmental hazards

A representative but incomplete list of London's major environmental hazards is provided in table 7.3. It includes a wide selection of extreme events or processes. Before considering these hazards in greater detail, it is important to take note of three factors that have broadly influenced the city's hazard-management system. First is a tendency toward public reticence about the existence or seriousness of environmental hazard. Second is the vulnerability of an increasingly complex urban infra-structure to short-term disruptions. Third is the difference in post-disaster recovery rates between rich and poor communities.

Factors influencing London's hazard-management system

Public information and attitudes about hazards

The tendency of many Britons to regard the ambient physical environment as benign has already been noted above. In combination with typical mass media reporting of disasters as "foreign news," this attitude has a dampening effect on levels of awareness and concern about domestic hazards. However, concern about different phenomena also varies widely as a consequence of information availability and information biases. Historically, British government agencies have not encouraged high levels of hazard awareness among the country's residents and have been wary of releasing public information about hazards.[3] Several reasons explain this stance. Officials are sometimes concerned about the economic impacts of releasing inaccurate information; thus, the dimensions of flood-hazard zones are often not publicized because survey information is lacking or incomplete. Defensiveness about the limitations of public policy responses is also evident. One example is the reluctance of London officials to admit that the costly and much-touted Thames Tidal Flood Barrier does not eliminate all significant risks of catastrophic flooding. There is also a widespread perception among civil servants and other public officials that laypeople will be needlessly alarmed by information about risks and will call for unnecessary hazard-mitigation efforts. A highly politicized discourse about nuclear radiation risks is rife with governmental accusations about irresponsible behaviour on the part of environmental pressure groups. Industrial leaders and managers, too, are usually reluctant to disclose risks, and often cloak them with arguments about the need for confidentiality in respect to commercial secrets. Conversely, well-timed releases of information about risks have sometimes been used to further partisan agendas of government, industry, and environmental groups.

When hazard information is available, the role of the mass media often becomes crucial. Sometimes, they have been a pivotal force for change, as in the case of London's smog-control legislation of the 1950s. Mounting media publicity provided much of the ammunition that was used by the National Society for Clean Air to force a reluctant government to investigate air pollution. Conversely, reporting of hazards and disasters in London may sometimes be misleading. During October 1994 the risk of disease being spread in London by visitors returning from plague-ravaged communities of India was much exaggerated in media reports. Usually, the level of reporting of hazard is mismatched with the seriousness of actual threats. Severe localized flooding occurs frequently throughout Greater London but receives remarkably little media attention; the aggregate social costs of the city's road accidents are also significantly underreported.

Despite a history of relative neglect by government and some distortion by the mass media, the role of public information about hazards is undergoing changes in contemporary London. One of these is a consequence of Britain's membership in the European Union (EU). In response to various experiences with unprecedented industrial disasters, the EU's so-called "Seveso Directive" requires governments to publicize information about certain industrial risks and hazards (De Marchi et al., 1996). The trend towards increased use of hazards information is also fed by growing use of detection and warning systems. Precautions taken in response to IRA bombing campaigns in London have been an obvious example. These are reinforced by police pronouncements about the need for public vigilance in a city that is increasingly concerned about rising rates of conventional crime. Public warnings about poor air quality, severe weather, and sunburn have also become more common in London. In summary, throughout the city and the United Kingdom, there is a growing belief that laypeople should be made more aware of environmental hazards.

Vulnerability of urban infrastructure

The degree of disruption caused by an extreme event depends among other things upon: its magnitude, timing, and location; types of activities affected; and the existence of feasible contingency plans. In a complex city such as London, the number of potential disruption scenarios is huge. However, it is possible to make several generalizations. Foremost among these is the fact that London's infrastructure is highly vulnerable to costly, short-term disruptions as a result of even mild environmental hazards such as light snowfalls and falling leaves, as well as more serious threats such as high winds and terrorist bombs. The city's various transportation systems are particularly susceptible to disruption. Although, in theory, they are characterized by a high degree of transferability (i.e. many alternative routes can be used if some are blocked), roads and railways tend to operate at capacity during commuting times, thereby reducing the availability of alternatives (Parker et al., 1987).

London's underground rail system is exposed to several types of risk. A major tidal flood could seriously affect many kilometres of track and about one-sixth of the city's underground stations. The probability of this is judged to be very low because it would require an extreme event that exceeded the design standard of London's tidal flood defences. Other types of underground disruption are more common. For example, during 1993 the Central underground line suffered several days of catastrophic electrical power failures that stemmed from a variety of different causes.

These closed the entire system and many passengers were trapped in stalled trains with limited ventilation. Fortunately, the thousands of casualties that were expected failed to materialize.

Improved contingency planning does not necessarily prevent infrastructure disruption, but it has reduced the effects and improved the overall survivability of London's businesses during disasters. This was well illustrated by the impact of a terrorist bomb in the City's Bishopsgate district in 1993. Although a large part of the neighbourhood was devastated, damage to financial institutions was limited because contingency plans had been developed both before and after previous bombings. Firms were able to transfer many activities elsewhere and specialist disaster recovery firms were on hand immediately after the blast. The rise of disaster contingency planning in the private sector is now contributing substantially to reducing disaster-vulnerability. Also, memories of how Londoners coped during the Second World War blitz have provided models for coping that have been emulated by people caught up in today's disasters. This was evident in a "business-as-usual" approach among many Londoners in recent IRA bombings.

Differential rates of post-disaster recovery

The pace of post-disaster recovery is mainly a function of three factors: magnitude of damage; opportunities for change created by disaster; and recovery priorities. At the end of the Second World War, these factors conspired to slow the pace of recovery from bombing destruction. Many working-class communities had been so devastated that they never fully recovered. In addition, planners and city officials perceived an opportunity to reduce inner-city densities by regrouping population in other parts of the metropolitan region. Because so many homes had been destroyed, priority was given to rebuilding houses in satellite towns; public services received less investment. In the city centre, only one new public building had been constructed even six years after the war and it was not until the 1960s that new inner-city housing and services were once again being provided in significant numbers. The pace of business recovery was similarly slow. Although reconstruction was being discussed in 1941, a plan for replacing the City's financial institutions was not approved until July 1947. Currency trading ceased in the City during the war, and the foreign exchange markets were not reopened until 1951. Indeed, a boom in post-war reconstruction did not really develop until the 1980s and early 1990s. It was only then that the last prefabricated (temporary) house was consigned to the Imperial War Museum, the derelict docks were cleared away, and large-scale investments flooded into London's financial district.

Unlike the past, the pace of recovery from more recent disasters has been dictated by two major variables: the speed with which utility services have been restored; and the workings of the private insurance industry. After the record windstorm of 15 October 1987, London's economic and social recovery was comparatively rapid, although it will take many decades for damaged woodlands to grow back (Mitchell et al., 1989). There was widespread, mainly minor, damage to buildings but most were quickly repaired using funds from insurance reimbursements. Within 72 hours most roads blocked by fallen trees had been cleared. There was an orderly and rapid restoration of power to many areas within 9 hours, except where local distribution lines had been lost. Ten days after the storm, about 50,000 (mainly exurban) consumers in the London region were still without electricity.

The role of insurance is increasingly important. Because there is now a relatively high level of private disaster insurance in the South-East of England, post-disaster recovery tends to take place quite rapidly. Although temporary shortages of some building materials, and competition for repairers, slowed recovery from extensive flooding in the Thames Valley in January 1991, this relatively affluent area bounced back rapidly and completely. In less affluent areas of London it is clear that the pace of post-disaster recovery is affected by several income-related factors: a relative lack of insurance; limitations on insurance policies that prevent "new for old" replacement of damaged property; and the extent of victims' alternative financial reserves (Tunstall and Bossman-Aggrey, 1988). As a general rule, recovery tends to be slowest in low-income areas where victims are either underinsured or uninsured. To a significant extent the preceding generalizations hold for all parts of London and all hazards. However, the trends that are affecting different hazards and different components of the same hazard (i.e. risk, exposure, vulnerability, response) are complex and require further elaboration. They are first examined separately for each hazard and then summarized in totality.

Natural hazards

Tidal flooding

London is the chief European city at risk of tidal flooding, and this has long been the city's most serious natural or quasi-natural hazard. Key government buildings at Westminster, the seat of British government, are among many thousands of structures threatened with inundation. Exceptional tides have caused deaths and/or substantial damage throughout the centuries, most recently during 1928, 1947, and 1953. In February 1953, the Thames came within a few millimetres of overtopping the flood defences of the Houses of Parliament. On this occasion, much of east coast Britain was severely flooded, causing approximately £1,000 million (*c.* US$1.5 billion) property damage at 1993 prices. Over 300 people died, about 100 of them in the Thames estuary. This flood also caused major property damage in London, and 13,000 people were evacuated (Summers, 1978).

The risk of tidal flooding is increasing mainly because mean sea level is rising in this part of England. Here regional sea level is rising at about 0.36 m per century, but the rate of rise is much faster in some places (e.g. 0.8 m per century at London Bridge). High-water levels are rising even faster than sea level (fig. 7.7); twice in 1978 the Thames reached levels similar to those of the highly unusual catastrophic floods of 1928. Several

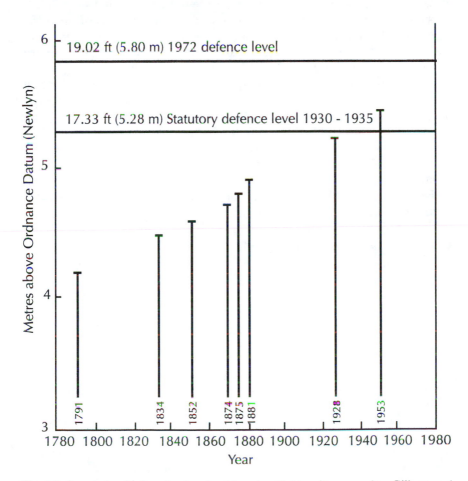

Fig. 7.7 Increasing high-water levels at London Bridge (Source: after Gilbert and Horner, 1984)

factors are responsible for these trends. They include geological processes, global warming, and human modifications of the hydrological system. For example, south-east England is gradually sinking to compensate for the continuing isostatic rise of northern Britain that began with the departure of Pleistocene glaciers; subsidence is also occurring in areas of excessive groundwater extraction. In addition, sea levels are thought to be rising because global warming is slowly melting the Antarctic ice sheet. The cumulative effects of dredging, embanking of former marshlands, and encroachment of buildings onto floodplains around the Thames have also increased the tidal range.

The traditional response to London's tidal-flood hazard was piecemeal and incremental: flood embankments were raised by small amounts to keep pace with increasing threats. By mid-century the limits of this strategy were at hand and new approaches were deemed to be necessary. Because of the increased frequency of high-tide surge levels, in the early 1980s the Greater London Council mounted a major public flood-warning awareness campaign and a flood emergency plan for London was devised. The public information campaign comprised mass distribution of flood leaflets to households in vulnerable areas, posters in prominent places, and graphic television advertisements. Unfortunately, on the rare occasions when awareness of flood warnings was subsequently investigated, the results were not encouraging. Surveys revealed that after receiving a one-hour warning many Londoners planned to take inappropriate actions, such as seeking to use the underground system (Penning-Rowsell et al., 1983).

The flood-warning campaign was mainly viewed as a stop-gap measure that would buy time pending the construction of an ambitious flood barrier across the lower Thames. A fixed flood barrage was considered during the nineteenth century and again during the 1930s. In the latter case, there was no follow-up because it was feared that the port of London would be vulnerable to closure if the barrage's lock gates were put out of action by enemy bombing. This concern disappeared after the Second World War, and a tidal exclusion barrier was finally opened in the early 1980s.

The Thames Barrier is the centrepiece of a £600 million, state-of-the-art, engineering flood-defence system that is intended to provide London with protection from extreme tidal surge events at least until the year 2030 (Gilbert and Horner, 1984). Normally, the barrier is open to permit shipping access to inner London. On receipt of a warning from the Storm Tide Warning System (established after the 1953 floods), gates can be raised off the bed of the river into a closed position. The barrier was designed not only to contain the currently estimated 1,000-year flood but to take account of a continuation in the upward trend of high-water levels plus an additional 1.5 m safety margin. The top of the barrier is 7.20 m above the present local mean sea level. Downstream, embankments along both sides of the Thames estuary have also been raised, and there are other smaller flood barriers on tributaries. One potential flaw in the system is its failure to take account of the possibility that global warming may accelerate the pace of sealevel rise. A faster rate of rise would reduce the effective useful life of the barrier. It is difficult to assess the added risk because of uncertainties about global warming. Several sealevel-rise scenarios have been considered by Turner et al. (1990). The "most likely case" assumes a net rise of 0.31 m between 1990 and 2100; whereas the "worst case" assumes a net rise of 1.1 m during the same period. Under some sealevel-rise scenarios, the probability of flooding may be 2 to 10 times more likely than at present.

Independent of any changes in flood risks, London's exposure to tidal flooding has increased dramatically. The Thames floodplain is the site of one of the greatest increases in flood loss potential in Europe. A serious flood would have major financial consequences for London (perhaps up to US$30 or 45 billion) and significant loss of life could occur. The flood-prone area includes the whole of London's docklands and riverside areas between the City (financial district) and Woolwich (fig. 7.2), both of which were the sites of major investment during the 1980s. Many of the existing riverside buildings, such as the Houses of Parliament, possess extensive basements and are susceptible to serious floods at street level. There has been a failure to prevent new building in the floodplain, although some recent structures have been elevated, thereby reducing their vulnerability. The area protected by the Thames Tidal Flood Barrier now contains approximately 1 million permanent residents; if daily commuters are included, the number rises to around 1.5 million.

The vulnerability of London's infrastructure to tidal flooding is high. Commuters in the exposed eastern part of the city are dependent on underground and surface rail lines. Electricity and telecommunication facilities are exposed and believed to be vulnerable, although the replacement of copper telecommunication cables with fibre-optic cables is reducing the flood-damage susceptibility of commercial linkages. The economic vulnerability of those living in the tidal-flood zone is highly variable, since it contains both high-income and low-income residents, the latter often composed of ethnic minorities. With the barrier in operation, London's tidal-flood hazard is considered to be a low-probability, high-consequence one. There exist various exceptional circumstances in which London might still be seriously flooded. Failure to close the barrier in advance of a major tidal surge through human error is one possibility. The tidal-flood defences that incorporate the barrier include 37 warning-dependent, moveable floodgates. Failure to close any floodgate could lead to serious tidal flooding in its immediate vicinity. Similarly, an exceptional tidal surge could cause the collapse of the north bank flood walls just downstream of the barrier, and this could lead to extensive flooding in Newham. Failure to synchronize the closing of the various subsidiary flood barriers along the Thames estuary could produce serious flooding, because closure of the barrier generates a reflected wave of water that could surge back downstream. Although it might be possible to warn of flooding caused by one of these failures, London's emergency planners are insufficiently aware of the areas at risk and by 1994 had been unable to test the effectiveness of the existing emergency response system to these kinds of threats.

During its existence, the barrier has been closed to prevent flooding on about 20 occasions and the frequency of closures is likely to increase. Its overall height can be raised by about 1 m, but this option is not available for many downstream flood defences because their foundations may not be able to bear additional loads. In summary, the future physical survival of a large part of London depends on a high-technology tidal-flood barrier that constitutes a fortress-like structural response to hazard. It involves very costly investment in a fixed line of defences that has a limited life-span. For most Londoners, the barrier has removed the perceived tidal-flood threat but, within a few decades at most, the structure must be replaced, perhaps by an even larger barrier on the outer estuary of the Thames. Meanwhile, a continuation of present land-use and development trends will progressively expose more and more of London to tidal flooding as the city's commercial centre migrates eastward (downstream) and the sealevel effects of probable global warming become more pronounced. Future tidal-flood problems will be even more difficult to resolve than those of the present.

Other types of flooding

Three other types of flooding are increasingly common in London: fluvial flooding, thunderstorm-related flooding, and sewer flooding. Fluvial flooding along the main stem of the Thames is a serious hazard independent of high tides or storm surges. There have been six major floods of this kind since 1894. They tend to occur after prolonged or heavy precipitation and mainly affect streamside areas from the river's tidal limit at Teddington westwards towards Reading. Localized flooding of tributaries such as the Lea, Brent, Crane, Wandle, and Ravensbourne (fig. 7.2) is also increasing. In central London, many tributaries were buried or canalized as the city grew and they now must carry increased volumes of runoff. Since the arrival of new housing and industrial estates there has been a threefold increase in flood peaks in the catchment of the Silk Stream, a tributary of the River Brent in north London (Hollis, 1986).

Outdated weir and bridge structures, which were designed for smaller flood flows, also contribute to tributary flooding, as does construction of riverside buildings that constrict water movement.

Throughout London there has been a marked increase in sewer-overflow floods, usually associated with intense convective rainfall (e.g. thunderstorms). Over 10,000 properties are at risk from sewage flooding at least once every 10 years; 8,300 were actually flooded during 1982–1983. Examples include severe flooding of Hampstead Heath (1975) and along the River Pinn (1977). The 1975 flood is estimated to have had a return period of approximately 1,000 years and caused over £1 million damage (Perry, 1981, p. 16). On 13 October 1993, London again came close to flooding disaster. Most of the Thames tributaries were bank-full and the ground was saturated following many weeks of heavy rain, but approaching storms veered away without adding to the problem.

During the past 30 years there has been a major increase in exposure to fluvial flooding. Among the newly constructed properties at risk are high-value residential and commercial developments and a large embanked water-supply reservoir. These developments have both raised the total value of properties at risk and displaced floodwaters onto adjacent land that was formerly not threatened. For example, between 1974 and 1984, 500 new properties were added to floodplain lands in Datchet near Heathrow Airport (fig. 7.8). The growth of flood-damage potential has continued to the present, and the community remains undefended against floods.

Vulnerability to rainfall-driven flood hazards is highly variable. Throughout the London region, vulnerability of road transportation to flash-flooding has increased because traffic volumes have increased and because the effects of congestion spread rapidly and widely, thereby reducing the network's ability to accommodate vehicles on alternative routes. For example, flash-flooding of London's inner ring road (North Circular Road) often brings the entire north-western sector of London to a complete standstill. The ageing of London's transportation infra-structure also helps to increase the city's vulnerability. On 11 August 1994, when two thunderstorms passed over central London (fig. 7.9), 28 Underground stations were closed because the Victorian-era drainage systems and pumps could not cope with floodwater. Train services were shut down for at least five hours and about 60,000 passengers were stranded in stations.

The typical response to fluvial flooding throughout London has been a combination of channel clearance, channel maintenance, and structural channel improvements. In some cases (e.g. the Lea Valley), flood relief channels were constructed early in the twentieth century. Flood retention reservoirs have also been employed and automatic flood-warning devices

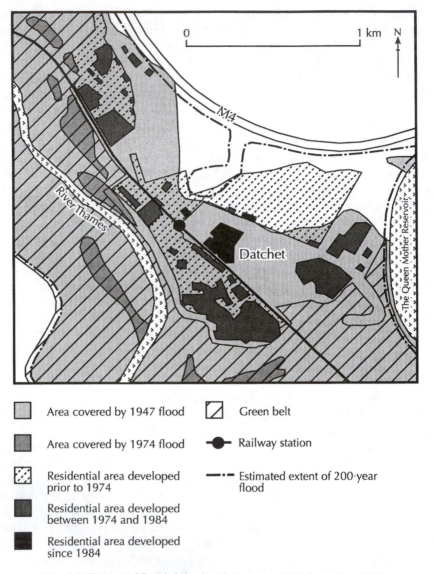

Fig. 7.8 Extent of floodplain development in Datchet since 1974

are used in a few locations. In theory, land-use controls offer the possibility of preventing further construction in flood-plains, but they are rarely effective because pressures to develop floodplains are intense in the London region. Only where a floodplain lies within the statutory green belt has development been resisted. Non-structural approaches to flooding are needed, and the National Rivers Authority is now pro-

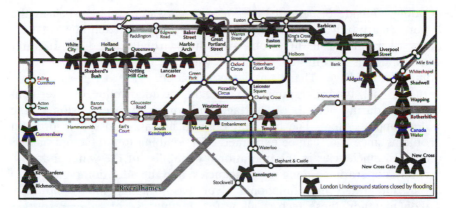

Fig. 7.9 London Underground stations closed by thunderstorm flooding, 11 August 1994 (Source: various newspapers)

moting openspace river corridors as conservation and amenity resources (Tunstall et al., 1993). Recognition is also growing that London's storm drains have an inadequate capacity for today's runoff .

Drought, land desiccation, and subsidence

The most notable recent drought in the London area occurred in 1976 and lasted for 165 days during the hottest summer since records began in 1727 (Parker and Penning-Rowsell, 1980). One consequence of anthropogenically accelerated climate change may be an increase in the risk of drought, but there is as yet no evidence that drought probabilities are changing in south-east England (Climate Change Impacts Review Group, 1991). Currently, water-resource planning is based upon the need to maintain reliable supplies to cope with a 1 in 50-year return period of low river flows and groundwater levels.

Exposure to drought and the incidence of drought-related subsidence through the drying out and cracking of clay soils have increased as London has grown. Following the 1976 drought there were over 20,000 claims in southern England for structural damage caused by subsidence. The main effect of drought on Londoners has been the transfer of drought risk to households through steeply rising subsidence insurance premiums and rising water charges. These have a disproportionate impact on people with low incomes. Per capita water consumption, and the population of the Thames catchment, have been growing steadily since 1945, and London's underground water resources were assessed as being fully exploited by the 1980s. Yet since 1970 reduced abstractions from aquifers have led to rising groundwater levels in central London, where underground structures and foundations are increasingly exposed to flooding. Outside of central London, surface water sources such as the River Lea are fully exploited and, until recent remedial action was taken, many rivers in porous limestone areas had virtually dried up. If water consumption is to continue growing, it will have to rely on further development of the River Thames as a supply source.

In the past, typical responses to drought included reduced water consumption during periods of shortage and longer-term extension of the water supply system. The fact that London has avoided severe drought problems since 1945 can be explained by a combination of factors such as improved efficiencies in the operational capability of the water industry, greater cooperation among water agencies, and the adoption (since 1974) of an integrated multifunctional river basin management approach. Taken together, these factors allow for flexible use of water resources and increased intra-basin transfers. The river basin management approach has significant untapped potential for further enhancing water supplies and reducing demand. For example, during the 1976 drought an intensive publicity campaign encouraged consumers to conserve water. Voluntary restrictions proved insufficient and unpopular mandatory hosepipe bans were introduced; these are now increasingly common in south-east England. During August 1976, Parliament introduced a Drought Act that extends the government's power to prohibit watering of parks, golf courses, and other facilities. A government minister was also appointed to coordinate responses to the drought.

Windstorm

Windstorm hazard is greatest on the west and south-west coasts of England, where the storm of record occurred in 1703 and killed about 8,000 (Perry, 1981, p. 24). Most windstorms are associated with intense depressions moving east or north-east off the Atlantic ocean. In autumn, intense depressions that were originally tropical storms or hurricanes sometimes reach the British Isles. The tornadoes that occasionally affect south-east England are not usually as intense and destructive as those of the United States. For example, tornadoes that affected Bexley on 25 January 1971 caused damage mainly to roofs and windows (Eaton, 1971). The possibility of increased storminess associated with climatic change cannot be ignored but, at present, there is no evidence of a worsening trend in the risk of windstorms for the London region.

Although London lies on the more sheltered east side of Britain, exposure to windstorm hazard is increasing as the city spreads over the confining hills on either side of the Thames valley and towards the coast. A particularly destructive storm characterized by high wind speeds affected south-east England on 15 October 1987 (table 7.2). This storm was the most severe event to strike the region since 1703, and Londoners were totally unprepared for its ferocity. Gusts of 94 mph were recorded in London and 100 mph in the Thames estuary. These gusts are estimated to have a return period of 120 years and more than 500 years, respectively (Mitchell et al., 1989). This storm produced insurance claims of £1.5 billion. The vulnerability of the London region to the 1987 storm was particularly high because nothing like it had occurred in living memory; people had not been prepared for such an event, and there was no public warning. As a result the region's infrastructure simply shut down. Deaths and injuries appear to have been distributed largely by chance. The most vulnerable people were those who had not heard early-morning news reports of the storm's developing effects, those who were standing beside walls or trees that collapsed or who were struck by flying debris, and those who were caught on journeys.

The Meteorological Office failed to forecast the storm and nationally televised weather forecasts neglected to convey the severity of the hazard. This subsequently led to severe criticisms of the Meteorological Office for failing to warn the public (Handmer and Parker, 1989). Controversy also surrounded an earlier decision to withdraw key weather ships from the area where the storm developed as part of a campaign to reduce public expenditures. A similar, but less intense, windstorm struck London and other parts of England in 1990. On this occasion the high winds were correctly forecast, but about 40 deaths occurred (some in London). Public knowledge about appropriate responses to severe weather warnings remains poor.

Wind-resistant building construction standards are the principal adjustment to windstorm hazard in Britain. Typically, buildings are designed and constructed to withstand wind loadings identified by the British Standards Institution code of practice. Unfortunately, enforcement of standards is uneven and sometimes lacking. Weather forecasting is the second most significant component of wind-hazard reduction strategy (Handmer and Parker, 1989). Improvements of forecasting and warning-dissemination systems have been implemented since 1987. For example, more attention is now devoted to explaining and displaying wind-speed information on the televised weather forecasts that most Britons rely on for information about wind hazards. Severe-weather warnings are now issued more frequently in London.

Fog

The Thames valley has always been fog prone, particularly during calm winter weather, and the historic record of fogginess is a long one. Brimblecombe (1987) demonstrates that since the seventeenth century the number of foggy days in London has fluctuated markedly but peaked in the 1890s (fig. 7.10). It is ironic that, just as the London Fog Inquiry of

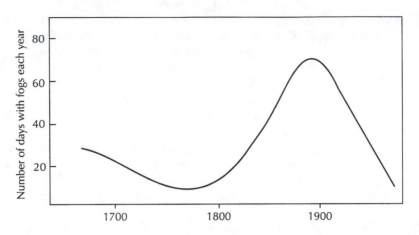

Fig. 7.10 Three centuries of fog in London (Source: Brimblecombe, 1987)

1902–1904 commenced, the number of fogs was decreasing! London fogs of the nineteenth century were closely related to levels of airborne particulate matter, especially fly ash derived from coal fires. A sharp decline in fogginess at the end of the nineteenth century has been attributed to actions promoted by the Coal Smoke Abatement Society, which was formed in 1899, but the possibility of climatic change cannot be ruled out. It is also possible that declining urban population densities, which accompanied the migration of London's population into expanding suburbs, may also have contributed to the trend by dispersing sources of pollution over wider areas.

Today, people living throughout the entire London region are exposed to fog hazards; damp low-lying areas are worst affected. Unfortunately, new high-speed roads such as the M25 have often been built through fog-risk areas, with predictable results in the form of major accidents. Vulnerability to fog hazards on the roads is also a function of journey times; early morning and late night travellers are most likely to experience fog. One of the most common responses to fog hazard is deferral of scheduled travel. Electronic fog-warning signs are installed on most high-speed roads, and fog warnings are a routine component of weather forecasts and severe-weather warnings. For air travellers, automated landing systems are also available at London's Heathrow Airport and elsewhere.

Seismicity

Earthquake risk in London has recently been assessed by the Department of the Environment (1993b). The annual probability of earthquakes with an MSK Intensity of V is 1:500 and for those of magnitude VII it is 1:10,000.[4] This means that London's seismic risk is small compared with places such as California, Japan, and Greece (Ambraseys and Jackson, 1985). None the less, significant earthquakes have occurred in the past (tables 7.1 and 7.2). For example, the following description of Londoners' reactions to a 1692 earthquake reveals a thoroughly disturbed population:

"All the People were possessed with a Panick Fear; some Swooning, others Aghast with Wonder and Amaze; the Houses were deserted, and the Streets thronged with such confused Multitudes, that there was no passing, nor could any Body give a true Relation of what had passed, for a considerable time.... so Great was the Confusion thereof (of the merchants), to the considerable Damage and Detriment of a great many Families.... for in houses in divers Parts, the Pewter and Brass were thrown from Shelves... several weavers, as they were at work, in Spittle-Fields, and various other Parts, had their work spoiled in their Looms; a House on Southwarkside, sunk several Foot into the Earth ... many Persons were taken with Giddiness in their Heads, and swooning fits ... these, and other Calamities, are occasioned by the Sins of this Nation." (Department of the Environment, 1993b, p. 6)

The Colchester earthquake of 1884 was the most severe for many centuries. Near its epicentre, on the edge of the contemporary London metropolitan region, brick buildings exhibited many cracks and chimneys collapsed; further away in east London and the lower Thames valley the earthquake merely rattled windows and doors.

Today, the entire population of the London metropolitan area is exposed to a small but significant earthquake risk. Few buildings have been constructed to take account of potential seismic loadings, but it is assumed that contemporary buildings are more robust than older ones. The Department of the Environment has used the 1989 earthquake that affected Newcastle, Australia, as a yardstick for estimating likely losses in a British quake. (There are many similarities of age and urban construction types between the two countries, although Newcastle's population densities are perhaps half those of London.) Assuming an earthquake of magnitude 5.5 Ms, with a focal depth of 10 km, the potential damage in Newcastle was estimated at between £1.2 and £1.7 billion. Adjusting for London's higher densities would double the economic losses and – given the very high value of many London buildings – this is still likely to be an underestimate of total losses. However, the annual probability of such an earthquake in London is about 1 in 50,000. Estimated fatalities for a Newcastle-type earthquake in the United Kingdom might be in the hundreds (Department of the Environment, 1993b).

Response to the seismic risk in London has so far been negligible. A government-funded research report on seismic hazards (Department of the Environment, 1993b) recommends that new building regulations should be considered, and that basic seismic requirements for conventional structures need to be developed. The national government is currently considering alternatives for introducing seismic building regulations. New buildings may be subjected to basic seismic checks, with buildings of national and historical importance subject to more stringent inspection. Seismic loading factors may also be incorporated into the design of potentially hazardous structures such as fuel and chemical storage installations, dams, and large sports arenas.

Technological hazards

Air pollution

Air pollution has been noted in London for 700 years. For example, problems associated with coal-burning prompted the creation of an investigatory commission in 1285. A subsequent ban on the use of "sea-coal" appears to have been largely ignored, although offenders were apparently tortured, hung, or decapitated – an early demonstration of the polluter-pays principle (Brimblecombe, 1987)! Use of "The Big Smoke" as a synonym for London has become part of folk history and few detective stories set in Victorian London would be complete without the backdrop of fog. During the nineteenth century, many Londoners became convinced about a connection between air pollution and worsening fogs. Fogs became thicker and darker coloured. The term "day darkness" was adopted to describe periods when the sun became totally obscured and lights had to be used. The number of polluted fogs in London climbed steeply and peaked in the 1890s. Subsequently they decreased in frequency and severity. There were also significant attempts to improve air quality. Smoke abatement acts were introduced. London's factory owners built taller chimneys and reduced black smoke emissions. Although the police were diligent in tracking down offenders, fines were not enough to discourage air pollution.

In the twentieth century, the mix of factors that contributed to London's air quality changed as the city became more dispersed and heating fuels became more diverse. Nevertheless, as recently as the decade after the Second World War domestic coal-burning was still the principal cause of air pollution. The 1950s and early 1960s witnessed a series of serious smog pollution episodes colloquially known as "pea-soupers." By far the worst occurred between 5 and 8 December 1952; in this one there were 4,000 extra deaths of mostly elderly and sick people. Large numbers had died in earlier London smogs (table 7.1), but the scale of the 1952 disaster finally triggered a major governmental response. This took the form of a Clean Air Act (1956), which granted local authorities the power to designate "smoke control areas" in which only smokeless fuels could be burned. Grants for heating appliance conversions were also provided. A subsequent act granted central government powers to force local authorities to take action. During the 1970s and 1980s millions of domestic and industrial properties were so regulated and this policy now forms part of broader European Union environmental controls. Continuing high concentrations of sulphur dioxide in London have prompted further legislation to prohibit the burning of high-sulphur fuels.

Smoke pollution in London has decreased dramatically since the 1950s. Legislation and regulation have had a hand in this, but other factors may have been just as important (Elsom, 1987). Beginning in the early twentieth century – and especially after the Second World War – an increasingly affluent society was already rejecting coal fires for more efficient heating systems that used cleaner-burning oil and gas. The history of air pollution episodes in London demonstrates the dynamism of risk. Smoke and sulphur dioxide concentrations have declined during the past 30 years but pollution from nitrogen dioxide, carbon monoxide, and particulate matter has increased – largely because of a dramatic increase in road vehicles (fig. 7.11) (Quality of Urban Air Review Group, 1993a). Increased vehicle emissions combined with recent hot, sunny summers have led to rising concern about photochemical smogs. These are now regular features of the summer atmosphere in London and they appear to have contributed to a dramatic increase in hay fever and asthma during the past 20 years. Periodic high concentrations of nitrogen dioxide have also begun to occur in winter (e.g. 12–15 December 1991), usually in combination with low wind speeds, low temperatures, high humidity, and

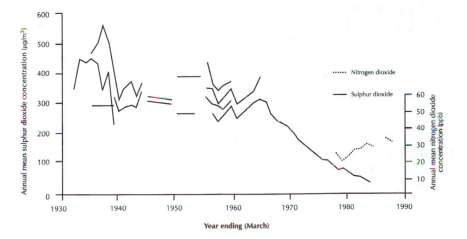

Fig. 7.11 Changes in air pollutants in London (Note: different lines relate to different measurement methods and slightly varying locations)

fog. An anticipated increase in summer temperatures associated with climate change, coupled with the nitrogen dioxide pollution trends, could significantly increase the risk of air pollution episodes in London (Lee, 1993). Rising numbers of diesel-fuelled cars may pose additional problems because they produce larger emissions of nitrogen oxides, particulate matter, and black smoke than catalyst-equipped petrol cars.

All of London's 13 million people are exposed to some air pollution, but there are significant spatial variations in exposure. Central London is particularly at risk, as are areas near heavily travelled roads. Vulnerability is differentiated by age, sex, state of health, and other factors. For example, children who take daily kerbside walks to school are likely to suffer disproportionately. London's infrastructure is also vulnerable to smog. The exteriors of most of London's historic buildings have been damaged by air pollutants.

Responses to air pollution are many and varied. Clean-up and restoration programmes are in effect for many pollution-damaged public buildings in central London. Technological responses include, among others, automobile exhaust controls. These may be jeopardized by changes in public preferences. It is feared by some that the beneficial effects of catalytic converters on conventional (petrol-powered) car exhausts could easily be wiped out by increased use of diesel engines (Quality of Urban Air Review Group, 1993b). Other technological responses include fuel-shifting to less polluting alternatives such as compressed natural gas, but this is still at a development stage.

Radical solutions to London's air-quality problems were recommended by the Royal Commission on Environmental Pollution, Transport and Environment (1994). This observed that the city's central problem is lack of an integrated approach to land-use and transport planning. Well-conceived and well-executed plans could modify the demand for travel. Non-technological options now being considered include establishing tighter statutory air-quality standards, and reducing vehicle use through increased planning controls, road pricing, and higher fuel taxes. Further developments in European Union legislation are promised. Within the next few years, attention is likely to turn to fuel quality, in-use testing, and on-board vehicle diagnostic systems. However, the benefits of introducing catalytic converters will not be as great as hoped unless the projected increase in vehicle numbers between now and 2010 can be avoided. Overall, London has been slow to recognize and respond to the largely predictable transformations of its ambient air quality.

Nuclear radiation

Unlike most of the other hazards considered here, the history of nuclear radiation is brief, beginning with the invention of nuclear weapons during the Second World War and the subsequent development of commercial power reactors. There are no nuclear generating stations or other large nuclear facilities in the immediate vicinity of London, but this does not mean that the risks of radiation are negligible. Radionucleids can be transported in the atmosphere, so the entire population of London is exposed and vulnerable. Although risk-assessment data are rarely made public, the UK government insists that the probability of nuclear accidents and subsequent contamination is very low. Potential sources of radioactive contamination have increased markedly in the past 50 years. They include: nuclear weapons production establishments to the west of London; trucks that transport nuclear warheads through the London region; and trains that carry nuclear wastes across London to a reprocessing plant at Sellafield in Cumbria. Coastal nuclear reactors in the United Kingdom and in northern France are another source of risk. There are now 14 commercial nuclear reactors within 160 km of London.

Measures for protecting the public against nuclear hazards are heavily reliant on the various safety systems that are employed by the nuclear industry. Safety audits are an essential part of the regulatory process. Public confidence was particularly shaken by the UK government's failure to plan for the kind of contingency that occurred at Chernobyl in 1986 and the initial feeble response to that event (Marples, 1996). Although the accident in Ukraine occurred on 25 April and the main radioactive cloud passed over the United Kingdom on 2 and 3 May, a central government lead department for coordinating government response was not formally designated until 6 May. Since then the central government has overhauled its civil emergency response strategy. None the less, there is a significant opposition in Britain to the whole concept of nuclear technology and frequent calls for phasing it out as the only reliable method of controlling these hazards.

Other industrial hazards

A range of other hazardous industrial installations are located in the Greater London area. Among them are plants that manufacture or store chlorine gas, sulphuric acid, and natural gas. In the parlance of British legislation, 19 of these are classified as "top-tier" major industrial hazard sites (fig. 7.12). Two concentrations of sites are particularly noteworthy: one stretches downriver from east London and the other is at Canvey. Although statistics are not available, the exposed population of London is relatively large, because many of the sites are located close to residential areas or upwind of them. A majority of nearby populations are members of low- to middle-income groups. The concentration of risks on Canvey Island was considered sufficiently hazardous for the Health and Safety Executive to undertake a risk assessment in 1978 (Health and Safety Executive, 1978). This investigation revealed that the annual probability of a local resident being killed in a major accident was about

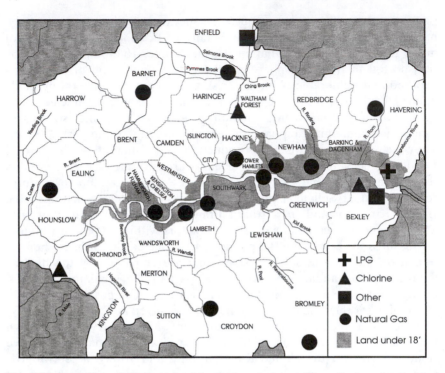

Fig. 7.12 Location of the principal hazardous industrial installations in Greater London

1 in 2,000. A number of recommendations made after this study were subsequently implemented (Health and Safety Executive, 1981).

Protective responses to industrial hazards mainly include safety audits and "fail-safe" technologies. Most of these are prescribed under the EU's Seveso Directive (1982) or the UK's Control of Major Industrial Hazards Regulations (CIMAH) of 1984. On-site and off-site emergency plans must be produced for installations above a certain size threshold, and local authority emergency response plans are rehearsed and publicized. Designated plants are regularly inspected by the Health and Safety Executive. Small plants are not designated and their on-site plans are confidential. CIMAH regulations are widely recognized as being innovative and comprehensive but their application remains controversial.

Social hazards

Terrorism

Terrorism has been an intermittent but generally limited threat to the lives and livelihoods of Londoners. Various political protest groups have

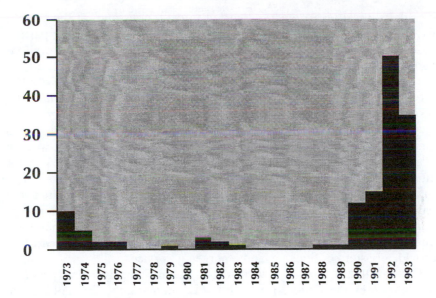

Fig. 7.13 Number of terrorist incidents in the London metropolitan area, 1973–1993 (Note: Records exclude incidents which have not been publicly disclosed. Source: Metropolitan Police Service, London)

employed violent tactics in the past that have resulted in deaths and property destruction. In recent years, the latest phase of the long-running antagonism between members of the Irish Republican Army (IRA) and the British authorities erupted into a terror bombing campaign that continued – with the exception of a truce in 1995–96 – from 1973 until 1997 (fig. 7.13). The number killed or injured was highest during the mid-1970s, with a secondary peak during the 1990s. Economic targets were preferred by the bombers; their objectives are believed to have been to undermine London's position as the premier global banking and finance centre, and to gain international publicity.

Bomb damage inflicted on property increased dramatically during the 1990s as large vehicle bombs became more common. The destruction of the Baltic Exchange in April 1992 was unprecedented in scale, but it was soon followed by a devastating bomb in Bishopsgate, and other attempted bombings were foiled. Terrorist bombings and incidents have affected locations throughout the London region, not just central London. Exposure to the risk has been widespread. Vulnerability has not been particularly differentiated; people are often killed or injured because they happen to be passing a bomb when it explodes. Members of the armed forces and rail commuters have been slightly more vulnerable than others because the IRA has targeted them.

Responses to IRA terrorism in London have been multifaceted and include political arrangements that led to cease-fires and, at the time of writing (September 1997), the opening of talks among opposing parties. Apart from the gathering and use of intelligence, anti-terrorist operations and improved security systems, public assistance was also mobilized. At the height of concern on the London underground system, items of "lost property" became "suspect packages," and London Underground Ltd. responded to around 90 suspect packages or hoaxes per day, often resulting in the evacuation of trains and stations. Commuters and employers have developed contingency plans, such as alternative ways of working or alternative routes, to minimize terrorist disruption. A major effort was made to find alternative premises for bombed-out firms. Firms such as the Deutsches Bank have constructed bomb-proof facilities and have arranged for data-processing outside of London. Disaster contingency planning is now built into many firm's plans. Back-up emergency financial offices have been established, with new firms being created to provide disaster contingency and emergency services.

Faced with the threat of vehicle bombs and steeply rising property insurance premiums, the Corporation of London and the City Police have responded with a package of measures designed to minimize the problem of terrorist attacks. These include controls on traffic access to the "Square Mile" (the central part of the City). Some roads are now completely blocked, with the effect that others can be policed more intensively. Random vehicle searches, closed-circuit television, and paging communication systems are also being employed. These measures have had a number of advantageous and unanticipated side-effects. For example, air pollution and noise have been reduced in the controlled areas. As a result, these "emergency controls" may become permanent.

Crime

London's social hazards include crime, civil disturbance, and problems involving crowds. Those most vulnerable are non-whites. Recent responses to crime take many forms and are highly controversial. Government has been promoting neighbourhood watch schemes, citizen monitoring, and a tightening of criminal justice legislation. The proposed changes in legislation are partly a response to civil disturbances and riots; ironically, during October 1994, demonstrations against these proposals themselves generated a riot in central London.

Following major loss of life at soccer stadiums in Sheffield and Brussels during the 1980s, the risk of crowd-related disasters appears to be high. Police have been developing crowd-management strategies to cope with large events including the annual Notting Hill Carnival, the New Year's Eve celebrations in Trafalgar Square, and large open-air concerts.

Complex emergencies and unprecedented hazards

Complex emergencies involve combinations of natural, technological, and social risks. London presents numerous opportunities for such emergencies, although few have occurred. The city's electricity supply system is so tightly coupled to its industrial, commercial, and residential systems that power failures can quickly spin off other problems such as paralysis of underground transportation, shut-down of subsurface ventilation, and crowd control crises. Many hazardous industrial installations are in or close to flood zones (fig. 7.12). Sudden flooding of London's many underground tunnels could cause major problems. This could happen as a result of ships grounding on tunnels beneath the Thames as well as exceptional flood flows. A number of vessels have collided with the Thames Tidal Flood Barrier, and vessels colliding with other flood defences could cause a complex emergency. Terrorists have so far refrained from threatening trunk sewers and flood defences, but there have been repeated attempts to disable London's mass transit systems. London is also at risk from new and so far unexperienced hazards such as a spatially concentrated crash by a major aircraft into the central city or by the widespread scattering of debris from a high-level aircraft disintegration. Few of these scenarios would surprise London's emergency planning community, and responses are being developed. Most of them focus on the reduction of tight-coupling and the preparation of emergency plans for large-scale evacuations.

Summary assessment of London hazards

In contemporary London, risks from most historic environmental hazards are either stable or increasing (table 7.4). Moreover, old hazards are being transformed into new variants. This is clearly demonstrated by the changing salience of different air pollutants and by the tendency of what were previously separate risks (e.g. floods, chemical emergencies) to metamorphose into hybrid ones. The number of low-probability, high-consequence hazards has increased.

The most pervasive and uncertain of the changes in natural hazards affecting London stem from climate modifications and the enhanced greenhouse effect. Climatic change is potentially the driving force behind possible changes in many of the natural hazards affecting London, including flood, windstorm, drought, and subsidence. Recent research has made progress in assessing the impacts of changes in mean climate upon economic and social systems in Britain, although most research on global climate change ignores urban areas (Climate Change Impacts Review Group, 1991). London is likely to be vulnerable to both shifts in mean

Table 7.4 Estimated changes in major hazard factors for hazards affecting London, 1945–2045

Hazard	Risk →50yrs	Risk 50yrs→	Exposure →50yrs	Exposure 50yrs→	Vulnerability of infrastructure →50yrs	Vulnerability of infrastructure 50yrs→	Vulnerability of people →50yrs	Vulnerability of people 50yrs→	Response →50yrs	Response 50yrs→
Sealevel rise and tidal flood	+	+	+	+	+	+	-	●	+	-
Fluvial flood	+	+	+	+	+	+	-	●	+	●
Sewer flood	+	+	+	+	+	+	-	●	-	●
Drought	●	●	+	+	●	●	-	-	+	+
Subsidence	●	●	+	+	+	-	●	-	-	+
Wind	●	●	+	+	+	-	+	●	-	+
Fog	●	●	+	+	+	+	-	●	●	●
Snow	●	●	+	+	+	●	+	-	●	●
Air quality	-	+	+	+	+	+	-	+	+	-
Water quality	-	-	●	●	●	●	●	-	+	+
Seismicity	●	●	+	+	+	+	●	●	●	+
Nuclear radiation	+	⊗	+	+	●	●	●	●	●	⊗
Wastes	+	⊗	+	+	●	●	●	●	●	⊗
Industrial sites	+	●	+	+	●	●	●	●	+	●
Transport	+	⊗	+	⊗	+	+	●	●	-	⊗
Crime	+	+	+	+	+	+	●	+	+	+
Terrorism	+	⊗	+	⊗	+	⊗	●	●	-	⊗
Complex	+	+	+	+	+	+	●	●	●	+

+ = increasing
● = stable
- = decreasing/improving
⊗ = not known, depends on socio-economic political events

climate and the frequency and magnitude of weather extremes. Because climate modelling is currently insufficiently advanced, considerable uncertainty remains about the links between a slowly changing mean climate and changes in the frequency or intensity of rainfall, winds, and storminess. A warmer Britain could bring changes in storm patterns and windiness. Global warming could precipitate storms nearer to Europe in the eastern Atlantic and storms might then arrive in Britain in a far more powerful and dangerous state (Simons, 1992).

Increased exposure to risks is one of the most important trends in contemporary London. This is occurring in part because urbanization is spreading into previously unoccupied hazard zones, in part because of reinvestment in older hazardous areas, and in part because more and more people are dependent on a few easily disrupted infrastructure systems. The overall vulnerability of residents to environmental risks has declined and probably will continue to do so, mostly because levels of living have risen and appear likely to continue to improve. But a widening vulnerability gap has appeared between rich and poor groups during the past 20 years. The increased vulnerability of ageing urban infrastructure is also a major contributor to hazard potential.

Responses to hazard are highly variable. In some cases there have been major – often successful – efforts to suppress risks; in others much less has been attempted or achieved. Lessons from the past are slow to be learned, and the concept of peacetime disaster management has been slow to gain official acceptance or public support. Many of the disasters that occurred in London during the past 50 years had similar counterparts in the nineteenth century.

Intervention to reduce hazards: Limits and opportunities

Constraints

Despite a long history of deliberate attempts to reduce environmental hazards, London's responses have often been ineffective. This suggests that opportunities for human intervention have been either lacking or unexploited and that the successful reduction of losses has occurred by chance or as a consequence of circumstances that are not directly connected with hazard. For example, comprehensive land-use planning controls have been in existence for almost 50 years, but floodplain development has not been prevented and flood losses continue to rise. Hazard reductions that have occurred were often the result of fortuitous changes in technologies and lifestyles. Thus a decline of fog and smog hazards after the 1890s probably stemmed more from the dispersal of population and industry in burgeoning suburbs than from special attempts to reduce air pollution by controls on the burning of fossil fuels. Even after smoke control laws became tougher in the 1950s, air-quality improvements owed still more to improved heating technologies.

Natural extremes have also failed to generate responsive public policies for other reasons. Some have been outranked by competing events. For example, Mitchell et al. (1989) demonstrate how a stock market crash that occurred four days after the storm of 15 October 1987 successfully diverted public attention away from the shortcomings of British policies for natural disaster response. London authorities made a delayed and confused response to the 1986 nuclear reactor fire in Chernobyl because they had judged that such an event was improbable and did not warrant planning for. This is still the dominate ethos among London's emergency managers with respect to earthquakes, accidental releases of nuclear radiation, and breaches of tidal-flood defences.

None the less, modest successes have been achieved by some deliberate actions – albeit often long delayed and adopted "just in time." Such actions have usually been taken in response to sudden high-intensity hazards rather than gradual low-intensity ones. For example, once it became appreciated that the Thames was lapping at the top of existing flood walls with distressing frequency, the political will to build a tidal flood barrier materialized. In retrospect it can be seen that the response was made not a moment too soon; since its creation the barrier has been closed on approximately 20 occasions! The rapid adoption of techniques for coping with large vehicle bombs provides another example of effective adjustment under the pressure of extreme events. On the other hand, responses have been less effective when hazards were perceived to be low-intensity events with difficult-to-identify impacts and diffuse costs. Repeated events that produce low to moderate losses have often spurred ineffective public responses. Despite considerable investment in physical measures for controlling riverine floods in London, such floods have intensified and are now widespread. Much of the problem is centred in the city's ageing and deteriorating infrastructure, which simply deepens vulnerability to flooding of all kinds. Failure to learn from previous hazards is strongly influenced by the "social matrix" in which hazards occur. Opportunities for improving hazard management in London have been constrained by a number of factors (Parker and Handmer, 1992b). A brief summary of these follows.

British hazard management has been buffeted and degraded by a series of apparently unconnected or loosely connected changes. Throughout London's history, prevailing doctrines of government have provided an important context for hazard management. Sometimes government policies favoured hazard-management initiatives such as the great nineteenth-century public hygiene reforms. At other times, such initiatives fell on barren ground; the persistent refusal of central government to give priority to peacetime emergency planning over wartime emergency planning during the 1980s is a trenchant example. In Britain, hazard management has often been assigned a low priority, partly because disasters are typically viewed as isolated non-recurring events. This leads to low levels of hazard awareness and the lack of a "safety culture" within organizations as well as governmental complacency. In the late 1980s, media coverage of post-disaster inquiries raised public awareness of London's disaster potential, but this impetus has not been sustained. Since then, media coverage of environmental disasters has rapidly given way to reports of economic events or political stories. For example, during early 1994, following successive moral scandals in the British government, news about serious flooding and landslides in Britain and Western Europe was trivialized by the media caricatures of a prime minister marooned by floods.

In most Western democracies, information about public issues is more or less freely available except in highly unusual circumstances when governmental restrictions are invoked. British governmental and industrial culture tends towards non-disclosure; officials and managers frequently limit information flow. For example, Spooner (1992) criticizes British industry for allowing information on known hazards to remain "buried in corporate files." This approach is in marked contrast to the "right to know" law of the United States. However, the recent success of programmes to increase public vigilance against terrorism now provides London with a public information and education model that can usefully be extended to other hazards. Though the United Kingdom is surely not alone in its preference for technological fixes to hazards, its heavy reliance on these measures has encouraged misconceptions about their effectiveness and about the availability of alternatives. The country has invested heavily in structural flood defences and flood forecasting technologies to the neglect of flood-warning dissemination and emergency response. Similarly blinkered responses are typical of other hazards.

A rapidly growing and diversifying mega-city such as London may require continuous changes of institutional and organizational arrangements if it is to keep pace with evolving public problems. In this kind of setting, information about hazards that reach crisis proportions only infrequently is apt to be lost in the absence of special measures to ensure continuity and visibility. Institutions such as the Royal Commission on Environmental Pollution have played a valuable role by drawing government's attention to long-term trends in risk, exposure, and vulnerability, but they rely upon other agencies routinely to collect, store, and disseminate hazard data. Initiatives within the UK Environment Ministry, which include hazard assessments, creation of a Government Office for London, and policy shifts to improve the coordination of transport and land-use policy, currently offer the greatest promise for the future.

Opportunities for creative hazard management

Debate about hazards in London has tended to ignore opportunities for integrating hazard management into broader fields of urban policy-making and urban management (Fainstein et al., 1992; Hall, 1989; Hoggart and Green, 1991; Self, 1957; Simmie, 1994; Thornley, 1992). Some urban policy interest groups have addressed the topic of global warming, but these have generally been bodies that lack executive authority (e.g. the Association of London Authorities, the London Boroughs Association (1991), the London Planning Advisory Committee). Even if the main urban decision-making entities were cognizant of hazards issues, the scope for improved hazard management would be limited by the lack of a coherent urban management strategy (Robson, 1994).

In London, opportunities for creative hazard management need to be explored in the context of both of the policy paradigms that have dominated Britain's political economy since 1945 (i.e. government-directed approaches; market-directed approaches). In the sphere of hazards, these policy paradigms are associated with contrasting, but not necessarily contradictory, values (table 7.5). Sometimes a lack of value consistency in the development or application of such policies produces confusion. For example, the practice of denying individuals risk information rests uncomfortably alongside an emphasis on encouraging self-reliance among hazard managers. Likewise, in heterogeneous late-twentieth-century democratic societies, market-based approaches may be possible only where the political power to enforce them is highly centralized. Successful hazard management will depend on adopting a consistent set of values and policies before looking for new ways of translating them into responses that reduce risk, exposure, and vulnerability.

Transportation planning and hazard management

There is a very large potential for hazard reduction in London's evolving transport system. Most observers have used terms such as "muddle," "failure," and "crisis" to characterize the existing system (Adams, 1986; Hall, 1989; Pharoah, 1991). Public policy has conspicuously failed to address important linkages between transport and land-use planning, safety, pollution, health, and climatic change.

Weakening dependence on motor cars is the most important goal because it will reduce congestion, air pollution, and the dispersal of development that contributes to the invasion of hazardous areas. Higher

Table 7.5 Contrasting values affecting hazard management in London

Values often associated with the market-based paradigm	Values often associated with the planning paradigm
Individuals should be self-reliant, responsible for their own actions; there is no such thing as "society".	The state should provide protection for the individual and society, and should act as a "collective memory" for extreme events.
Risk information is "confidential," risk communication is restricted, emphasis upon commercial interests.	Risk information should be publicly available and communicated. The public have a right to know.
The market can be used to deliver hazard reduction, and can be modified through taxation and subsidy to achieve hazard reduction.	Market failures make the market largely inappropriate for delivering effective hazard-reduction measures.
Hazard management should be delivered by both state and private sector, with the emphasis upon private solutions and hazard reduction as a private good.	Hazard management should be delivered by the state and the public sector with controlled employment of private sector. Hazard reduction is principally a public good.
Deregulation is necessary to encourage enterprise and economic growth; planning and zoning restrictions are relaxed or removed.	The regulatory approach is one of the principal responses to hazard creation and perpetuation; planning controls and zoning restrictions should be tighter.
Urban regeneration should be used to facilitate private sector development and growth, with the minimum of restrictions.	Urban planning should be used to control and guide development, with restrictions to control development of hazards.
Public services should be made more responsive to customers by target-setting and customer charters.	Public services can be made responsive to public need by placing them under local democratic control.
Disaggregated approach to public sector, separately managed units, rejection of metropolitan government.	Aggregated approach and enthusiasm for metropolitan government or strong coordinating metropolitan-wide agencies.

fuel taxes, road rationing, road pricing, and reduced vehicle trips are needed. The scale of long-distance car commuting into London might be reduced by creating park-and-ride stations and other incentives for long-distance car commuters to transfer to mass transport. Hall (1989) proposes a Paris-style "Reseau Express Regional" for London, mainly using existing rail lines. Such changes might require institutional change, possibly through the revival of a planning authority for London. Increased sharing of experience with other countries would also be valuable. London was early to adopt traffic restraint measures, but it has been slow to learn about the benefits of Dutch or German traffic management. Pedestrianization has been limited, and attempts to encourage "soft" modes of travel (walking and cycling) could also be encouraged. Finally, the health benefits of reduced car use are likely to become more salient. Evidence is emerging in London about the possible links between urban air pollution and diseases such as asthma, and research on these issues is being stepped up. (In Britain, 1 in 5 schoolchildren are now asthma sufferers.)

Improving development planning and urban sustainability

Early intervention in the development planning process and adoption of an extended time-horizon for planning could make important contributions to hazard reduction and urban sustainability. Unfortunately, there is no statutory provision for regional planning in England and Wales, although advice about planning is currently prepared by several voluntary groupings of London and South-East planning authorities. Single-municipality plans are being encouraged by the UK government, and these offer opportunities to influence the plan-making process at an early stage. For example, at present the National Rivers Authority must devote significant resources to advising local planning authorities about applications for building in floodplains. By seeking to influence the development plan-making process in its formative stages, the Authority has an opportunity to reduce future workloads and flood losses. This requires the NRA to formulate catchment management plans that can interact with development plans to ensure long-term sustainability. By formulating model flood-hazard reduction policies and communicating them to developers and local planning authorities, the NRA can indirectly constrain the growth of exposure to flood hazards (Tunstall et al., 1993).

Standards of service and hazard management

Better hazard management could be encouraged by improving the standards of services that are delivered to the public. One way of doing this is to make services (including hazard protection) more responsive to customer needs. This may be done by establishing and publishing target standards, and then progressively improving them. One means that is currently being used by central government to raise standards for governmental services is the Citizen's Charter (Major, 1991). Under the Charter, public agencies are expected to publish explicit standards of performance as well as data on standards achieved. The Charter approach has now been applied to British Rail, London Underground Ltd., and the NRA. Safety and other aspects of hazard management currently have a low profile within existing Charters, but there is potential for setting and raising standards of hazard management. The NRA is developing standards of service for flood forecasting and warning-dissemination systems. Consideration is being given to improving standards for a range of parameters, including warning lead-times and the proportion of those subsequently flooded who receive prior warnings (Parker and Handmer, 1992b).

A standards-of-service approach could be used to deliver a range of hazard-management services to the public more effectively, by either the public or the private sector. These services could include information on risks, warning systems, self-protective measures, technical advice, and community responses. Some initiatives of this sort include the increased provision of information on air pollution levels in London via help lines. Preparedness and risk communication for natural disasters could be further enhanced by extending the model of the Seveso Directive (European Community, 1982), which addresses industrial hazards. The Seveso Directive is the most coherent and systematic civil emergency planning legislation originating from the European Union. The focus of the Directive is information about risks and safety measures. Information must flow from employers to employees, from industries to competent authorities, and from government to citizens.

Emergency planning and response

Within the London region, responsibilities for civil emergencies are now highly fragmented amongst central government ministries, the Metropolitan Police, the City of London Police, the British Transport Police, the London Fire Brigade, the London Fire and Civil Defence Authority, 33 London boroughs, county councils, the London Ambulance Service, health authorities, and other agencies such as the NRA. There are also numerous private emergency service groups. Coordination among all of these groups is voluntary and there is no comprehensive administrative organization. Every borough is authorized to undertake emergency planning, but there is no statutory obligation to plan for peacetime emergencies. Despite over 20 years of voluntary liaison between London's emergency planners and service providers, there is disagreement over roles. Statutory duties and powers for civil emergency planning and response are far from clear. A trend towards further fragmentation and the dismantling of central emergency planning is continuing, with unfortunate consequences. For example, in the Metropolitan Police, responsibilities for natural-disaster response are being devolved to 63 separate police divisions. Since there is a low annual probability that the territory within any single division will experience a natural disaster, the salience of emergency planning within the police force as a whole is likely to decrease. In the event of an emergency, these trends will leave much of the London region in a highly vulnerable position. London's predicament reflects a larger disaster planning and management problem that faces all of Britain. A fixation with "civil defence" (i.e. war planning) constipated the development of civil disaster planning in Britain for many years. Although more attention is now being given to civil emergency planning, funding remains limited (Handmer and Parker, 1991).

Institutional design and development

For London to respond more successfully to many of its hazards, mega-city-wide strategies and actions are required. This is because many of the natural and social processes that generate increased risk, exposure, and vulnerability in London do not stop at the city's boundaries. A mega-city government with strong powers to plan and implement a London-wide transportation strategy, and to link this to land-use planning, would clearly assist hazard management. However, this is an unpopular prescription in the Britain of the 1990s because large-scale regional planning during the 1960s and 1970s proved to be largely ineffective and was firmly rejected by recent UK governments. However, as Hall (1989, p. 180) points out, there may well be a return to regional strategic plans in the future as "planning by the roulette wheel" proves unsatisfactory.

Under the prevailing philosophy of centralized government, a parliamentary advisory Metropolitan Planning and Transport Commission for the London region is a feasible model (Hall, 1989). Such a Commission would have responsibility for producing a regional framework for sustainable development into which hazard management could be fitted. Other possibilities include extending the role of the national government's Health and Safety Executive to embrace natural and social hazards. This might enhance the overall disaster learning process because the Executive has taken a strong lead in addressing technological and industrial hazards.

There are also important opportunities to bring together "top–down" intervention and "bottom–up" strategies. Urban regeneration initiatives provide a basis for community-based hazard management. Community action groups have been successful in influencing the location of hazardous activities (Blowers, 1994). The partnership approach that is being developed and funded by central government offers opportunities for urban regeneration and job creation, as well as environmental improvement. It might be extended to include pollution controls and the reduction of dangers to property, health, and life. Already there are signs of movement in this direction. A new urban village in West Silvertown incorporates the concept of a pleasant and non-threatening environment; more explicit connections could be made to the goal of reducing various specific hazards.

The precarious position of London's civil emergency planning requires the creation of a new lead or coordinating authority. Proposals have been made for the London Fire and Civil Defence Authority to take charge of civil emergency planning powers, thereby assuming the leading role in disaster prevention and planning that was formerly taken by the GLC. An alternative is for the police authorities to be given these planning powers.

Reducing social vulnerability

In the next decade, urban regeneration programmes could make an important contribution to reducing the vulnerability of Londoners to hazards. Currently, in London there is little recognition of the fact that differential vulnerability to risk is an important factor that affects patterns of losses. This is partly because data on vulnerability are poorly developed but also because vulnerability reduction is rarely perceived by policy makers as a realistic response to hazards.

Developing alternatives to the technological fix

Finally, it should be noted that most of the opportunities for creating better hazard management involve non-technological alternatives. These are particularly compatible with the goal of sustainable development that the British government has officially embraced (Department of the Environment, 1990).

Conclusions

The record of successful responses to environmental hazards in London is a modest one. Failure to act on the lessons of its own history is probably the single most important characteristic of hazard management in this mega-city. Here the process of adjustment to environmental risk has typically been slow, troubled, and closely reflective of the city's complex and changing social matrix. Lack of scientific knowledge has hampered effective hazards policy-making to some degree (especially with regard to environmental health hazards during the pre-Victoria era), but it is rarely the crucial variable today. Institutional inertia and cultural, economic, and political factors that favour countervailing public policies are more important barriers to hazard reduction.

Failure to take a broad systems view of hazard – including its creation, perpetuation, and intensification – has also hampered effective policy-making, planning, and management. Public leaders and administrators have tended to ignore the complex linkages between national economic policy and urban management on the one hand and the forces that create hazards and disasters on the other. In other words, a wide range of public policies with indirect effects on hazard have been developed and implemented in a compartmentalized manner, without regard to their interactive or synergistic effects.

In London, improved responses to hazard typically followed a series of mounting losses that served as a continuing spur to action. Sudden, intense, episodic events (e.g. tidal floods) most often exhibited this pattern. Slow-developing, low-intensity hazards were usually ignored – even if they were widespread and repetitive – at least until an upward surge in losses occurred (e.g. air pollution). Yet, however faltering London's response to familiar hazards has been, the city's government and people have also shown that they are sometimes capable of rapid and robust action in the face of unprecedented "surprises" (e.g. large vehicle bombs).

Hazard responses in London have nearly always been imperfect – sometimes glaringly so. Attempts to control air pollution provide a number of salutary examples. The history of partial and uneven compliance with London's smoke control laws is well known and continues today. Technological fixes for air pollution have been no more effective. The recent introduction of super-unleaded petrol in London has proved to be an embarrassing failure because it is now recognized that the new fuel is no less harmful to health than the regular unleaded fuel that it replaced. Moreover, even as the city struggles to develop more effective ways of coping with hazards, the very nature of the problems that are faced changes. Like disease bacteria that produce new drug-resistant strains, the hazards themselves sometimes "bounce back" as new variants of their former selves. Thus, problems caused by airborne lead or hydrocarbons emerge as problems associated with particulate matter recede. The salience of different types of flood (e.g. tidal, riverine, and sewer flooding) exhibits comparable shifts.

As London's experiences clearly demonstrate, policies and programmes of urban hazard management require continuous readjustment. In the next decade the primary stimulus for readjustment will probably come, not so much from increased natural, technological, or social risks, but from increased exposure to risk and increasing differential vulnerability – particularly as this is manifest in widening gaps between white and non-white racial groups, between males and females, and between rich and poor. These trends will constitute potent challenges for London's urban hazards managers of the twenty-first century.

Acknowledgements

I wish to acknowledge the helpful comments on drafts of this chapter made by Ken Mitchell, Roger Leigh, and John Handmer. I am particularly grateful for the assistance of Sue Tapsell in obtaining information.

Notes

1. When "city" is used in the text it generally refers to Greater London. To Londoners, "The City" refers to the ancient core of London – a municipality known by its formal title as the City of London.
2. This definition of London matches the so-called "Outer Metropolitan Area" identified by planners (fig. 7.1). However, the name London has been applied to many different areas (Hall, 1989). The ancient City of London consists of 1,690 hectares within a 15 km^2 zone commonly termed "Central London." This area contains one of the largest concentrations of employment in the world. Central London is surrounded by "Inner London," a region of mainly older and denser housing corresponding roughly to the built-up area that existed in 1914. In turn, Inner London is contained by "Greater London" – the area occupied by 33 London boroughs. During the 1960s, London's planners also defined an "Outer Metropolitan Area." Continued outward growth led in the 1980s to the recognition of what Hall (1989, p. 2) terms "Roseland" (Rest of South East). This is an extension of London characterized by fast growth of population and urban forms. Finally, both Hall and Warnes (1991) recognize a further zone characterized by a complex set of commuting relationships, which Hall terms the Greater South East (fig. 7.1).
3. It remains very difficult for the public to obtain information about many types of hazard, especially nuclear radiation and on-site emergency plans for responding to industrial accidents. Gould (1990) relates how government agencies fought a stubborn rearguard action over more than 30 years to restrict information about radiation risks in one British community. The seriousness of an initial graphite reactor fire at Windscale in 1957 was not revealed for three days and additional details were not published until the late 1980s because the facts were considered to be too sensitive. Windscale was subsequently renamed Sellafield in an attempt to reduce the public visibility of the disaster.
4. The MSK (Medvedev–Sponheuer–Karnik) scale is a Central European variant of the Modified Mercalli scale, which is widely used in the United States.

REFERENCES

Adams, J. 1986. "Transport dilemmas." In H. Clout and P. C. Wood (eds.), *London: Problems of change*. New York: Longman, pp. 75–82.

Ambraseys, N. N. and J. A. Jackson. 1985. "Long-term seismicity of Britain." In *Earthquake engineering in Britain*. Proceedings of a Conference Organised by the Institution of Civil Engineers in Association with the Society for Earthquake and Civil Engineering Dynamics, University of East Anglia, 18–19 April, pp. 49–66.

Appleyard, B. 1993. "A city stretched to its limits." *The Independent*, 24 November, p. 19.

Bennett, R. J. 1991. "Rethinking London government." In K. Hoggart and D. R. Green (eds.), *London: A new metropolitan geography*. London: Edward Arnold, pp. 205–219.

Blowers, A., ed. 1994. *Planning for a sustainable environment*. A report by the Town and Country Planning Association. London: Earthscan Publications.

Breheny, M. and R. Rookwood. 1993. "Planning the sustainable city region." In A. Blowers (ed.), *Planning for a sustainable environment*. A report by the Town and Country Planning Association. London: Earthscan Publications, pp. 150–189.

Brimblecombe, P. 1987. *The Big Smoke. A history of air pollution in London since medieval times*. London: Methuen.

Brownhill, S. and C. Sharp. 1992. "London's housing crisis." In A. Thornley (ed.), *The crisis of London*. London: Routledge, pp. 10–24.

Champion, A. G. 1987. "Momentous revival in London's population." *Town and Country Planning* 56: 80–82

Climate Change Impacts Review Group (UK). 1991. *The potential effects of climate change in the United Kingdom*. First report of the United Kingdom Climate Change Impacts Review Group. London: Her Majesty's Stationery Office.

Clout, H. and P. Wood, eds. 1986. *London: Problems of change*. London: Longman.

Coleman, A. 1985. *Utopia on trial: Vision and reality in planned housing*. London: Shipman.

Coupland, A. 1992. "Every job an office job." In A. Thornley (ed.), *The crisis of London*. London: Routledge, pp. 25–36.

De Marchi, B., S. Funtowicz, and J. Ravetz. 1996. "Seveso: A paradoxical classic disaster." In James K. Mitchell (ed.), *The long road to recovery: Community responses to industrial disaster*. Tokyo: United Nations University Press, pp. 86–120.

Department of the Environment. 1990. *This common inheritance. Britain's environmental strategy*, Cmnd. 1200. London: Her Majesty's Stationery Office.

———— 1993a. *East Thames Corridor: A study of development capacity and potential*. London: Her Majesty's Stationery Office.

———— 1993b. *Earthquake hazard and risk in the UK*. London: Her Majesty's Stationery Office.

Department of Transport. 1988. *Investigation into the Kings Cross Underground fire* (The Fennell Inquiry), Cmnd. 499. London: Her Majesty's Stationery Office.

Doornkamp, J. C. 1995. "Perception and reality in the provision of insurance against natural perils in the UK." *Transactions of the Institute of British Geographers* 20(1): 68–80.

Downing, T., E. J. A. Lohman, W. S. Maunder, A. A. Olsthoon, R. Tol, and H. Tiedmann. 1993. *Socio-economic and policy aspects of changes in the incidence and intensity of extreme (weather) events*. Workshop at Instutuut voor Milieuvraagstukken, Vrije Universitiel, Amsterdam, 24–25 June.

Eaton, K. J. 1971. *Damage due to tornadoes in SE England 25 January 1971*. Building Research Station Current Paper 27/71. Watford: Building Research Station.

Elsom, D. 1987. *Atmospheric pollution*. Oxford: Blackwell.

European Community. 1982. *Council Directive of 24 June 1982 on the major accident hazards of certain industrial activities*, 82/501/EEC. Brussels.

Fainstein, S. S., I. Gordon, and M. Harloe, eds. 1992. *Divided cities: New York and London in the contemporary world*. London: Blackwell.

Gilbert, S. and R. Horner. 1984. *The Thames Barrier*. London: Thomas Telford.

Gordon, I. and S. Sassen. 1992. "Restructuring the urban labour markets." In S. S. Fainstein, I. Gordon, and M. Harloe (eds.), *Divided cities: New York and London in the contemporary world*. London: Blackwell, pp. 105–128.

Gottmann, J. 1961. *Megalopolis: The urbanised north eastern seaboard of the United States*. New York: Twentieth Century Fund.

Gould, P. 1990. *Fire in the rain. The democratic consequences of Chernobyl*. Cambridge: Blackwell.

Government Office for London. 1994. *East London and the Lee Valley: A programme for economic regeneration in London's key industrial zone*. London: Government Office for London.

Hall, P. 1989. *London 2001*. London: Unwin Hyman.

———— 1993. *Cities of tomorrow*. London: Blackwell.

Handmer, J. W. and D. J. Parker. 1989. "British storm-warning analysis: Are customer needs being satisfied?" *Weather* 44(5): 210–214.

———— 1991. "British disaster planning and management: An initial assessment." *Disasters* 15(4): 303–317.

Harloe, M., P. Marcuse, and N. Smith. 1992. "Housing for people, housing for profits." In S. S. Fainstein, I. Gordon, and M. Harloe (eds.), *Divided cities: New York and London in the contemporary world*. London: Blackwell, pp. 175–202.

Health and Safety Executive. 1978. *Canvey: An investigation of potential hazards from operations in the Canvey Island/Thurrock area*. London: Her Majesty's Stationery Office.

———— 1981. *Canvey: A second report*. London: Her Majesty's Stationery Office.

Hebbert, M. 1992. "Governing the capital." In A. Thornley (ed.), *The crisis of London*. London: Routledge, pp. 134–148.

Hoggart, K. and D. R. Green, eds. 1991. *London: A new metropolitan geography*. London: Edward Arnold.

Hollis, G. E. 1986. "Water management." In H. Clout and P. Wood (eds.), *London: Problems of change*. London: Longman, pp. 101–110.

Hood, C. and M. Jackson. 1992. "The new public management: A recipe for disaster?" In D. J. Parker and J. W. Handmer (eds.), *Hazard management and emergency planning: Perspectives on Britain*. London: James & James, pp. 109–126.

Horlick-Jones, T. 1990. "The nature of disasters." *Procs of 2nd Disaster Limitation Conference: Emergency Planning in the 1990s*. Bradford: University of Bradford.

Jones, E. L., S. Porter, and M. Turner. 1984. *A gazetteer of English urban fire disasters, 1500–1900*. Historical Geography Research Series No. 13. Norwich: Geo Books.

Lee, D. 1993. "Climate change and air quality in London." *Geography* 78(1): 77–79.

Lewis, M. 1989. "How a Tokyo earthquake could devastate Wall Steet and the world economy." *Manhattan, inc.*, June: 69–79.

Logan, J., J. Taylor-Gooby, and M. Reuter. 1992. "Poverty and income inequality." In S. Fainstein, I. Gordon, and M. Harloe (eds.), *Divided cities: New York and London in the contemporary world*. London: Blackwell, pp. 125–150.

London Boroughs Association. 1991. *Global warming: How London boroughs can respond*. London: London Boroughs Association.

Major, J. 1991. *The Citizen's Charter. Raising the Standard*. London: Her Majesty's Stationery Office, July.

Marples, D. 1996. "The Chernobyl disaster: Its effect on Belarus and Ukraine." In James K. Mitchell (ed.), *The long road to recovery: Community responses to industrial disaster*. Tokyo: United Nations University Press, pp. 183–230.

Mitchell, J. K., N. Devine, and K. Jagger. 1989. "A contextual model of natural hazard." *Geographical Review* 79(4): 391–409.

Parker, D. J. and J. W. Handmer. 1992a. "Improving hazard management and emergency planning." In D. J. Parker and J. W. Handmer (eds.), *Hazard management and emergency planning: Perspectives on Britain*. London: James & James, pp. 261–274.

———— eds. 1992b. *Hazard management and emergency planning: Perspectives on Britain*. London: James & James.

Parker, D. J. and E. C. Penning-Rowsell. 1980. *Water planning in Britain*. London: Allen & Unwin.

Parker, D. J., C. H. Green, and P. M. Thompson. 1987. *Urban flood protection benefits: A project appraisal guide*. Aldershot: Gower Technical Press.

Penning-Rowsell, E. C., D. J. Parker, D. Crease, and C. Mattison. 1983. *Flood warning dissemination: An evaluation of some current practices within the Severn Trent Water Authority area*. Flood Hazard Research Centre, Middlesex Polytechnic, Enfield.

Perry, A. H. 1981. *Environmental hazards in the British Isles*. London: Allen & Unwin.

Pharoah, T. 1991. "Transport: How much can London take?" In K. Hoggart and D. R. Green (eds.), *London: A new metropolitan geography*. London: Edward Arnold, pp. 141–155.

Quality of Urban Air Review Group. 1993a. *Urban air quality in the United Kingdom*. London: Quality of Urban Air Review Group.

———— 1993b. *Diesel vehicle emissions and urban air quality*. London: Quality of Urban Air Review Group.

Reddaway, T. F. 1940. *The rebuilding of London after the Great Fire*. London: Jonathan Cape.

Robson, B. 1994. "No city, no civilization." *Transactions of the Institute of British Geographers* 19(2): 131–141.

Royal Commission. 1994. *Royal Commission on Environmental Pollution, Transport and Environment*. London: Her Majesty's Stationery Office.

Self, P. 1957. *Cities in flood: The problems of urban growth*. London: Faber & Faber.

Simmie, I., ed. 1994. *Planning London*. London: University College London Press.

Simons, P. 1992. "Why global warming could take Britain by storm." *New Scientist* 136(1846): 35–38.

Spooner, P. 1992. "Corporate responsibility in an age of deregulation." In D. J. Parker and J. W. Handmer (eds.), *Hazard management and emergency planning: Perspectives on Britain*. London: James & James, pp. 95–108.

Summers, D. 1978. *The east coast floods*. London: David & Charles.

Thornley, A., ed. 1992. *The crisis of London*. London: Routledge.

Tunstall, S. and P. Bossman-Aggrey. 1988. *Waltham Abbey and Thornwood, Essex: An assessment of the effects of the flood of 29 July 1987 and the benefits of flood alleviation*. Flood Hazard Research Centre, Middlesex University, Enfield.

Tunstall, S., D. J. Parker, and D. Krol. 1993. *Planning and flood risk: A strategic approach for the National Rivers Authority*. Bristol: NRA.

Turner, R. K., P. M. Kelly, and R. C. Kay. 1990. *Cities at risk*. London: BNA International.

Warnes, A. M. 1991. "London's population trends: Metropolitan area or megalopolis?" In K. Hoggart and D. R. Green (eds.), *London: A new metropolitan geography*. London: Edward Arnold, pp. 156–175.

Wise, M. J. 1962. "The London region." In J. Mitchell (ed.), *Great Britain: Geographical essays*. Cambridge: Cambridge University Press, pp. 57–85.

8

Lima, Peru: Underdevelopment and vulnerability to hazards in the city of the kings

Anthony Oliver-Smith

Editor's introduction

The impact of external economic factors on susceptibility to urban environmental hazard is a theme that appears in several chapters of this book but nowhere is it more strongly presented than in this case-study of Lima. Anthony Oliver-Smith underlines the city's role as a conduit for resources mobilized first by self-serving decisions of colonial masters in Spain and later by exploitative international financiers based in the United Kingdom and the United States. Here Lima implicitly stands in place of other Latin American mega-cities with similar origins and histories of dependency that have also experienced severe environmental hazards. On the other hand, Lima is distinctively different. It can be argued that the signal risks of this site would have posed serious difficulties for large-scale urbanization under any socio-economic regime. Moreover Peru has fostered links with Japanese hazards scientists and other specialists that have made a distinctive contribution to the national hazard-management system. Indigenous experiences and resources combined with external skills and modest investment in local hazard management could make significant inroads on the city's admittedly daunting hazard problems.

Introduction

Peru is Lima and Lima is the Street of the Union and the Street of the Union is the Palais Concert.

So went the ironic comment of turn of the century Peruvian writer, Abraham Valdelomar (1888–1929; as quoted in Ortega, 1986, p. 1). In a single sentence, Valdelomar captured the essence of one of Peru's most critical problems. Since its founding over 450 years ago, the citizens of Lima have considered themselves to be Peru and have always cast their gaze toward Europe rather than toward the country at their back (fig. 8.1). The strength of that vision has strongly affected Lima's social and economic development, its demographic growth, and its spatial expansion – all of which have in turn reinforced the city's vulnerability to natural and technological hazards.

Lima was founded in the middle of a broad valley that includes three separate rivers. Since pre-Colombian times the central river has been called the Rimac (literally, "he who speaks" in the native Quecha language). The city that grew up on its banks has unquestionably spoken the loudest of any in Peru's post-conquest history. As Lima continues the process of explosive growth that began early in the twentieth century, its voice threatens to drown all others. Homes and workplaces have expanded away from the Rimac into the adjacent Lurin and Chillon river valleys as well as the surrounding Andean foothills. What was once a single small municipality is now a complex fragmented urban region that contains 50 separate local governments and houses over 7.5 million people.

Lima's environmental hazards can be understood only in a wider context. The vulnerability of Lima to natural and technological hazards has been deeply influenced by historical processes of development and underdevelopment that shaped much of the world system (Ortiz de Zevallos, 1986). In Peru these processes have not just created an unequal distribution of wealth and benefits, but also given rise to powerful symbolic images. These include polarized visions of society that define and reinforce identity and power: colonizer/colonized, white/Indian, coast/highlands, civilized/barbaric, Lima/village (Jacobsen, 1993, p. 3). For many generations of urban migrants it is the contrast between Lima's image of utter centrality and primacy and the rest of Peru's image as exploitable resource base or barbarous frontier that has come to define concepts of progress, hope, and modernity. These polarized visions, forged from the very moment of Lima's founding, lie at the heart of the contemporary city's substantial vulnerability to hazards.

Fig. 8.1 Peru

The physical environment

The natural settings in which humans find themselves, whether by accident or by design, invariably contain advantages as well as disadvantages. Often, the same phenomenon constitutes a resource as well as a hazard. For example, settlements near rivers or coastlines can be places of both sustenance and danger. Indeed, human settlements are often initially sited in close proximity to resources that later turn out also to be hazards. Such was the case in sixteenth-century Lima. The natural resources of the site included plentiful water supplies from the valley's rivers, ample terrain for a settlement and supporting agriculture, a fine port, and a temperate climate. The Rimac is the richest and most extensive of all the valleys on the central Peruvian coast. On either side of the river's mouth are two long bays that shelter the best natural anchorage on the entire coast (see fig. 8.2). When Francisco Pizarro chose this location for Peru's new capital in 1535 he recognized natural advantages that were already well known to the indigenous peoples, who had practised a complex, intensive system of agriculture there for thousands of years (Pacheco Velez, 1982, p. 13). The place was also well regarded by its earliest European inhabitants. In the words of one chronicler of those times:

Its Fertility joyn'd to the Plenty of all things, never sensible of any intemperature in the air, which is so uniform, was it not interrupted by the frequent earthquakes, I do not think, says Frezier, that there is a fitter Place to give us an Idea of a terrestrial Paradise. (Lozano, 1755, p. 4)

But Lima was – and is – substantially at risk from several different kinds of natural extremes.

Natural systems and natural risks

Three vast interacting natural systems affect the environment of Lima: the Pacific Ocean, the coastal desert, and the Andean cordilleras. A fourth, the tropical Amazon forest, plays a more limited but still important hydrological role in coastal climates. These systems interact to create a unique set of geological, atmospheric, and hydrological hazards.

Lima lies within a broad zone of crustal plate convergence that is characterized by land deformation, orogenic uplift (mountain-building), seismic activity, and vulcanism. Subduction of the eastward-moving oceanic Nazca plate, as it slips under the rim of the continental South American plate, has given rise to the Andes mountains as well as conditions of high seismic instability and volcanic activity (Repetto et al., 1980). Extensive weathering and erosion of the mountains around Lima

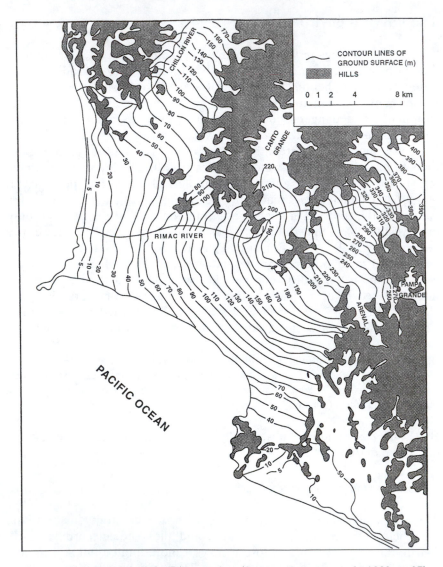

Fig. 8.2 Topography of the Lima region (Source: Repetto et al., 1980, p. 37)

have contributed large sediment loads to adjacent valleys, and the city's substrata are mainly composed of boulders, cobbles, and gravel in a mix of silty sand. Together these materials usually form a very hard and compact mass, although clays and loose sandy soils are also present in some areas at depths of 100 m (Moran et al., 1975, pp. 5–6).

The city sits on an elevated, gently sloping plain that covers about

3,500 km^2 and stretches inland along the Rimac valley towards steep mountains approximately 40 km to the east. Within this plain the local rivers are separated by low rolling hills. The availability of relatively flat land has permitted Lima to spread westward toward the coast and also – to some degree – into the hills that lie north and south. Hills such as the Morro Solar, el Agustino, San Cristobal, and San Geronimo have in recent years been occupied by migrant populations from rural districts. The built-up area has also begun to extend eastward into the narrower upper part of the Rimac valley. There, towns such as Chosica and Chaclacayo are still separated by farmland from metropolitan Lima, but it is likely that this physical separation will soon end (Ferradas, 1992, pp. 36–37). Much of metropolitan Lima is roughly 100 m above sea level, but the coastline varies from 80 m bluffs near Chorrillos to sea level at the port of Callao (located on the westernmost peninsula; fig. 8.2) (Kuroiwa et al., 1984, p. 801). Predictions of sealevel rise due to global warming indicate that large, now densely populated areas of Callao may be in significant danger of future flooding.

The Rimac valley cuts across a long narrow strip of extremely arid coastal desert that stretches from the Andes to the ocean between Ecuador and northern Chile. In the vicinity of Lima, annual precipitation is low and ranges from 0 to 50 mm. The scant precipitation is insufficient to support much flora beyond a few hardy species in the coastal desert. Although the appearance of the natural environment is often harsh and barren, when the soils of these desert valleys receive water from rivers or irrigation systems, they burst into florescence; they are some of the most fertile agricultural land in Peru, on which the vast majority of export crops are grown.

In many respects the climate is unusual for a low-latitude location (approximately 12 degrees N). From May to December the region is normally completely covered with dense clouds and fog at altitudes from 500 to 800 m. These render only a heavy mist, locally called "*garua*," which is described as "more bothersome than dampening" (Pulgar Vidal, 1987, p. 34). Thus, for as much as eight months of the year, when air temperatures rarely drop below 11 °C, the city of Lima is shrouded in a humid blanket that feels penetratingly chilly. On rare occasions, summer temperatures have reached heat wave proportions, as they did in early January of 1983 when 17 people succumbed to heat prostration and dehydration and over 2,700 people were affected (OFDA–AID, 1988, p. 154).

The unusual climate has evoked very different responses among those who experience it. In sharp contrast to Lozano's portrayal of Lima as an earthly paradise, Herman Melville described it in quite different terms:

Nor is it, altogether, the remembrance of her cathedral toppling earthquakes; nor the stampedoes of her frantic seas; nor the tearlessness of arid skies that never rain; nor the sight of her wide field of leaning spires, wrenched copestones, and crosses all adroop (like canted yards of anchored fleets); and her suburban avenues of house-walls lying over upon each other, as a tossed pack of cards; it is not these things alone which make tearless Lima, the strangest, saddest city thou can'st see. For Lima has taken the white veil; and there is a higher horror in this whiteness of her woe. Old as Pizarro, this whiteness keeps her ruins for ever new; admits not the cheerful greenness of complete decay; spreads over her broken ramparts the rigid pallor of an apoplexy that fixes its own distortions. (1961, p. 194)

The bizarre climatic combination of heavy cloud cover, moderate temperatures, extremely arid terrain, high humidity, and low rainfall is closely connected with the juxtaposition of very humid cold oceanic air, which overlies the Peruvian and Humboldt offshore currents, and very dry stable air over the land. Although these conditions regularly reproduce the city's most characteristic weather, the region is also extremely sensitive to perturbations in the ocean-to-atmosphere energy transfer system (Moseley et al., 1981, p. 234). At irregular intervals, there occur natural perturbations that heat up the normally cold ocean currents and overlying air masses, reduce the biological productivity of important marine fisheries, and affect the onshore precipitation regime. These perturbations are collectively known as the El Niño Southern Oscillation (ENSO). In addition to disruptions of the marine food chain and subsequent reductions of the coastal fishing economies, El Niños bring torrential rains to the western slopes of the Andean cordilleras and devastating flash-floods in the rivers that descend from them (Caviedes, 1981, p. 288). Even without El Niños, coastal Lima is subject to variations in the climate of the neighbouring Andean region. Rainfall, or the lack thereof, in the central Andes affects water supplies and hydroelectric power for the Lima metropolitan area, as well as the threat of flooding in eastern districts of the city. The possibility of global warming via the greenhouse effect – and the consequences of its impact on El Niño – are difficult to determine, but may include accelerated desertification of coastal valleys such as the Rimac (Antunez de Mayolo, 1986, p. 61) or increased flooding.

Lima: The city of the kings

Population projections indicate that Lima may soon host over 9 million people. For most of its existence the city grew slowly; it did not exceed 100,000 for almost three and a half centuries and required a further 80

Table 8.1 Population growth of Lima, 1614–1990

Year	Population
1614	26,441
1700	37,259
1836	55,627
1857	94,195
1862	89,434
1876	100,156
1881	80,000
1891	103,956
1898	113,409
1903	130,089
1908	140,844
1920	223,807
1931	373,875
1940	662,885
1961	1,632,370
1972	3,002,043
1980	4,410,000
1990	6,500,000
2000[a]	9,140,000

Sources: 1614–1972 – Dobyns and Doughty (1976), p. 298; 1980–2000 – Wilkie and Contreras (1992), pp. 119–120.
a. Projection.

years to reach 1 million. Since the end of the Second World War population numbers have accelerated dramatically (table 8.1). This recent demographic shift has placed enormous burdens on Lima. Each migratory wave from the rural hinterland has exerted overwhelming demands on employment, services, infrastructure, and land. In other words, Lima provides a truly classic illustration of a third world city where life-support systems are drastically out of balance with population numbers.

The pre-Colombian role of natural hazards

The part of Peru in which Lima is located has a history of human settlement that stretches back more than 10,000 years, including at least 4,000 years of complex civilizations. Although tectonic uplift, earthquakes, floods, and droughts appear to have influenced large-scale culture change throughout Peru, archaeological and ethno-historical records suggest that the pre-Colombian population had developed a set of relatively effective adaptations to such threats; there is little evidence of massive mortality and destruction owing to sudden-onset disasters. All indications are that hazard preparedness and mitigation were important components in the prevailing cultural perspective of ancient Peruvians. Evidence of sensitivity to natural hazards shows up in the sites that were selected for settlements, urban design, building technologies, and building materials. But these features of Peruvian life were altered significantly by the Spanish conquest (Oliver-Smith, 1994).

Conquest of Peru and foundation of Lima

When the Spaniards toppled the Inca empire in 1532, Pizarro at first wanted to place his new capital in the highlands near Jauja (Miro Quesada, 1982, p. 13) (fig. 8.1). He was forced to reconsider because rival conquistadors were plotting incursions into the new territory from coastal bases in Central America. If Pizarro was to retain hegemony over the new domain, he required a capital that had ready access to maritime communication. The Rimac valley was situated at the approximate midpoint of an important maritime route between Panama and Santiago in Chile. In addition to the other natural advantages of this site that have been described above, it was located at the seaward end of the primary Inca road out of the southern sierra – from which much gold and silver would eventually issue.

Thus, for reasons of both political economy as well as access to important natural features and resources, on 18 January 1535 Pizarro founded Lima – the City of the Three Kings – in the lower Rimac valley. Some historians claim that the title "City of the Three Kings" derives from the fact that Lima was established just after the Christian festival of Epiphany (6 January), which commemorates the delivery of gifts to the infant Jesus from three kings (Miro Quesada, 1982, p. 15). The name Lima is interpreted as deriving from Hispanization of the local word for the Rimac River, pronounced "Limac" in Indian dialect (ibid.).

The city was sited just 12 km from the coast, not far from the Inca temple and pilgrimage site of Pachacamac. An expedition led by Pizarro's brother, Hernando, had been greeted by a strong earthquake there almost exactly two years before (Giesecke and Salgado, 1981, p. 11), but this does not appear to have deterred the founders. Although it was not immediately apparent, the Rimac and other coastal valleys were also prone to mosquito infestation, which was later to lead to malaria and yellow fever epidemics in the city (Dobyns and Doughty, 1976, p. 125). The site was hazardous in another important way. Pizarro rejected Cusco, the inland Inca capital around which pivoted the internal social and economic networks of the vast conquered territory, in favour of a coastal location that looked outward towards the colonial power overseas. By placing his capital on the coast he established the basis for a form of urban and national development that is vulnerable to various external forces that have produced – and now affect – the enormous mega-city of contemporary Lima. The act of selecting a coastal site for the colonial capital – repeated elsewhere in many other European colonial settlements – cast a mould that has compounded the vulnerability to natural hazards that is inherent in Lima's location.

The colonial model of development

Lima was established with Spain's needs in mind. Spain's incursion into Latin America occurred simultaneously with the expansion of mercantile trading in Europe. The newly conquered possessions, particularly Peru and Mexico, both of which had an abundance of precious metals, were developed to enhance Spanish economic power. The Spanish towns and cities of Peru were either gold and silver production centres in the mountains (e.g. Arequipa, Ayacucho, Potosi [now in Bolivia]) or administration and transhipment points on the coast (e.g. Lima, Trujillo, Lambayeque, Tumbes) (Wilson, 1987, p. 200). Owing in large measure to the transportation hazards presented by bandits, pirates, and other brigands, it was decided to concentrate military protection on a single city. Because Lima had also been designated the capital, it became both the administrative centre in touch with the Spanish metropolis as well as the principal location through which surplus wealth could be extracted from the colony (Wilson, 1987, p. 200). The stage was set for the city's rapid ascension to primacy, initially for all of the continent and later for the country of Peru.

Early Lima

Settlements in the early Spanish colonies of the New World were thoroughly regulated by the Laws of the Indies, which included "Royal Ordinances for the laying out of new cities, towns or villages" (Griffin and Ford, 1980; Nuttall, 1922). These ordinances recommended sites that provided fresh water, included land for farming and pasture, and were elevated so that diseases would be less prevalent. They also contained very specific requirements that cities be laid out in a grid pattern, with regular east–west and north–south streets radiating from a central plaza. The principal church, the town hall, and other official structures were to be constructed around the plaza, and nearby blocks were designated for élite residential development. (Proximity to the plaza became a marker of social status in the colonial city that continues to this day in many smaller and medium-sized cities of Latin America – Griffin and Ford, 1980, p. 399). Lima was begun before the ordinances were adopted but its layout anticipated them because such guidelines flowed from widely held

Spanish perceptions of ideal urban settlements. All the implicit guidelines were followed regarding the location of the city, water sources, pasturage, wind currents, and altitude as well as the geometric urban plan.

Lima was located with its back approximately 100 paces west of the Rimac river bank. This close proximity permitted irrigation from the river and the disposal of wastes, including those from a slaughterhouse. Indeed, the river was one of the primary formative elements of the city's development (Ortiz de Zevallos, 1986, p. 42). A second formative element was the chessboard design of streets and urban lots (ibid.). The municipal offices, the cathedral, and the palace of the governor would all eventually be located on the plaza in the eastern sector of the city close to the river. Around this plaza the primary conquistadors were allocated plots of land ranging from 4 blocks for Pizarro down to 1 block for less important Spaniards (fig. 8.3) (Pacheco Velez, 1982, p. 18).

The social structure of the colony in its earliest years consisted of a very thin layer of dominant Spaniards grafted onto an enormous indigenous population of various ethnic identities. As the colony matured, Spanish women joined the original male settlers and a Creole population of American-born Spaniards emerged. Unions between Spanish males and Amerindian women were also established almost immediately by force, purchase, gift, or, in some cases, choice, resulting in a population of people of mixed Spanish and indigenous heritage that grew steadily through the colonial period. So rapid was the growth of the mestizo population that Magnus Moerner (1967) has referred to the conquest of the New World as the conquest of women. At the same time, the number of indigenous peoples diminished catastrophically because of disease, exploitation, and outright murder. It has been calculated that in the first 100 years of the colony, the indigenous population was reduced by 98 per cent from its estimated pre-conquest level of 14 million to approximately 600,000 (Cook, 1981, p. 114). Nevertheless, Indians still constituted an overwhelming majority of the population of colonial Peru and would continue to do so into the twentieth century. Enslaved and free Black people, from Africa as well as the Iberian peninsula, had participated in the conquest as soldiers and servants. They also figured importantly among the growing number of artisans of colonial Lima (Lockhart, 1968). By the eighteenth century the social and economic base of the viceregal city became dependent on Black labour (Dobyns and Doughty, 1976, p. 124).

Though the population of Lima was composed of Spaniards, mestizos, Indians, and Blacks, the city's identity was bound up with its status as the Spanish capital and the residence of the colony's élites. Here the white population differentiated itself from the Indians; mestizos were essentially unrecognized by the élite, although Blacks were valued more highly

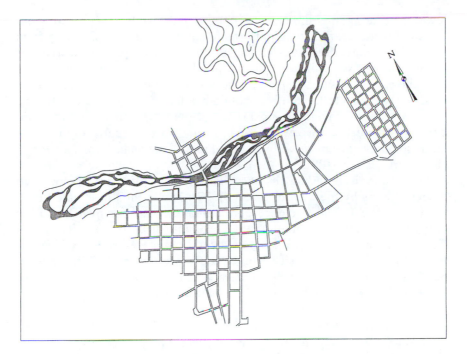

Fig. 8.3 Lima, 1613 (Source: based on a historical reconstruction by Juan Bromley in Gunther Doering, 1983)

for valuable skills, particularly in carpentry, masonry, and construction trades. However, the colony's economic structure was built on rights to Indian labour, enshrined in an institution known as the *encomienda*. It was this that formed the basis for the initial Spanish extraction of surplus and accumulation of wealth.

In its earliest years, Lima was merely a more or less orderly collection of large, poorly built adobe houses with woven split-cane roofs. Indian rebellions and seditious revolts by Pizarro's rivals limited the city's growth (Miro Quesada, 1982, p. 17). Later, the provisional buildings were replaced by more substantial structures. Large pretentious residences began to be constructed for members of the Spanish élites (Lockhart, 1968, p. 108). Monumental churches were also created. The task of building such grandiose edifices was carried out by drafted Indian labourers, in effect compounding their defeat by forcing them to contribute to the construction of the most visible symbols of their physical and ideological subordination (Fraser, 1990).

Although basically Spanish in style and inspiration, early Lima was also influenced by the cultural and environmental features of its Peruvian

location. The temperate climate permitted use of light, cheap construction materials following patterns familiar to the conquistadors from
their Caribbean expeditions (Pacheco Velez, 1982, p. 19). Scarcity of
locally quarried stone dictated the use of adobe and wood, and the lack of
rain allowed for easily constructed flat roofs. The experience of frequent
earthquakes quickly limited most buildings to two floors (ibid.). By the
mid-1540s the architectural appearance of the city, which was to remain
fixed for centuries, was established – long lines of bare adobe walls
interrupted by elegant carved wooden doors, some as large as 7 by 9 feet
(Lockhart, 1968, p. 108).

Shortly after its founding, the once inauspicious settlement was awarded
a coat of arms by the king of Spain (1537). It was then designated both
the site of the Royal Audience and capital of the Viceroyalty of Peru
(1543); its bishop was later (1545) given jurisdiction over dioceses in
Ecuador, Colombia, Nicaragua, and several other countries. In 1551 the
New World's oldest university, later to become known as the University of
San Marcos, was founded (Miro Quesada, 1982, p. 17). With the influx of
these functions the city began to take on a more impressive appearance.

"very good houses and some very elegant with towers and roof terraces, and the
plaza is large with wide streets and through most of the houses pass irrigation
channels that are of no little advantage for from the waters of these are served
and irrigated their many fresh and delightful gardens and vegetable plots."
(Pedro Cieza de Leon, as quoted in Miro Quesada, 1982, p. 16)

Sixteenth-century Lima evolved within the grid pattern focused on a
central plaza; it also possessed many open spaces occupied by gardens
and groves. Water was drawn from the efficient pre-Colombian irrigation
complex constructed by the Maranga Indians. The city was also situated
at the crossroads of a number of Incaic roadways, guaranteeing a steady
flow of travellers (Pacheco Velez, 1982, p. 15). By the early seventeenth
century, Lima had evolved into the form that it maintained almost to the
beginning of the twentieth century (fig. 8.3). Instead of the city expanding
into new territory, its gardens and groves became urbanized and existing
lots were redeveloped at higher densities. Defensive walls were also constructed to provide protection against marauding pirates – first around
Callao in 1639 and later around Lima in the years between 1684 and 1687
(Gunther Doering, 1983, p. 9). El Cercado, the largest Indian sector, was
located outside the walls, but significant numbers of Indian servants and
concubines lived in Spanish homes inside the walls (Butterworth and
Chance, 1981, p. 15). Small settlements such as Carabayllo, Magdalena,
Surco, and Cieneguilla also began to appear in the countryside at some
distance from the walled city.

A flourishing colonial capital: Seventeenth and eighteenth centuries

Early in the seventeenth century there was an upsurge in the construction of major buildings, among them convents and monasteries, public buildings, and private palaces. These often adopted monumental baroque styles of architecture and layout (Pacheco Velez, 1982, p. 29). The rigid quadrangular grid of streets was cut across by diagonals, some following the line of old Indian roads, and numerous small plazas of varying shapes were added. A hierarchy of urban spaces also appeared in the form of ethnic and socio-economic neighbourhoods (*barrios*): Black neighbourhoods in San Lazaro, Pachacamilla, and Malambo; Indian neighbourhoods and towns in the Cercado and Magdalena; Chinatowns; the English avenue; commercial streets, silversmiths' streets, swordmakers' streets. By 1630, when Lima's population reached 30,000, the city had taken on the essentials of its modern personality. Its political power and economic privilege were confirmed by the number and variety of its offices and public facilities. Furthermore, it held unquestioned status as the site of Euro-American high culture in this part of Latin America; poets, architects, painters, musicians, philosophers, and theologians pursued intensely artistic lifestyles, despite the prevalence of poverty and exploitation among a majority of the population.

Lima had also become an educational centre, owing in part to a high concentration of religious institutions and clerics (Dobyns and Doughty, 1976, p. 127). Clerics and viceregal bureaucrats displayed their finery in masses, processions, *autos-da-fé*, and journeys to and from the ships docked in Callao (ibid.). Lima of the élites in the years before the great earthquake of 1746 was well summed up by Lozano:

Lima had arrived to as great a Degree of Perfection as a City situate at such a distance from Europe, and discouraged by the continual dread of such Calamities was capable of. For although the Houses were of but moderate Height, being confined to one story only, yet the streets were laid out with the exactest regularity, and adorned with all that beauty which a nice Symmetry could give: So that they were equally agreeable to the Sight as commodious to the inhabitants; and display'd as much elegance as if all the ornaments of the best Architecture had been bestowed upon them ... It may be affirmed, that the Magnificence of these Edifices, if it did not exceed, at least might rival that of the grandest Fabricks in the whole World: for the beauty of their design, their Profiles, their Cemeteries, the largeness of their Naves, their Cloisters and Stair cases, was such as they had no Cause to envy any such for Size or Elegance. (1755, p. 5)

In contrast to the ostentation and display of the members of the élites, the general populace lived in conditions of insecurity and privation. Much of the city was extremely unsanitary. Refuse littered the streets, most of which were uncobbled dirt with open ditches for domestic water supplies and sewage disposal. Not surprisingly, gastro-intestinal diseases doomed half of the city's newborn and many adults each year. Malaria and yellow fever were constant preoccupations. Every ship that put into Callao constituted a potential disease vector of major proportions; epidemics of smallpox, influenza, typhus, typhoid, or plague were frequent (Dobyns and Doughty, 1976, p. 127). Earthquakes also shook the city. In 1687, one of these ushered in the use of *quincha* walls in preference to adobe. Quincha is a form of wattle and daub that greatly reduces weight and enhances structural flexibility (Gunther Doering, 1983, p. 10). Many quincha buildings from this and later eras still exist in present-day Lima. Unfortunately, quincha is vulnerable to fires, which are easily ignited from cooking surfaces (Dobyns and Doughty, 1976, p. 127).

Real and imagined Indian rebellions posed other threats to the well-being of Limenos. Like peasant uprisings in the Old World – which generally occurred when abuse and exploitation reached intolerable levels – Indian uprisings in the sierra were not uncommon. The apprehension of Limenos about what was essentially a distant highland phenomenon, frequently resulted in the panicked discovery of plots to burn the city, followed by summary executions of purported Indian conspirators (Dobyns and Doughty, 1976). However, fears of rural uprisings directed against Lima were ultimately realized when the Shining Path guerrilla movement *(Sendero Luminoso)* emerged in the late twentieth century.

During the eighteenth century, Lima experienced crises and other changes that undermined its power and monopolies. The nation's capacity for food self-sufficiency was seriously affected by a plague that destroyed wheat production and required the importation of food. At the same time, the seemingly inexhaustible sources of silver and mercury, on which Peru's economy depended, began to decline. Finally, the 1746 earthquake reduced Lima to rubble and created a tsunami that virtually obliterated Callao and killed 4,000 of its inhabitants (Gunther Doering, 1983, p. 11; Lozano, 1755, p. 9). At that time, Lima contained roughly 3,000 houses in 150 blocks, occupied by approximately 40,000 people. Although the earthquake killed just over 1,100 people in Lima, it left scarcely 20 houses standing there (Lozano, 1755, p. 4).

The Earth struck against the Edifices with such violent repercussions, that every Shock beat down the greater part of them; and these tearing along with them vast Weights in their Fall (especially the Churches and high houses) complicated the destruction of every Thing they encountered, even of what the Earthquake had spared. (Lozano, 1755, p. 3)

The natural catastrophes were compounded by a series of political and economic changes. Many élite Limenos lost their wealth when the Spanish monarch abolished the *encomienda* system. The creation of competing viceroyalties in Buenos Aires and Santa Fe reduced the extent of Lima's administrative hegemony (Gunther Doering, 1983, p. 11). Finally, the convoy and armada system, which routed all South American merchandise (and export duties) through Callao, was also discontinued. Although this was initially seen as a calamitous blow to Lima's economy, the liberalization of trade that resulted actually proved beneficial.

Lima and Peru benefited from other changes that were sweeping the market structure of Europe. There, increasing urban populations generated demands for new agricultural products and this resulted in the emergence of a new form of colonial productive relations called *latifundios*. These were large landed estates, initially in the highlands and later on the coast, that utilized economies of scale to produce an exportable surplus (Wilson, 1987, p. 201). Lima and other coastal cities that were involved in the shipment of this surplus saw the emergence of a new class of wealthy merchants, brokers, financiers, and transporters. The tertiary sector of the economy developed significantly, adding to the cultural and political life of the city and decidedly altering consumption patterns toward European luxury goods (Wilson, 1987, p. 202).

Notwithstanding the calamities and other changes, Lima continued growing slowly but still at a rate that outstripped its closest competitors. According to the census of 1796, it had a population of 52,627, roughly 6 per cent of the total within the viceroyalty, and only double that of 1614 (Dobyns and Doughty, 1976, p. 124). The area encompassed by the city's wall was a total of 506 hectares, divided into 358 ha urbanized, 130.9 ha in gardens and groves, and 16.8 ha in military fortifications (Barbagelata, 1971, p. 7). Despite the massive destruction that it caused, the earthquake of 1746 provided an opportunity for the thirty-first viceroy, Manuel Amat y Junient, to enact significant changes as Lima was rebuilt. Public lighting was installed on the main streets and the previously unadorned plazas and parks were embellished. Churches, promenades, and a formal bullring were built. New streets and promenades were also put in place outside the city wall along the eastern bank of the Rimac (Dobyns and Doughty, 1976, p. 129). These initiatives added 58.8 ha to Lima's total area (Barbagelata, 1971, p. 7). Thus, at the end of the eighteenth century, the primacy of Lima was unchallenged in the nascent nation of Peru. Such was the city's pre-eminence that Spaniards charged with crimes (or official displeasure) were often exiled 5 or 50 leagues from Lima for varying lengths of time (Dobyns and Doughty, 1976, p. 126). Banishment from Lima meant banishment from the life of European luxury goods, from cafe society, and from the intellectual centre of the colony.

Republican Lima

Peru proclaimed its independence from Spain in 1821 but did not finally achieve control over its own affairs until 1824. The independence movement was led by colonial mercantilists and latifundistas who were strongly influenced by new political ideas and fresh capital inputs from Europe. Enlightenment France contributed ideas about the state, natural liberty, and the rights of man; capital came mostly from England. However, neither independence nor the shift from a hybrid form of feudalism and mercantilism to capitalism did anything to alter the social system of Peru or the commercial hegemony of Lima, although the outward flow of goods was increasingly directed to British markets (Wilson, 1987, p. 202). Chief among the new export products from the 1840s onwards was a fertilizer composed entirely of bird dung. Known as *guano*, this came from enormous deposits on offshore islands. The demand from European countries innovating in agricultural technology and practices created a virtual boom economy, providing a windfall with which Peru hoped to pay its revolutionary war debts.

None the less, the face of Lima did not change much for decades after independence. Population grew slowly and the city continued to experience a wide variety of communicable illnesses, including yellow fever and malaria epidemics, as well as endemic diuretic diseases fostered by the still crude water and sewage-disposal system. Even the improved water systems that were installed in Callao (1846) and in Lima (1856) failed to affect urban death rates significantly (Dobyns and Doughty, 1976, p. 170).

The demolition of Lima's city walls between 1868 and 1871 marked the beginning of significant alterations to the city. These were initiated in large measure by the North American entrepreneur and contractor Henry Meiggs, who was originally hired to construct a rail line from Arequipa to the coast but quickly adopted more ambitious plans (Barbagelata, 1971, p. 1). Although the walls' demolition did not result in any immediate expansion of the city, it paved the way for a later comprehensive redevelopment scheme. This came in the form of the 1873 plan for an area adjacent to the old city, designed along French lines, that included wide avenues, promenades, sidewalks, and multiple circular street intersections occupied by small parks and monuments, reminiscent of many European cities (Dobyns and Doughty, 1976, p. 185; Pacheco Velez, 1982). This plan called for more bridges over the Rimac, the river's canalization, a beltway following the line of the demolished wall, the extension of existing streets, and the creation of new broad avenues toward the coast, first in the direction of Callao and later toward Chorrillos. The greater attention to Callao was stimulated by completion of the Panama Canal, which increased the level of commerce in the port. In effect, the coast became a third formative element in the development of the city (Ortiz de Zevallos, 1986, p. 42). The 1873 plan accurately projected and shaped the course of Lima's urban growth for the next half-century. Indeed, some of the projects that it called for were not completed until 1960. Unfortunately, some aspects of the plan were never carried out, with substantial costs to the city. These included the recommended urbanization of peripheral areas and lands gained by rechannelling the river; instead of being systematically and carefully developed, these places were later invaded and occupied by the informal settlements of migrants (Barbagelata, 1971, p. 9).

The development of Lima in this period was flawed in a way that was to prove particularly pernicious for the country; it was financed entirely on credit obtained mostly from abroad by mortgaging future production of guano and other domestic resources. This strategy cemented the neo-colonial condition of the country, solidified its dependent status, and mortgaged national economic sovereignty for over a century (Dobyns and Doughty, 1976, p. 192). Peru's acute contemporary debt crisis is thus part of a pattern of dependent development that began by borrowing to pay off debts from the war of independence and increased by borrowing to finance national and urban growth.

The War of the Pacific (1879–1883) was fought between Chile, Peru, and Bolivia over mineral resources in ambiguously demarcated territory near their mutual frontiers. The Chilean navy began bombarding Callao in 1880, invaded Lima in 1881, and occupied the city until 1883. Living conditions plummeted and death rates rose as food became scarce. People fled the city and the population fell from 100,156 to around 80,000 (Dobyns and Doughty, 1976, p. 197). The results of this war were disastrous for Peru and Lima. The state's bankruptcy was compounded by severe damage to infrastructure, and its major resources were controlled by foreign interests. The cost in casualties, money, and damage to infrastructure was far worse than experienced in any earthquake (Dobyns and Doughty, 1976, p. 203). Faced with economic collapse, successive governments attempted to raise monies by taxing the Indians, but this merely provoked rebellions during 1885 and 1892. None the less, reconstruction was begun using money raised by British bond holders in exchange for long-term control over resources. The rail system was rebuilt and expanded, linking settlements in both the sierra and along the coast to Lima. These and other infrastructural improvements in the sierra helped to stimulate migration to the capital. There living conditions were also improving; the incidence of waterborne diseases diminished after an improved sanitation system was introduced in 1884, and electricity was installed throughout the city in 1886, providing light for homes and power for emerging small industries (Dobyns and Doughty, 1976, p. 202).

Larger industries such as shoes, matches, bricks, and ceramics also emerged and power lines reached out to supply expanding sugar mills on the coast north of Lima (Dobyns and Doughty, 1976, p. 218).

On the eve of the new century, Lima was modernizing. Educational institutions, from primary schools to universities, were available. Up-to-date sanitation and medical facilities had finally made significant inroads into the abominable health conditions that had previously kept population in check and discouraged in-migration. For the first time in its history, Lima's death rate fell below its birth rate. Electricity, clean water, theatres, schools, medical care, churches, trolley cars, automobiles, bicycles, fire companies, police protection, all became available in Lima in the period between 1885 and 1895 (Dobyns and Doughty, 1976, p. 213). The city was now ever more attractive to rural migrants at a time when population growth, agricultural stagnation, economic injustice, and other problems were making life in the rural areas less and less viable. In 1821, at independence, Peru's population stood at roughly 2 million people, with Lima hovering around 50,000 inhabitants. By 1895 the national population had more than doubled to 4.5 million. The combination of rural problems (push factors) and urban attractions (pull factors) set the stage for a demographic avalanche that would begin to encroach upon Lima in the early years of the twentieth century and later engulf it by the end of the century.

Lima metropolitana: Evolution of a mega-city

The turn of the century brought a change in Peru's economic relationships: the dominance of British industrial capitalism was replaced by that of US monopoly capitalism. Major US investments in mining and export agriculture strengthened enclaves of production and further solidified the financial and commercial roles of Lima (Wilson, 1987, p. 203). The large-scale rural migration to Lima that began in the late nineteenth century increased steadily during the first two decades of the twentieth. Beginning in 1908, migration-induced changes in Lima occurred in three stages: 1908–1940; 1940–1960; and 1960–present.

The first stage was marked by accelerated population growth and urban expansion well beyond the perimeter of the old medieval city. An ambitious national programme of road-building increased the mobility of highland residents and the accessibility of the capital. As a result, the urban population growth rate jumped from 2.5 per cent per year in 1920 to 6.1 per cent in 1930 (Enriquez and Ponce, 1985, pp. 13–14). The city continued expanding south toward the coast, spawning middle-class suburban communities such as Magdalena Nueva and San Miguel, upper-class suburbs such as San Isidro and Miraflores, and beachside areas including Barranco and Chorrillos.

During the second stage of twentieth-century urban changes, the zone between Lima and Callao became fully urbanized. Urban industrialization remained sluggish, but in-migration continued at rates in excess of 9 per cent per year. However, the nature of the migrants changed: what had been primarily a movement of members of the rural élite and people from provinces near Lima became a massive and generalized phenomenon involving migrants from the whole country (Enriquez and Ponce, 1985, pp. 13–14). By 1954, the entire region between the old city centre and the southern coastline became urbanized, and informal squatter settlements began to appear in desert areas to the south and north (fig. 8.4).

During the third and latest stage, Lima has become a city of provincial migrants, and the character of the city has changed utterly (Enriquez and Ponce, 1985, pp. 13–14). To place this growth in perspective, Lima in 1984 contained the same number of people, roughly 6 million, as the entire nation had in 1940. Or, further, in 1940 Peru was a predominantly rural nation with somewhat less than 35 per cent of the population living in urban settlements. Overall rural–urban migration to all cities has turned Peru into a nation that was 70.2 per cent urban in 1990 (Economic Commission for Latin America and the Caribbean, 1991, p. 8). The city of Lima contained 29.1 per cent of the total national population in the same year (Wilkie and Contreras, 1992, p. 120). The overwhelming primacy of Lima by 1989 is evidenced by table 8.2.

Hazards and vulnerability

Over the past 450 years Lima has experienced a variety of hazards and disasters. Some of these are outgrowths of the city's location in an area of known risks. Others are the direct result of human alteration of the environment in ways that make its inhabitants more risk prone. And still others, perhaps now the vast majority, are a function of socio-economic inequities that force the most vulnerable to bear the greatest risks. Table 8.3 provides a chronology of the more noteworthy disasters.

Because this table focuses on large-scale devastating events, it significantly understates the hazardousness of Lima. A full accounting of Lima's hazards would include many past outbreaks of epidemic diseases that are now less threatening, a wide range of more or less chronic seasonal hazards, and a growing list of low-intensity risks that have serious cumulative effects. For example, malaria, yellow fever, smallpox, and plague were regular visitors to the city before the mid-nineteenth cen-

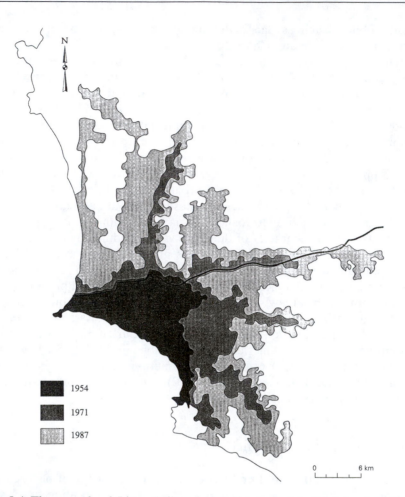

1954

1971

1987

0   6 km

Fig. 8.4 The growth of Lima–Callao during the second half of the twentieth century (Source: Burga and Delpech, 1988)

tury. Peru has also experienced approximately 50 major earthquakes and countless smaller ones since historical records were begun (Gieseke and Salgado, 1981, pp. 65–66). Even the smaller tremors, which do little damage, constitute a continuing source of human stress. Seasonally, Lima is subject to floods and *huaycos* (mudslides), landslides, water scarcity, food shortages, dehydration, and heat prostration. There are also the daily dangers of waterborne diseases, malnutrition, fires, toxic waste exposure, air pollution, and, since 1980, guerrilla warfare of varying intensity.

All these hazards are borne unequally, and mostly by the underprivileged. The middle and upper classes can afford anti-seismic construction

Table 8.2 The primacy of Lima in Peru, 1989

Indicator	Lima's share (%)
National population	28
Gross national product	69
Private investment	98
Banks	83
Commercial bank deposits	77
Informal businesses	70
Tertiary product (services)	80
Tax revenues	87
Public employees	51
University population	53
Educators	39
Doctors	73
Economically active population	32
Workers	36
Employers	50
University professors	62
Telephone hook-ups	76
Hospital beds	48

Source: Vargas Chirinos (1989), pp. 101–102.

and can live above floodplains on solid terrain in uncongested areas with good escape routes. They can also obtain nourishing food, purified water, and good medical care. High quality and accessibility to other services are considered to be part of the bundle of rights to which the affluent are entitled and that they can normally expect to receive. The poor of Lima, more than half the citizenry, can neither avoid the hazards nor purchase the means that might mitigate them. Vulnerability to hazards in Lima, like poverty, is stratified by race, ethnicity, age, gender, and income. Moreover, the pattern of vulnerability in Lima is closely tied to the structure of the nation's political economy and the role of the city in that structure.

Economic context

Since the Second World War, economic vulnerability has grown in Lima as a consequence of government policies and the working of international market forces. After the war, Peru's government – like most in Latin America – adopted a strategy of industrialization that involved heavy investment of borrowed overseas capital. Foreign borrowing accelerated after the global oil price rises of the early 1970s (Portes, 1989, pp. 9–10). At the same time, multinational corporations began to reorganize production by shifting manufacturing towards lower-cost developing coun-

Table 8.3 Disasters in Lima

Date	Agent	Impact
1524	Smallpox epidemic	16 million deaths in Peru
1531	Smallpox epidemic	Generalized in Peru
1546	Plague or typhus epidemic	
1552	Earthquake	Moderate damage
1555	Earthquake	Significant damage
1558–60	Influenza and smallpox epidemic	
1578	Earthquake	Severe damage to homes, churches, viceroy's palace
1584	Earthquake	Structural damage
1586	Earthquake	Significant damage; 14–22 deaths
	Tsunami	Major damage in Callao
1585–91	Influenza epidemic	
1609	Earthquake	Structural damage; Cathedral damaged
1618	Measles epidemic	
1630	Earthquake	Structural damage; deaths undetermined
1655	Earthquake	Heavy damage
	Tsunami	Damage in Callao; undetermined number of deaths
1678	Earthquake	Severe damage
1687	Earthquake	Heavy damage in Lima; wide regional damage in adjacent coastal valleys
1690	Earthquake	Structural damage
1746	Earthquake	Almost total destruction of the city; 1,141 deaths
	Tsunami	Total destruction of Callao; 4,000 deaths
	Epidemics	Post-earthquake conditions
1806	Earthquake	Significant damage
1828	Earthquake	Significant damage; 30 deaths
1867	Yellow fever epidemic	4,445 deaths
1897	Earthquake	Significant damage in Callao
1904	Earthquake	Moderate damage
1932	Earthquake	Damage: Lima – moderate; Callao & Rimac – severe

270

1926	Earthquake	Moderate damage
1940	Earthquake	Heavy damage; 179 deaths; 3,500 injuries; regional impact
1966	Earthquake	Heavy damage (1 billion soles); 100 deaths; regional impact
1974	Earthquake	Heavy damage (2.7 billion soles); 78 deaths; 2,500 injuries
1991	Cholera epidemic	76,190 cases; 196 deaths

tries. Peru experienced some gains in average worker salaries as a result of this process, but total employment did not increase because much of the new investment was capital intensive. In addition, many of the domestically consumed goods that would previously have been produced in Peru began to be supplied from specialized assembly points elsewhere (Wilson, 1987, pp. 203–207). As a result, during the period in which migration to Lima intensified, the urban employment market actually began to shrink.

Between 1970 and 1984 the percentage of the labour force in manufacturing fell from 18.4 to 15.6 and the urban real minimum wage dropped from 107.3 in 1976 to 53.3 in 1985, using a 1980 value of 100 as a base. In the same period, urban formal unemployment rose from 8.4 per cent to 16.4 per cent, and the rate of informal market participation reached all-time highs in 1984 (Portes, 1989, pp. 26–27). Most of the informal sector involves so-called "subsistence commerce" in urban areas, but it also includes the (mainly rural) cocaine industry, which has not generated much new employment. For certain urban dwellers, cocaine sales support the consumption of imported luxury goods and allow them to engage in urban real estate speculation as well as some construction and service businesses. The cocaine industry has also stimulated a drug problem in the cities as well as in the production areas, adding yet another hazard to Peru's list (Morales, 1989).

Changes in the global economic system did not compensate for reductions in Peru's labour market; instead the problem was worsened by International Monetary Fund austerity measures that had profound recessionary effects and produced negative growth rates for the first time in 50 years. They also reduced the resources that were available for maintaining urban infrastructure (Portes, 1989, pp. 13–14). The outcome was catastrophic misery for millions of already desperately poor Limenos during the 1980s and 1990s. It was said that Lima, where abject poverty had never been far away, was becoming "Calcutta-ized."

Land-use context

The chronic and worsening imbalance between population growth and employment in Lima finds expression in two distinctive urban forms, each with its own pattern of vulnerability to hazards: (1) informal squatter settlements, which are proliferating around the fringes of the city; and (2) inner-city neighbourhoods that are becoming increasingly crowded and run-down.

Many post-war migrants dealt with the high price of urban space and urban housing by illegally occupying land on the city's periphery. Despite their somewhat ramshackle appearance, these settlements are the result of highly organized and purposeful behaviour. Traditionally, migrants who arrived in Lima from the highlands sought housing in the *tugurios* (slums) of the old inner city. Once living quarters and jobs were secured there, strategies for an "invasion" of vacant land on the urban periphery would be developed. Clandestine visits to chosen sites would be undertaken for purposes of mapping out lots for future houses, other structures, and community institutions. On an appointed day, usually in the earliest hours of the morning, rented trucks would transport the conspirators from the *tugurio* to the site. Before dawn brought discovery and the summoning of police, squatters would construct an entire community out of light timber and *estera* (woven split-cane) mats. Owners and officials were presented with the unappetizing task of forced removal or acceptance of a *fait accompli*. In the long run, the authorities have collaborated – at least passively – in this process by granting land titles as well as formally recognizing ownership rights and community charters (Butterworth and Chance, 1981; Mangin, 1967; Maskrey, 1989; Millones, 1978). In this way there have arisen urban communities that lack planned infrastructure or services and are vulnerable to a wide range of hazards.

During the 1950s, 56 of these settlements were located on the periphery of the city (some were then already more than 20 years old); in 1984 there were 598 such *barriadas*. Now called *pueblos jovenes* (young towns), they contained close to 40 per cent of Lima's population (approximately 2.5 million people) (Fuenzalida, 1986, p. 185). Older *barriadas* gradually evolved into permanent communities and grouped together to form separate municipalities (Burga and Delpech, 1988). For example, San Juan de Lurigancho, which was incorporated in 1967, is composed of over 200 different units: 113 *pueblos jovenes,* 31 urban developments, 40 building associations, and 25 building cooperatives (Poloni, n.d.). The process that gave rise to the *pueblos jovenes* may be self-limiting. The *tugurios* are no longer staging areas for invasions of vacant peripheral land. Instead their deteriorating buildings are overcrowded with job-seekers who can no longer afford the long commutes from peripheral *pueblos jovenes* to work in central Lima where most of the informal commerce is concentrated – even as such job opportunities are themselves declining (Portes, 1989, p. 32). More and more people are becoming trapped in the deteriorating buildings of old Lima.

Between 1940 and 1984 Lima's built-up area expanded from 3,966 ha to 31,255 ha. A great deal of the newly urbanized land was formerly desert, river beds, and hillsides; most of the rest is converted agricultural land that had for long provided much of the city's food (Fuenzalida, 1986, p. 184). Not all of the urban expansion is accounted for by *pueblos jovenes*. Upper-middle- and upper-class Limenos have also fled before the incoming wave of poor migrants, and many now live in large expensive new developments, particularly to the south-west of the old city. However, the areal differentiation of rich and poor was reduced during the 1980s, partly because the debt crisis impoverished middle-income groups and forced them to resettle working-class areas and partly because the poor tend to occupy open land in any location (Portes, 1989, p. 22).

Infrastructure

Vulnerability to improperly managed hazardous materials and other environmental contaminants has increased in Lima, while industrial infrastructure has remained inadequate because it is underfunded and poorly developed. This has led to more respiratory, gastro-intestinal, and malnutrition-related diseases and has contributed to increased morbidity and mortality among the city's disadvantaged populations. Vulnerabilities associated with the deterioration or absence of technological infrastructure interact with natural and technological hazard agents, each compounding the effects of the others. It has also become increasingly difficult to extend existing infrastructure to service the outer metropolitan populations because they are so thinly spread. However, the combination of lower population densities in the newer *pueblo jovenes* and improved building materials (including both lightweight temporary wood constructions and solidly built replacements) has reduced levels of seismic vulnerability among the populations of squatter settlements (Maskrey, 1989, p. 6).

Vulnerability to natural hazards

Earthquakes

In the past, earthquakes have caused Lima's most frequent and serious disasters. Vulnerability to earthquakes is clearly related to social class and construction type, while both of these factors are in turn correlated with location. For example, middle- and upper-income groups reside in the suburbs and the poor occupy *tugurios* or peripheral squatter settlements. Soil stability is also important. The soils of central Lima generally bear static loads well (Moran et al., 1975, p. 6), but they tend to lose bearing strength during earthquakes, especially in a belt of alluvial soils along the river banks (Kuroiwa, 1977, p. 13). Soils that underlie middle- and upper-class areas to the south of the city and the port of Callao are also subject to seismic instability (Moran et al., 1975, p. 6; Kuroiwa, 1977, p. 13).

Inner-city slums are the most seismically vulnerable areas of the city. Houses are built from adobe (dried mud), a material that readily cracks and crumbles during earthquakes if it is poorly mixed with binding agents, deteriorated, or weakened by previous tremors. Deterioration is common because rents are often controlled and landlords who cannot raise money by increasing rents often abandon their property. Moreover, many buildings are overcrowded, surrounded by narrow alleys, and not near open spaces that could act as refuge sites. In the economic crisis of the 1970s and 1980s, building costs became prohibitive and the budgets for self-help housing programmes were cut back. Slum-clearance projects further reduced the availability of rental housing in the inner city. More and more people have found themselves trapped permanently in the *tugurios* (Maskrey, 1989, p. 10). It has been estimated that a repeat of the 1940 earthquake (8.2 on the Richter scale) would destroy more than 26,000 dwellings, leave 128,000 people homeless, and kill or injure large numbers of people (Maskrey, 1989, pp. 7–11).

Bricks and concrete have been used as building materials in Lima for about 80 years (Moran et al., 1975, p. 5). Because they are more expensive than adobe, quincha, and the temporary materials employed in squatter settlements, their use has been restricted to middle- and upper-class residences and the engineered structures of commerce, government, and industry. However, brick and concrete are considered by the poor to be "noble materials" and their use for housing is much aspired to. Indeed, the houses in many older squatter settlements have evolved from *estera* mat huts to adobe and quincha and ultimately to brick and concrete (Burga and Delpech, 1988; Poloni, n.d.; Degregori et al., 1986).

Most brick and concrete buildings are less than three storeys high. Roofs and intermediate floors are made of reinforced concrete slabs, sometimes with hollow brick infill to decrease weight. Over the past 30–40 years, there have been some changes in construction practices that affect seismic stability. Reinforced concrete columns have been placed at corners and a ring beam of concrete has been poured where walls contact floors or ceilings; this makes brick structures quite rigid (Moran et al., 1975, p. 5). Buildings higher than three storeys are generally of reinforced concrete frame design. Anti-seismic design to resist lateral force effects became mandatory in the 1950s, but compliance with the code was uneven. In essence, earthquake resistance has become a significant part of design and construction practice only since the 1960s (Moran et al., 1975, p. 5).

Tsunamis

Together with the seismic vulnerability of adobe construction in older parts of Lima, the threat of tsunami (earthquake-generated waves) to the port of Callao is considered to be the most important natural hazard faced by the metropolis (Kuroiwa et al., 1984, p. 801). Callao occupies a transition zone between the north coast of Peru – where significant tsunamis have not occurred in the past – and the south coast, where they are common (Kuroiwa et al., 1984, p. 804).

Most of the major earthquakes (1687, 1746, 1940) caused tsunamis that inflicted severe damage on Callao, and recent urban expansion has greatly exacerbated the problem. A repeat of the 1746 tsunami, which devastated a 6.5 km^2 area, would inundate 150,000 people, 67 schools, the Peruvian Naval Academy, the Maritime Academy, and the nation's primary port and industrial area. Since 1940, 13 tsunamis have produced varying degrees of damage in the low-lying coastal areas of Lima. Recent calculations underscore the difficulties of evacuating the growing population in the event of a tsunami generated by a nearshore earthquake. As little as 20–30 minutes of warning may be all that is possible (Kuroiwa et al., 1984, p. 804).

Huaycos *and floods*

Over a distance of about 120 km, the Rimac River descends 5,000 m between its source and Lima. On the way it passes through several zones that are characterized by sharply different rainfall regimes. For example, between December and April the upper valley receives 865–1,025 mm, while the foothills rarely get more than 30 mm per year and the coastal desert is in effect dry (Antunez de Mayolo, 1986, p. 61). During the December–April period, a considerable number of floods and *huaycos* take place along the Rimac. According to Maskrey (1989), a *huayco* occurs when heavy rainfall saturates steep slopes and runs off into adjacent streams. Long-term deforestation, overgrazing, and inappropriate management have reduced vegetative cover so that runoff is typically charged with soil and boulders. This load is eventually deposited in broad fans at points where the valley widens out. In the lower valley, the net effect is erosion of river banks and elevation of the river bed as it becomes choked with deposited materials.

Huaycos often block or destroy the highway and the railway between Lima and the central Andes, causing large losses. The city is deprived of food and the country of foreign exchange earned by exporting minerals. *Huaycos* can deposit solids in the city's main reservoir, thereby cutting water supplies by 25 per cent and forcing additional expenses for cleaning the water treatment plant. Finally, the homes of the migrant poor occupy floodplains, steep slopes, and alluvial fans in the Rimac's tributary valleys. When *huaycos* occur, these homes are destroyed and occupants are killed. Vulnerability to *huaycos* is in large measure determined by where one lives, and that is itself determined by socio-economic status (Maskrey, 1989, pp. 13–18).

People are moving in increasing numbers into the upper Rimac valley. At these slightly higher altitudes, residents are located above Lima's six- to seven-month cloud cover. Indeed, the year-round sunshine has attracted middle- and upper-class Limenos to weekend homes since the nineteenth century. Increasing occupancy of the upper Rimac valley has implications for future hazard losses. The river is the principal source of both water and electricity for the city. Its valley is also the route followed by roads and railways that connect Lima with the central Andes and the jungle.

Water shortages, water pollution, and public health

It is estimated that only the most affluent quarter of Lima's population receives constant water service (Reyna and Zapata, 1991, p. 36). Occupants of the *pueblos jovenes* are particularly vulnerable to various water-related hazards. Inadequate supplies of potable water are one of these. Approximately 55 per cent of Lima's water comes from surface sources and the rest from boreholes. In both cases, the water originates primarily from the Rimac River (Puri et al., 1989, p. 290). Lima's water supply and disposal system was constructed piecemeal and a good part of it is now 90 years old (Reyna and Zapata, 1991, p. 34). Leaks are endemic; almost 30 per cent of supplies are lost from the system owing to broken municipal pipes and leaking plumbing fixtures in homes (ibid., p. 37). Customers have little incentive to conserve piped water because only a third of the existing water meters function properly and charges are small (ibid., p. 38). Moreover, a new system would be prohibitively expensive. In 1991, the lowest estimate for a minimally adequate new system was US$470 million (ibid., p. 39). Workable and relatively inexpensive plans exist to pipe rainwater and spring water directly from the mountains, but they are not implemented because Lima's water authority does not have enough money to undertake these projects (ibid., p. 45). In the meantime, the city continues to drill new boreholes and drain even more subterranean water (Puri et al., 1989). Without adequate recharge, subterranean reserves will eventually be drained dry (Reyna and Zapata, 1991, pp. 39–40).

Unacceptable water quality is an even greater problem than water supply (Reyna and Zapata, 1991, pp. 35, 44–46). The Rimac is considered to be one of the world's most highly polluted rivers. More than 300 pipes dump domestic and industrial wastes in it and there are 14 large solid-waste dumps along its banks. Purification using chlorine alone costs more than US$0.5 million per year. But bacterial contamination is not the only problem. The head of Peru's Society of Engineers has stated that neither chlorination nor boiling is sufficient to ensure potable water because the Rimac is dangerously contaminated with heavy metals from mining and industrial wastes (Reyna and Zapata, 1991, p. 35).

The homes of 20–25 per cent of Lima's residents, primarily occupants of *pueblos jovenes,* lack water and sewer services. For these people water comes from tank trucks, but it has not been subject to quality controls (Reyna and Zapata, 1991, p. 35). There are 13 suppliers of water to the trucks, all of them connected with SEDEPAL (Servicio de Agua Potable Acantarrillado), but administered by Lima's various municipal councils. Customers pay up to 30 times more for water than do those with pipes, and they receive inferior water. This segment of the city's population uses only 2 per cent of Lima's fresh water (ibid., p. 36). In 1987, samples of trucked water were found to be contaminated with faecal coloforms and other bacteria (Gilman et al., 1993, p. 1556). Another study of the *pueblos jovenes* focused on infection by *Helicobacter pylori*, a bacterium that appears to be endemic in Lima (Klein et al., 1991, p. 1503) and that is "associated with gastric and duodenal ulcers [and] may also contribute to an increased risk of gastric carcinoma." The bacterium was three times more common among children whose homes were dependent on trucked water than among those in homes with piped water from the central water system. However, the risks are not solely a function of income. Children from higher-income families with piped water had 12 times the infection rates of those from higher-income families that received their water from community wells, which are entirely separate from Lima's municipal system (Klein et al., 1991, p. 1505).

A study of 70 of Lima's 300 boreholes showed that, although water quality was generally good, water retrieved after pumping had stopped and was then restarted tended to have high counts of *E. coli* (Puri et al., 1989, p. 293). This suggests that the cessation of pumping creates negative water pressures in the system and that *E. coli* enters because contaminated water and sewage are then sucked into water pipes (Committee on Foreign Affairs, 1991, p. 35; Reyna and Zapata, 1991, p. 34). Of Lima's 27 municipalities, 26 pour untreated sewage into the ocean (Committee on Foreign Affairs, 1991, pp. 52–53). As a result of oceanic pollution, many of Lima's beaches are unsafe for swimming and fishing (cited in Reyna and Zapata, 1991, p. 46). Another study, in 1988, has determined that swimmers face an increasing risk of contracting contagious gastro-intestinal and skin diseases (cited in Reyna and Zapata, 1991, p. 47).

Peru's poor suffer severely from from a variety of at least 20 waterborne diseases, including cholera (Witt and Reiff, 1991, p. 461). For example, between 1981 and 1991 gastro-intestinal disease increased sevenfold, while influenza and respiratory diseases rose twelvefold (Committee on Foreign Affairs, 1991, p. 4). Hepatitis is endemic and unvaccinated dogs and cats pose a constant threat of rabies. However, the most visible and acute water-related crisis emerged during the cholera epidemic of the early 1990s (Witt and Reiff, 1991, p. 452; PAHO, 1994, p. 16).

Cholera

The cholera epidemic of 1991–1993 has had the widest impact, if not the highest mortality, of any Latin American disaster during the same period. One US political leader perceived a direct connection with the global economic policies pursued by external governments and agencies: "The cholera epidemic is, I suggest, the human face of the debt problem, a repudiation of the trickle-down theories that have guided development policy for more than a decade" (Rep. Robert G. Torricelli in Committee on Foreign Affairs, 1991, p. 1).

The first case of cholera in Latin America was reported in Peru on 23 January 1991. The government confirmed the presence of cholera on 5 February (PAHO in Committee on Foreign Affairs, 1991, p. 86). Peru then established a Committee of Epidemiologic Surveillance (Staff, 1991), and the Peruvian Ministries of Health and Housing both initiated programmes to purify community water supplies (Reiff, 1992, p. 19). Owing to the rapid response, the death rate was held to fewer than 1 per cent of those who contracted the disease (PAHO, 1994, p. 13). This is particularly impressive because 30–50 per cent of untreated cholera victims are usually expected to die (ibid., p. 14). Peruvians who died during the 1991–1993 epidemic appear to have received no medical treatment or to have received it late (ibid.). Cholera tends to be a disease of adults, but it responded well to oral rehydration therapy (ORT), the treatment of choice for severe childhood diarrhoea (Committee on Foreign Affairs, 1991, p. 12). Fortunately, Peru's Public Health Service had an excellent ORT programme already in place, developed with funds from the US Agency for International Development, and it received considerable international assistance during the epidemic (ibid., pp. 13–16).

Ironically Lima had the lowest percentage of cholera deaths – despite its questionable water – because Lima has far better health facilities than most other parts of Peru (Reyna and Zapata, 1991, p. 26). Cholera victims in Lima generally received prompt treatment, owing to the ready availability of ORT and of people trained in its use (Committee on Foreign Affairs, 1991, p. 80). As a result, of the first 55,000 people with cholera in Lima, only 150 died, whereas in Cajamarca, 320 out of 6,000 cholera sufferers died (ibid.). From the start of the epidemic until May 1991, Lima had 76,190 cases of cholera, of which 26,326 were hospitalized and 196 (0.26 per cent) died. In the same period, Cajamarca had 8,602 cases, of which 4,538 were hospitalized and 376 (4.37 per cent) died. Lima's 76,190 cases represented 34.84 per cent of Peru's total cases (Reyna and Zapata, 1991, pp. 109–111).

Most of Lima's cholera victims were poor people who lived in *pueblos jovenes* (Reyna and Zapata, 1991, p. 26). Cholera throughout Latin America is primarily – but not exclusively – a disease of the poor

(PAHO, 1994, p. 14). Highly contaminated water from decaying municipal water systems appears to have been a major reason for the spread (ibid., pp. 14, 16). In Peru, "cultural reasons" for the spread included the belief that freezing would kill the cholera pathogen, which led to the use of ice made from impure water (Reiff, 1992, p. 18). Cholera also appears to have been spread through dish-washing water used (and reused) by street vendors, and ingestion of raw seafood (Glass et al., 1992). On 12 February, Lima's mayor prohibited further street vending of food (Reyna and Zapata, 1991, p. 126). During the epidemic there was tremendous controversy over the eating of *ceviche*, Peru's national dish, which is made from marinated raw seafood. President Fujimori was shown on television eating *ceviche* after health recommendations against eating it had been publicized. This was followed by a rise in new cholera cases after each of Fujimori's appearances (ibid., pp. 125–132).

The cholera epidemic provoked a number of secondary stresses also. Securing sufficient fuel to boil all drinking water for 10 minutes, as recommended by Peru's Minister of Health, could be costly, particularly for those in *pueblos jovenes* who used kerosene for cooking (Reyna and Zapata, 1991, p. 79). On 25 February, Peru's Minister of Health requested that the government subsidize the price of kerosene so that poor people could afford to boil their water (ibid., pp. 79, 127). On 13 March, just before the start of a new school term, a group of sanitary engineers revealed that approximately 70 per cent of Lima's public schools had insufficient sanitary facilities and should not be opened for the new school session (ibid., p. 128). Many parents kept their children out of school (Committee on Foreign Affairs, 1991, p. 63). In the end, President Fujimori ordered that all of Peru's schools remain closed until 15 April (Reyna and Zapata, 1991, p. 130).

The degradation and contamination of the rivers that supply most of the drinking water for many Latin American communities can lead to the endemic presence of cholera. In addition, the contamination of the ocean has aided in keeping the cholera pathogen viable. It was once thought that cholera was carried only by humans, and spread solely through contamination by human faecal material. However, recent studies demonstrate that the pathogen can survive in oceanic plankton and algae that thrive on industrial wastes, fertilizer runoff, and raw sewage, even in the absence of faecal material (Emmett, 1993, p. 13). It appears that this can result in continued re-contamination of drinking water in coastal cities such as Lima.

Fires

Fires still constitute a major hazard in the squatter settlements, much as they did in colonial times. In some cases the causes of fires are the same.

The provisional huts of *estera* mats, although never seen as permanent, are frequently lived in for extended periods while resources are accumulated for more durable housing materials of adobe or brick. The *estera* mats are flammable and kitchen mishaps with kerosene stoves frequently ignite these fragile dwellings, quickly consuming the hut and all its contents and spreading to other huts built close by. Similar fires break out in overcrowded slum dwellings (Kuroiwa, 1977).

In other cases, however, the rapid expansion of both squatter settlements and other forms of formal and informal housing and industry has created circumstances of extreme danger from industrial fire and explosion. In Callao, for example, there is considerable risk in the port and industrial zone because petroleum, gasoline, and natural gas storage tanks are surrounded by homes and their accompanying cooking facilities as well as mobile fast-food vendors and their carts. Furthermore, the historic zone of Callao, consisting of the Fort Real Felipe and many wood-frame buildings, is also close by (Preuss and Kuroiwa, 1990, p. 278). In fact, the entire greater port area suffers from severe congestion, with storage facilities for combustible and toxic substances surrounded by dense, low-quality residences of the *pueblo joven*, Puerto Nuevo. In addition, between the navy base in Callao and the commercial port is the area's fishing fleet. The conjuncture of all these uses constitutes a high risk of fire starting either inside the industrial area from accident, inadequate maintenance, or impact or outside by the cooking carts, home accidents, terrorism, or an accident involving the many trains and trucks entering and leaving the area. In the event of such a fire, there would be little to keep the blaze from becoming a major conflagration (ibid., p. 280). The danger of fire caused by earthquakes or accidents is compounded by this mixed use pattern in this area as well as other parts of the city. Furthermore, the system of streets and roads within both *pueblos jovenes* and the old city very often presents an obstacle in itself for the access of emergency services of any kind. In the older parts of the city, dating from the colony and early republic, streets are narrow and extremely congested. The *pueblos jovenes* are honeycombed with narrow alleys and passageways, sometimes ascending the steep hillsides to the east of the city and virtually impossible to traverse for emergency vehicles. Compounding the problem of fires is the fact that Lima has only a volunteer fire department that is woefully underequipped for any fires of an industrial or chemical nature (Preuss, personal communication, 1994).

Industrial and technological hazards

Although, nationally, Peru's level of industrialization is low, almost 70 per cent of all industrial establishments are located in metropolitan Lima. In 1985 the city also accounted for 57.6 per cent of the nation's gross

Table 8.4 The share of industrial value, by sector, produced in metropolitan Lima

Sector	Lima's share (%)
Foodstuffs	57.7
Beverages	63.0
Tobacco products	100.0
Textiles	83.3
Clothing	94.3
Leather	70.7
Shoes	95.0
Wood products	26.0
Paper products	75.7
Printing and publishing	83.0
Chemical processing	77.5
Other chemicals	92.4
Petroleum	60.1
Oil and coal derivatives	35.8
Rubber	98.8
Plastics	90.4
Ceramics	98.6
Glass	91.8
Non-metallic minerals	52.6
Basic industrial iron and steel	29.0
Non-ferrous basic industries	12.4
Metals	77.7
Non-electric machinery	76.8
Electric machinery	99.0
Transport	86.5
Optical and photographic industries	98.5
Other industries	86.9

Source: *Atlas del Peru* (1989), p. 191.

national product (*Atlas del Peru*, 1989, p. 192). The city's dominance is a function of two factors. Because Lima contains almost one-third of the national population, it provides a huge market. Secondly, Lima is the only city in Peru that can satisfy the demand of industry for infrastructure and services (energy, water, transportation networks, communications networks, banking, commercial and distribution chains) at competitive prices. The city's nodality is well seen in the Peruvian road system, which radiates from it. Nodality is even more exaggerated in an economic sense (ibid.; table 8.4).

By itself the concentration of industry would pose a significantly increased threat to health and safety in Lima compared with other communities. When industrial concentration is combined with other factors, the potential for disaster is even higher. Such factors include: inadequate infrastructure; deteriorating facilities; insufficient maintenance; lack of enforcement of industrial regulations; and mixed land-use patterns that feature intermingling of housing and industrial plants. To these can be added the propensity of developed countries to ship hazardous wastes to already burdened poor states. In these circumstances, reported negotiations between a Peruvian company and a US company that wished to dispose of toxic wastes from American industries (e.g. solvents, burnt oils, and chemical wastes) take on added importance (Hardoy and Satterthwaite, 1989, p. 188).

The combination of industrial emissions with emissions from poorly tuned vehicles using fuels with a high lead content generates severe air pollution problems in Lima. According to available data this is not yet on a par with some other developing metropolises, but this may be due to lack of research as much as to actual pollution levels. Impressionistic data indicate that the situation is worsening and represents a grave enough potential to be discussed as a threat in civil defence literature (Comite Nacional de Defensa Civil, n.d.).

The juxtaposition of industrial and residential uses is characteristic of a number of areas in Lima. It is also important to recall that the figures cited above represent only that portion of the total industrial output that is produced by the formal sector. Admittedly, almost 60 per cent of the informal economy is dedicated to tertiary activities, but there is evidence that small-scale industrial production and maintenance activities go entirely unregulated, many of them located in the *tugurios* and the *pueblos jovenes*. Although many of the *pueblos jovenes* often plan an industrial zone apart from residential areas, mechanical and electrical workshops are frequently set up in or adjacent to dwellings (Poloni, n.d.; Burga and Delpech, 1988).

Social unrest and terrorism

In the colonial era, residents of Lima frequently regarded the oppressed Indian population as a menace to the city. Indian rebellions certainly occurred in the highlands, from the early days of the colony to the nineteenth century, but Lima was never seriously threatened. Apparent conspiracies were uncovered and supposed plotters were executed to assuage the anxiety of the members of Spanish élites. But these anxieties were not realized until the recent emergence of the members a political movement known as the Peruvian Communist Party of the Shining Path (*Sendero Luminoso*). In many ways, Shining Path was an expression of rage and frustration felt by some sectors of Peruvian society against conditions of injustice and poverty that afflicted the majority of Peruvians (Palmer, 1992). Founded in 1970 by a philosophy professor from the University of San Cristobal de Huamanga in the highland city of Ayacucho, the party adopted a Maoist interpretation of Peruvian society, and emerged in 1980 to wage a guerrilla war that is unparalleled in Peruvian history for its success and cold-blooded brutality. Its leader, Abimael Guzman, interpreted Peruvian underdevelopment as deriving from the colonialist, semi-feudal system, which subjugated and exploited the Indian masses for the benefit of urban élites (Granados, 1992). He believed that urban élites were assisted by international allies in the capitalist system and therefore he prescribed a Maoist-style war on the cities. Initially this targeted provincial and departmental capitals, but the conflict gradually worked its way toward and into Lima.

During the first year of hostilities (1980) there were over 1,000 attacks on government offices, high tension towers, and police posts in highland areas (Kirk, 1991). At first the guerrillas were dismissed as gangsters and fanatical thugs, but gradually the government found itself fighting a protracted war against opponents who were often indistinguishable from the peasant population. Counter-insurgency efforts were soon marked by numerous human rights violations against the peasantry as government forces attempted to combat the growing effectiveness of *Sendero Luminoso*. The guerrillas replied in kind, executing peasant leaders and often whole families or neighbourhoods. Such was the brutality that large numbers of refugees began fleeing the highlands in 1983, most of them ultimately taking up residence in the shanty towns of Lima. Many of the *displazados* (displaced or refugees), in effect, began to establish their own shanty towns, invading unoccupied land at the edge of other settlements. The *displazados* have been estimated to have contributed approximately 200,000 people to Lima's already enormous shantytown population (Kirk, 1991).

As the *displazados* began pouring into the city, *Sendero Luminoso* and another group – the Tupac Amaru Revolutionary Movement (MRTA) – initiated campaigns of urban terrorism. The first actions taken against residents of Lima involved dynamiting transmission towers carrying power to the city from hydroelectric plants. These blackouts (*apagones*) became an unmistakable indicator of the city's vulnerability to terrorism. Shortly after the first one, car bombs and assassinations also began; these targeted public functionaries, neighbourhood and *barriada* leaders, as well as public offices, embassies, and other facilities. After 1986, sounds of gunfire and bombings became commonplace in the city. Electricity and water were rationed and supplies were unreliable as attacks on public utilities escalated. Assassinations and bombings continued until the mid-1990s. Urban terrorism forced some better-off residents to emigrate to the United States, but most Limenos could not avail themselves of this option. Parts of the city where *Sendero* support was strong became known as "liberated" zones. Often these were areas of strategic significance that controlled access to vital resources. For example, Ate-Vitarte was a major focus of *Sendero* activity because it is a nerve centre for both water and electrical services, as well as being the site of many large *pueblos jovenes* (Gonzalez, 1992; Rojas Perez, 1992, p. 127). The great mass of Lima's population endured what became known as the "Beirutization" of their city. The capture of Abimael Guzman and many of the leadership cadres in September of 1992 has diminished *Sendero*'s effectiveness, if not its brutality. Conflict continues, but more sporadically; the bombings and the assassination of public officials and organizational leaders are now less frequent. A high-profile four-month-long seige of the Japanese embassy by the Tupac Amaru ended in April 1997 with the death of all the guerrillas in the building. The war thus far has cost over 25,000 lives and US$10 billion in terms of destruction and damage (Doughty, 1993).

Opportunities for intervention to reduce urban hazards

Today Peru is beset by a vast range of problems, including hazards and disasters. The country suffers profoundly from severe underdevelopment and "maldevelopment" (Amin, 1990), both of which undergird vulnerability to hazards. From the standpoint of urban hazards, Lima even appears to be regressing; this is best indicated by the resurgence of epidemic diseases that had not occurred since the early nineteenth century. Now cholera (and AIDS) have reinstated epidemics on the list of burgeoning urban environmental hazards.

The national and international dimensions of underdevelopment should be addressed, as well as the more immediate manifestations of hazard. For example, the hazards of waterborne diseases such as cholera, hepatitis, and dysentery could be substantially mitigated by addressing deficiencies in the water system, particularly with regard to issues of purity, maintenance, and delivery as well as adequate sewage and waste disposal. Fire hazards could be substantially reduced by separating industrial and residential areas, widening access roads, improving housing stock, and professionalizing fire services. Vulnerability to *huaycos* and floods would be substantially reduced by appropriate land-use planning and enforcement as well as by land acquisition and housing programmes for low-income people. Earthquake vulnerability in inner-city slums could be reduced by renovating buildings and reducing densities via provision of alternative housing. These steps are fundamentally contributions to development as well as hazard mitigation. Indeed, by itself, the provision of basic services to all of the city's population would go a long way toward mitigating many of Lima's hazards. Sadly, given the enormous numbers of people without basic services and the resources available for such a task, this form of hazard reduction will be uneven and slow at best.

Unfortunately, the structure, circulation, and accumulation of international capital have created an economic system that consigns Peru to the role of raw materials supplier and processor. The current production system will not generate sufficient capital to alleviate Peru's perennial international deficits, much less realistically address the multiple hazards that face the population of Lima on a daily basis. Adequate responses to these problems would require enormous transfers of capital and an escalation of the country's indebtedness to unimagined heights. Such a goal is to be hoped for, but there seems little basis for believing that it will happen in the foreseeable future.

That fact notwithstanding, it must be recognized that numerous cities in the developed world are experiencing similar difficulties coming to grips with problems of air and water quality, fire service, land use, sanitation, toxic water disposal, and other aspects of environmental hazard. And, however overwhelming the challenges facing Lima may be, Peru is not without significant resources that can be deployed to reduce them. Important steps to address many disaster-related problems are being taken. This is partly because hazard awareness is high in Lima. It is impossible to live there and not be aware of the potential of natural, social, or technological hazards. Limenos have joked to me that they are the most adaptive people in the world because they have developed a coping system for chaos! They are extremely serious about coming to grips with hazards, despite their lack of economic resources to resolve them. They have learned from the experience of the great 1970 earthquake and have embarked on a wide range of hazard-reduction research projects.

The earthquake of 1970 killed approximately 65,000 people and devastated a vast area in the north-central coastal and Andean regions (Oliver-Smith, 1992, 1994). Although the shocks were strongly felt in Lima, they did little damage there. However, the earthquake performed a valuable service for non-affected areas by stimulating a programme of research on hazard and vulnerability that continues to the present (Casaverde and Vargas, 1984; Gieseke and Salgado, 1981; Kuroiwa, 1977, 1983; Kuroiwa and Alva, 1991; Kuroiwa and Tanahashi, 1989; Kuroiwa, Sato, and Kumagai, 1992; Kuroiwa et al., 1984; Torres Cabrejo and Huaman Egoavil, 1992). Key individuals in the National Engineering University (UNI), the National Civil Defence Committee, the Catholic University of Peru, and other national and international organizations assumed responsibility for investigating the earthquake vulnerability of Lima and the rest of Peru. Secondary disasters such as tsunamis and fires were also studied. Initial work relied heavily on graduate student theses that investigated vulnerability of structures (e.g. housing, hospitals, schools, industries) and public services (e.g. water, sewerage, energy, transportation, communications). A map of probable seismic intensities was constructed, based on past events and soil and geological analysis. This was combined with maps that showed probable damage based on the prevalence of different construction materials (Kuroiwa, 1977).

Analyses of these data revealed several critical problems in Lima. A primary finding was the debilitated condition of overcrowded adobe and quincha housing in older parts of the city, specifically Barrios Altos, Cercado, Rimac, Callao, Barranco, and Chorrillos. In the event of a serious earthquake, such dwellings remain likely to collapse, trapping victims, filling streets with rubble, and impeding rescue. Hospitals and schools in these neighbourhoods are in similar weakened condition. Moreover, the soils that underlie the older parts of Lima tend to amplify seismic movements. The same soil conditions also affect Callao, whose vulnerability to tsunamis constitutes the second most critical problem facing the Lima metropolitan area.

Newer middle- and upper-class residential neighbourhoods tend to be located on stable soils and to employ better materials and building technologies. Minimal damage is predicted for these areas. However, several factories have been found to be vulnerable to seismic damage and public services are also at risk. Water pipes are particularly vulnerable because of either age or soil instability. Electricity supplies are less vulnerable because of good maintenance. Traffic circulation in the older areas of Lima would be subject to much worse congestion in the event of a disaster and the communication system, radio, TV, telegraph, etc. were found to lack disaster preparation.

Most of the findings for the 1977 earthquake vulnerability study remain valid today, with the exception of significant improvements in the communication system (Maskrey and Romero, 1986; Maskrey, 1989). There have been continuing efforts to establish hazard-research organizations and projects, some in cooperation with international bodies and other countries. Principal among these is the Centro de Investigacion Sismica y Mitigacion de Desastres (CISMID – Centre for Earthquake Engineering Research and Disaster Mitigation). CISMID was formed under an agreement between the UNI and the Japanese government. Its facilities include a structural lab, a geotechnical lab, a computer centre, a national disasters data bank, a disaster mitigation department, a library, and educational, training, and diffusion services. The Japanese International Cooperation Agency (JICA) has send expert advisers to Peru, and 25 CISMID staff members were trained in Japan during the late 1980s (Kuroiwa and Tanahashi, 1988). One major focus of investigation has been the transferability to Lima of disaster-mitigation techniques that are employed in Tokyo (Kuroiwa, 1977).

The process of political reorganization that is under way in Peru has broad implications for hazard reduction in Peru as a whole as well as in Lima. A basic law for political regionalization and implementing a National Plan for Regionalization were important innovations included in the new constitution of 1979. This legislation has subsequently been modified and complemented by amendments and additional laws promulgated in 1984, 1987, and 1988. It is a substantial attempt to come to grips with the overwhelming centralization of Lima and to develop an integrated policy of economic and social development among the country's 12 regions (Vargas Chirinos, 1989; Kuroiwa and Tanahashi, 1989). Important progress has been made, including the election of regional assemblies in some regions, but the process – and its intended effects on decentralization – is far from complete.

Regionalization has also provided a structure and a focus for a national programme of disaster reduction. The technical knowledge and skills of hazard-management individuals and organizations are being mobilized to develop a national disaster-reduction plan that is based on regional components. Currently, the principles and procedures of micro-zonation – a widely used tool of earthquake engineering – are being refined for use in urban and regional land-use plans throughout the country. Micro-zonation techniques are normally applicable to small areas, usually of a few square kilometres. At the regional scale, they are assigned to high-priority locations such as those with rapidly growing populations, areas with high disaster frequency and vulnerability, and areas where important engineering projects are to be located (Kuroiwa and Tanahashi, 1989). The separate regional plans will eventually be combined in a single National Programme for Disaster Prevention and Mitigation, which will also double as Peru's main contribution to the International Decade for Natural Disaster Reduction. A pilot case-study of the Grau district in the far north of Peru was carried out and produced important findings about soils, land-use planning, appropriate construction techniques and materials, expansion areas for urban development, and disaster probabilities and vulnerability (Kuroiwa, Sato, and Kumagai, 1992).

Because micro-zonation techniques were originally developed for urban planning, their application for hazard analysis for Lima is relatively straightforward and will constitute a major effort in the near future. Peru's newly established regions provide the organizational structure for a hazards database. Supporting research has been funded by a greatly expanded budget of the National Council of Science and Technology (CONCYTEC). An education, training, and diffusion initiative is also encouraged by an accord between the Ministry of Education and the UNI (Kuroiwa and Tanahashi, 1989). Eventually it is intended that all national construction in Peru will factor in disaster-mitigation measures and all citizens, regardless of place of residence, will be informed about disaster preparedness (Kuroiwa and Alva, 1991). The level of technical understanding of environmental hazards is high and there has been a continuous process of refinement of emergency and mitigation plans for the city and the nation (Kuroiwa and Tanahashi, 1989; Kuroiwa, Sato, and Kumagai, 1992). Notwithstanding the progress that is being made in formulating national and regional policies for natural hazards and disasters, unless the Peruvian government can implement and enforce these policies, their value is limited. This has been the fate of previous efforts to modify Lima's urban development by either planning controls or building codes.

In any case, national policies cannot reduce urban hazards by themselves; they must be complemented by hazard-mitigation and preparedness measures that incorporate high levels of local community participation. During the past decade in Peru, important non-governmental organizations (NGOs) dedicated to a community-based hazard reduction have emerged in Lima. They include organizations such as the Centro de Estudios y Prevencion de Desastres (PREDES) – a locally based NGO; and the Intermediate Technology Development Group (ITDG) – an international NGO with offices in Lima. Both of these and other groups are attempting to link disaster preparedness and mitigation to locally perceived existing needs. They seek to build awareness of hazards and risks at the community level by means of pilot projects that reinforce, rehabilitate, and rebuild slum dwellings or improve soil- and water-management techniques in huayco-prone communities. In so far as such strategies require the exercise of political pressure for expropriation of land, transfer of capital resources, or alternative financial mechanisms, they help to build the basis for future agendas aimed at general transformation of underdeveloped communities (Maskrey, 1989, pp. 45–46). The immediate goals of NGOs such as ITDG and PREDES are local, but their long-term implications are far-reaching. None the less, they realize that, at the most basic level, vulnerability to hazards is deeply embedded in broader political economic structures that are based in both national and international systems of capital circulation and accumulation. In that sense, they work not only for disaster mitigation in specific local contexts but also for forms of political empowerment to address broader issues.

Conclusions

Since the beginning of the twentieth century, and especially during the past 30 years, Lima has been transformed. Once it was a bastion of criollo élite culture where the privileged few presided over a vast multi-ethnic population. Many Limenos used to take pride in the fact that they had travelled widely throughout Europe and North America but had never been to the highlands of their own country. They can no longer practise such evasions, for the highlands have now come to Lima. All the polarized metaphors and visions that élite Limenos once employed to define their identity are becoming irrelevant in what has become a massive city of migrants. The problems of exploitation and dependence that Lima's dominance created for the rest of Peruvian society have now been replicated in the capital. Demographically at least, the nation has become unified. If Lima was the springboard from which Peru was conquered, it is now the chief prize of the reconquest of Peru by its native peoples. Instead of Lima "being" Peru, as Valdelomar joked, Lima finally "belongs" to Peru.

Although efforts to address environmental hazards at both national and community levels are significant, given the monumental dimensions of the problem and the complexity of the tasks required there is little reason to expect a significant reduction in Lima's vulnerability to hazards in the near future. Both national policy implementation and community action may reduce the vulnerability of specific local populations, but an overall reduction in hazard vulnerability requires a series of actions that would alter the structure of national and international markets, debt–credit relationships, demographic movements, and political and administrative institutions. In the unlikely eventuality of such changes, efforts to come to grips with hazard vulnerability in Lima and in Peru as a whole take on real significance. The fact that such efforts are really unable to address the root causes of that vulnerability, so embedded in the historical development of the nation and its capital city during its over four and a half centuries of existence, does not mean that they will fail completely; only that they will never succeed completely. Given the similar situation of many mega-cities in the developed world, such a goal is worth the effort.

Acknowledgements

I would like to acknowledge and thank Ms. Suzanne Autumn and Ms. Lisa Perry for their invaluable assistance in the research for this chapter.

REFERENCES

Amin, Samir. 1990. *Maldevelopment: Anatomy of a global failure*. London: Zed Books.

Antunez de Mayolo, Santiago. 1986. "Sequias e inundaciones." *Boletin de Lima* 46(8): 61–64.

Atlas del Peru. 1989. Lima: Instituto Geografico Nacional.

Barbagelata, Jose. 1971. *Un siglo del acontecimiento historico precursor del desarrollo urbano de Lima moderna.*

Burga, Jose and Claire Delpech. 1988. *Villa el Salvador: La ciudad y su desarrollo, realidad y propuestas.* Lima: Centro de Investigacion, Educacion y Desarrollo (CIED).

Butterworth, Douglas and John K. Chance. 1981. *Latin American urbanization.* Cambridge: Cambridge University Press.

Casaverde, Lucia A. and Julio Vargas N. 1984. "Seismic risk in Peru." In *Proceedings of the Eighth World Conference on Earthquake Engineering.* Englewood Cliffs, NJ: Prentice-Hall.

Caviedes, Cesar. 1981. "Natural hazards in Latin America: A survey and discussion." In Tom L. Martinson and Gary S. Elbow (eds.), *Geographic research on Latin America: A benchmark.* Conference of Latin Americanist Geographers. Muncie, IN: Ball State University, vol. 8, pp. 280–294.

Comite Nacional de Defensa Civil. n.d. *Manual de seguridad.* Lima: Comite Nacional de Defensa Civil.

Committee on Foreign Affairs (CFA). 1991. *The cholera epidemic in Latin America.* Hearing before the Subcommittee on Western Hemisphere Affairs of the Committee on Foreign Affairs, House of Representatives, One Hundred Second Congress, First Session, May 1. Washington, D.C.: US Government Printing Office.

Cook, Noble David. 1981. *Demographic collapse: Indian Peru, 1520–1620.* Cambridge: Cambridge University Press.

Degregori, Carlos Ivan, Cecilia Blondet, and Nicolas Lynch. 1986. *Conquistadores de un Nuevo Mundo.* Lima: Instituto de Estudios Peruanos.

Dobyns, Henry and Paul L. Doughty. 1976. *Peru: A cultural history.* New York: Oxford University Press.

Doughty, Paul L. 1993. "War as another disaster: The 'earthquake' in Latin American life." Unpublished paper delivered at International Congress of Anthropological and Ethnological Sciences, Commission on the Study of Peace, Mexico City, 30–31 July.

Economic Commission for Latin America and the Caribbean. 1991. *Statistical yearbook for Latin America and the Caribbean.* Santiago, Chile: United Nations, ECLAC.

Emmett, Ariele. 1993. "Pollution in the time of cholera." *Technology Review* 96 (May/June): 13–14.

Enriquez, Narda and Ana Ponce. 1985. *Lima: Poblacion, trabajo y politica.* Lima: Pontificia Universidad Catolica del Peru.

Ferradas, Pedro. 1992. *Quirio: Prevencion de desastres, tradicion, y organizacion popular en Chosica.* Lima: Predes Centro de Estudios y Prevencion de Desastres.

Fraser, Valerie. 1990. *The architecture of conquest.* Cambridge: Cambridge University Press.

Fuenzalida, Fernando. 1986. "Lima: Ciudad abierta." In Augusto Ortiz de Zevallos (ed.), *Lima a los 450 anos*. Lima: Centro de Investigacion de la Universidad del Pacifico (CIUP), pp. 183–196.

Giesecke, Alberto and Enrique Salgado. 1981. *Terremotos en el Peru*. Lima: Ediciones Rikchay Peru.

Gilman, Robert H., Grace S. Marquis, Gladys Ventura, Miguel Campos, William Spira, and Fernanto Diaz. 1993. "Water cost and availability: Key determinants of family hygiene in a Peruvian shantytown." *American Journal of Public Health* 83: 1554–1558.

Glass, R. I., M. Libel, and A. D. Brandling-Bennet. 1992. "Epidemic cholera in the Americas." *Science* 256, 12 June, pp. 1524–1525.

Gonzalez, Jose E. 1992. "Shining Path's urban strategy: Ate-vitarte." In David Scott Palmer (ed.), *The Shining Path of Peru*. New York: St. Martin's Press.

Granados, Manuel. 1992. *El PCP Sendero Luminoso y su ideologia*. Lima: Eapsa.

Griffin, Ernst and Larry Ford. 1980. "A model of Latin American city structure." *Geographical Review* 70(4): 397–422.

Gunther Doering, Juan. 1983. *Planos de Lima*. Lima: Municipalidad de Lima Metropolitana.

Hardoy, Jorge and David Satterthwaite. 1989. *Squatter citizen: Life in the urban third world*. London: Earthscan Publications.

Jacobsen, Nils. 1993. *Mirages of transition*. Berkeley: University of California Press.

Kirk, Robin. 1991. *The decade of Chaqwa: Peru's internal refugees*. Washington D.C.: American Council for Nationalities Service, US Committee on Refugees.

Klein, Peter D., David Y. Graham, Alvaro Gaillour, Antone R. Opekun, and E. O'Brian Smith. 1991. "Water source as risk factor for *Helicobacter pylori* infection in Peruvian children." *Lancet* 227, 22 June, pp. 1503–1506.

Kuroiwa, Julio. 1977. *Proteccion de Lima metropolitana ante sismos destructivos*. Lima: Secretaria Ejecutiva del Comite Nacional de Defensa Civil.

——— 1983. *Tsunamis: Efectos sobre las costas de Lima metropolitana*. Lima and Geneva: UNDRO.

Kuroiwa, Julio and J. Alva. 1991. "Microzonation and its application to urban and regional planning for disaster mitigation in Peru." *Proceedings of the Fourth International Conference on Seismic Zonation*. Stanford, CA: Earthquake Engineering Research Institute, pp. 771–793.

Kuroiwa, Julio and Ichiro Tanahashi. 1989. "A national plan of hazard reduction." In *Proceedings of the Ninth World Conference on Earthquake Engineering*. Kyoto: Japan Association for Earthquake Disaster Prevention, vol. VII, pp. 1057–1062.

Kuroiwa, Julio, J. Sato, and Y. Kumagai. 1992. "Peru's program for disaster mitigation." *Proceedings of the Tenth World Conference on Earthquake Engineering*. Rotterdam: Balkema, pp. 6203–6208.

Kuroiwa, Julio, E. Alegre, V. Smirnoff, and J. Kogan. 1984. "Urban planning for disaster prevention in the low coastal area of metropolitan Lima." In *Proceedings of the Eighth World Conference on Earthquake Engineering*. Englewood Cliffs, NJ: Prentice-Hall, pp. 801–808.

Lockhart, James. 1968. *Spanish Peru 1532–1560*. Madison: University of Wisconsin Press.

Lozano, Pedro. 1755. *A true and particular relation of the dreadful earthquake which happen'd at Lima, the capital of Peru, and the neighboring port of Callao on the 28th of October, 1746*. Boston: D. Fowle.

Mangin, William. 1967. "Latin American squatter settlements: A problem and a solution." *Latin American Research Review* 2(3): 65–90.

Maskrey, Andrew. 1989. *Disaster mitigation: A community based approach*. Development Guidelines No. 3. Oxford: Oxfam.

Maskrey, Andrew and Gilberto Romero. 1986. *Urbanizacion y vulnerabilidad sismica en Lima metropolitana*. Lima: Predes Centro de Estudios, y Prevencion de Desastres.

Melville, Herman. 1961. *Moby Dick*. New York: Signet Classic (Penguin Books).

Millones, Luis. 1978. *Tugurio: La cultura de los marginados*. Lima: Instituto Nacional de la Cultura.

Miro Quesada S., Aurelio. 1982 *Nuevos temas Peruanos*. Lima, Peru.

Moerner, Magnus. 1967. *Race mixture in the history of Latin America*. Boston: Little Brown.

Morales, Edmundo. 1989. *Cocaine: White gold rush in Peru*. Tucson: University of Arizona Press.

Moran, Donald, Freer Ferver, Charles Thiel, Jr., James Stratta, Julio Valera, Loring Wyllie, Jr., Bruce Bolt, and Charles Knudeson. 1975. *Engineering aspects of the Lima, Peru earthquake of October 3, 1974*. San Francisco: Earthquake Engineering Research Institute.

Moseley, Michael, Robert A. Feldman, and Charles R. Ortloff. 1981. "Living with crises: Human perception of process and time." In Mathew H. Niteck (ed.), *Biotic crises in ecological and evolutionary time*. Princeton, NJ: Princeton University Press.

Nuttall, Z. 1922. "Royal ordinances concerning the laying out of towns." *Hispanic American Historical Review* 5: 249–254.

OFDA–AID (Office of Foreign Disaster Assistance – Agency for International Development). 1988. *Disaster history: Significant data on major disasters worldwide, 1900–present*. Washington. D.C.: USAID.

Oliver-Smith, Anthony. 1992. *The martyred city: Death and rebirth in the Andes*. Prospect Heights, IL: Waveland Press.

——— 1994. "Peru's five hundred year earthquake: Vulnerability in historical context." In Ann Varley (ed.), *Disasters, development, and environment*. London: Belhaven Press.

Ortega, Julio. 1986. *Cultura y modernizacion en la Lima del 900*. Lima: Centro de Estudios para el Desarrollo y la Participacion (CEDEP).

Ortiz de Zevallos, Augusto. 1986. "Lima como expresion material de una civilizacion." In Augusto Ortiz de Zevallos (ed.), *Lima a los 450 anos*. Lima: Centro de Investigacion de la Universidad del Pacifico (CIUP), pp. 27–68.

Pacheco Velez, Cesar. 1982. "Lima: Tiempos y signos de Lima vieja." In *Lima*. Madrid: Instituto de Cooperacion Iberoamericana, pp. 10–56.

PAHO (Pan American Health Organization). 1994. "Cholera situation in the Americas." *Epidemiological Bulletin* 15(1): 13–16.

Palmer, David Scott (ed.). 1992. *The Shining Path of Peru*. New York: St. Martin's Press.

Poloni, Jacques. n.d. *San Juan de Lurigancho: Su historia y su gente*. Lima: CEP.

Portes, Alejandro. 1989. "Latin American urbanization during the years of crisis." *Latin American Research Review* 24(3): 7–44.

Preuss, Jane and Julio Kuroiwa. 1990. "Urban planning for mitigation and preparedness: The case of Callao, Peru." In *Proceedings of the Fourth National Conference on Earthquake Engineering*. San Francisco: Earthquake Engineering Research Institute.

Pulgar Vidal, Javier. 1987. *Geografia del Peru*. Lima: Promocion Editorial Inca, S.A. (PEISA).

Puri, S., J. L. Petrie, and C. Valenzuela Flora. 1989. "The diagnosis of seventy municipal water supply boreholes in Lima, Peru." *Journal of Hydrology* 106: 287–309.

Reiff, Fred M. 1992. "Cholera in Peru." *World Health* July–August: 18–19.

Repetto, Pedro, Ignacio Arango, and H. Bolton Seed. 1980. *Influence of site characteristics on building damage during the October 3, 1974 Lima earthquake*. Report No. UCB/EERC-80/41. Berkeley: College of Engineering, University of California.

Reyna, Carlos and Antonio Zapata. 1991. *Cronica sobre el colera en el Peru*. Lima: DESCO (Centro de Estudios y Promocion del Desarrollo).

Rojas Perez, Isaias. 1992. "Las barriadas de Lima: La batalla pendiente." *Ideele* 4: 45–46, 124–135.

Staff. 1991. "Cholera – Peru, 1991." *Journal of the American Medical Association*, 13 March, p. 1232.

Torres-Cabrejo, R. E. and C. E. Huaman-Egoavil. 1992. "Evaluation of seismic risk on water supply pipelines." In *Proceedings of the Tenth World Conference on Earthquake Engineering*. Rotterdam: Balkema, pp. 5609–5612.

Vargas Chirinos, Raul. 1989. *Regionalizacion del Peru y gobierno nacional descentralizado*. Lima: CONCYTEC.

Wilkie, James W. and Carlos Alberto Contreras. 1992. *Statistical abstract of Latin America, vol 29*. Los Angeles: UCLA Latin American Center Publications.

Wilson, Patricia Ann. 1987. "Lima and the new international division of labor." In Joe R. Feagin and Michael Peter Smith (eds.), *The capitalist city: Global restructuring and community politics*. Oxford: Basil Blackwell.

Witt, Vicente M. and Fred M. Reiff. 1991. "Environmental health conditions and cholera vulnerability in Latin America and the Caribbean." *Journal of Public Health Policy* 12(4): 450–464.

9

Social vulnerability to disasters in Mexico City: An assessment method

Sergio Puente

Editor's introduction

Mexico City is a testament to the durability of environmental hazards in many parts of the world. Successive cultures that occupied this site, from the pre-Aztec to the post-colonial, have faced demanding constraints of seismicity, vulcanicity, subsidence, aridity, flooding, and altitude and have adjusted with different degrees of success. All have either made careful allowances for these hazards in the design and operations of their communities or been forced to take them into account retroactively. But none has had to cope with such an explosive expansion of population and urban development as has occurred in the late twentieth century. As a result, instead of drawing on reliable mechanisms of urban society to buffer themselves from environmental risks, modern residents of Mexico City occupy an uncertain, often provisional, metropolis. Dizzying growth and deep-seated social dislocation challenge inherited ways of thinking and acting about environmental risk and foster the emergence of new hazard-management approaches. This is a situation that calls for sensitive handling and enlightened analysis, a challenge to which Sergio Puente's paper directly responds.

[T]he Aztecs and the tribes that preceded them did not believe that their valley had attained physical stability. Their religion was filled with prophecies that the world would be torn asunder by cataclysmic earthquakes. The Spanish conquistadors witnessed volcanic activity in the crater of Popocatepetl ... In 1985, thousands of Mexico City residents perished and hundreds of buildings were destroyed by a giant quake. And dozens of times every year, the capital's inhabitants experience the stomach-churning sensation of smaller tremors that cause no damage but remind the populace that the earth beneath their feet is unsettlingly alive.

(Kandell, 1988, pp. 10–11)

Mexico must stop acting as if Mexico City were some sort of natural disaster out of control.

(Vargas, 1995)

Introduction

Vulnerability can be defined as the propensity to incur loss. The disaster-vulnerability of urban areas is not exclusively determined by human factors but, during the twentieth century, the human contribution has come to dominate. This is partly a result of profound worldwide social and technological transformations that have concentrated population and industry in cities. It is also an outgrowth of other broad societal changes that raise serious questions about the proper role of humankind in a changing world. These include: a breakdown of existing relationships between society and nature; crises of economic development and world order; and pressures to redefine the societal roles of science and technology. To resolve problems of environmental hazard in cities it may be necessary to modify fundamental aspects of society, including the human axiological system.

The mega-city as locus of hazard

It has been argued that natural events become natural disasters because humankind is preoccupied with the goal of economic growth to the exclusion of sensible precautions for human safety. From this perspective, the city represents the maximum human appropriation of Nature (Ibarra, Puente, and Schteingart, 1984). By spatially concentrating many activities of social reproduction, the city also concentrates risks that are inherent in its location and the indifference of its residents to natural processes. Because mega-cities usually contain the largest concentrations of population and economic activities, they may well be the loci of highest risk and greatest exposure to natural disasters.

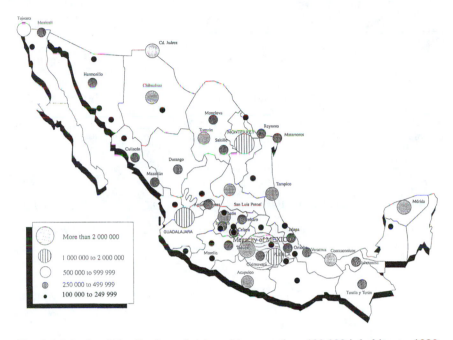

Fig. 9.1 Mexico: Distribution of cities with more than 100,000 inhabitants, 1990

Proneness to disaster also depends on many other factors, including: physical characteristics of the city's location; differences in the material fabric of buildings and infrastructure; variable flows of resources and wastes that shape the urban metabolism; and degrees of coordination among different social agents charged with responsibilities for protecting city residents – especially the state. The surplus of resources that is available for investment in improvements to the urban fabric and the socio-economic status of the population are further important determinants of mega-city vulnerability. Such factors vary considerably from country to country but tend to favour reduced vulnerability in the cities of industrialized states. Elsewhere, as in Mexico City, they do the opposite.

Mexico City in its regional setting

Mexico City is the dominant city of Mexico, where it stands at the head of a growing number of large urban centres (fig. 9.1). The Federal District comprises the heart of Mexico City and the two terms are often used interchangeably.[1] However, the mega-city has spread well beyond the District into the surrounding Basin (Valley) of Mexico. Many parts of the

Table 9.1 Population of the megalopolis of Mexico, 1970–1990

State	1970	% national	1980	% national	1990	% national
Federal District	6,874,165	13.56	8,024,498	12.00	8,235,744	10.14
Hidalgo	259,624	0.51	357,461	0.53	479,740	0.59
Mexico	3,285,288	6.48	6,044,871	9.04	8,973,925	11.05
Morelos	382,052	0.75	616,600	0.92	814,337	1.00
Puebla	896,286	1.77	1,342,305	2.01	1,689,789	2.07
Tlaxcala	319,850	0.63	456,729	0.68	629,885	0.78
Total Central Region	12,017,265	23.70	16,842,464	25.20	20,823,420	25.63
National total	50,695,000	100.00	66,847,464	100.00	81,249,645	100.00

Sources: Population and Housing Census, INEGI, Mexico, 1970, 1980, 1990, tables I, IX, X, and XI.

Mexico City Metropolitan Zone (MCMZ) or Central Region are now included within greater Mexico City. That Zone – which is the focus of this chapter – includes portions of six Mexican states: the Federal District, Hidalgo, Mexico, Morelos, Puebla, and Tlaxcala.

Mexico City has become a paradigm of fast and anarchic urban growth during the second half of the twentieth century (Ezcurra and Mazari-Hiriart, 1996). Between 1950 and 1975 its population expanded from 3 million to 11.9 million, while it advanced from fourteenth (1950) to third (1975) in the ranks of large cities. Today, with at least 18.5 million inhabitants – and perhaps as many as 20.8 million – Mexico City is among the world's three largest metropolitan areas (table 9.1). Similar meteoric urban growth has occurred in other developing countries, whereas the big cities of developed countries now generally have stable or declining populations. In 1950, 9 of the 30 largest cities were located in developing countries; by 1990 the balance had shifted – 17 of the top 30 cities belonged to developing countries.

According to some population projections, Mexico City will be the world's biggest urban place by the year 2000, with around 28.4 million residents. Whether these figures are correct is much disputed; recent United Nations estimates suggest that there may be as few as 15.6 million residents at the turn of the century. Compared with secondary cities such as Monterrey, the capital's "costs of admission" (i.e. for land, housing, employment, etc.) have risen prohibitively for poor in-migrants and may drive out some of those who are already there (Linden, 1996, p. 54).[2]

Mexico City clearly illustrates a distinctive feature of third world urbanization that has emerged in the years since the Second World War, namely the upsetting of more or less stable and complementary relations between city and countryside. On a wordwide basis this is reflected in the growing contribution of urban areas to gross domestic products and by the high concentration of development in a small number of large urban places. It is this process that has most challenged the state's capacity to respond to demands for urban investment and urban services. In 40 years Mexico City grew from 1,640,314 inhabitants (1940) to 13,354,371 (1980). During much of this period, annual growth rates varied between 5 per cent and 6 per cent, falling to around 4.4 per cent only in recent years, while the urbanized area increased from 117 km^2 to 1,000 km^2 (Ibarra, Puente, and Saavedra, 1988).

The rapidity with which mega-cities in developing countries are growing is particularly important for students of urban hazards. The big cities of the developed world (e.g. New York and London) took a long time to reach their present size – sometimes several centuries. Because the same process is occurring quickly in developing countries, the state's capacity to provide the necessary urban infrastructure is often exceeded. Prolonged conditions of scarcity, acute polarization, and social marginalization have become distinctive features of third world urbanization. In Latin America, the large urban agglomerations are what Bryant Roberts has labelled "cities of peasants," disproportionately made up of recent in-migrants from the countryside (Roberts, 1980). In Mexico City it has been difficult for migrants to adopt urban ways of behaving that involve cooperation, concentration, and subordination of the individual interest to the public interest. As a result, the effectiveness of existing public strategies for reducing urban vulnerability in the face of natural or social induced disasters is limited.

Socio-demographic profile

Expansion of urban territory

One important indicator of urbanization in Mexico City is the rate at which rural land is converted to urban uses. Before 1960, urbanized land was confined within the boundaries of the Federal District. Rates of incorporating new territory reached maxima during the 1920s (6.38 per cent per annum) and again during the 1940s (6.94 per cent). These growth spurts were triggered by broad changes in Mexican society. The 1920s were a time of recuperation after a decade of deep social conflict and urban paralysis during the Mexican Revolution. The 1940s witnessed an upsurge of industrialization that took advantage of a surplus of cheap and mobile labour. Population densities in the Federal District remained more or less constant throughout these decades. During the 1960s, municipalities of the State of Mexico added 22.1 per cent more people and 6.62 per cent more land. In the following decade, the increments were 10.29 per cent and 9.65 per cent, respectively. During the 1980s, however, population and territorial growth rates fell off and density declined; a period of reduced growth and spatial dispersion had begun (table 9.2).

The emergence of a dual city

Mexico City contains a mixture of dispersed and concentrated populations. The mega-city as a whole has been growing, but there have also been substantial shifts in the locus of population, and the State of Mexico is now growing fastest of all units of the Mexico City Metropolitan Zone (MCMZ). In 1970, the Federal District was twice as populous (6.9 million) as the State of Mexico (3.3 million), and the whole mega-city was home to a total of 12 million people. By 1990, the State of Mexico had almost tripled its population (8.9 million) and had overtaken the Federal District (8.2 million). At that time the Metropolitan Zone housed 20.8 million (later corrected to 21.6 million) (table 9.1).

During the 10 years between 1980 and 1990, although the natural population increase was substantial in the Federal District, almost as many people migrated out of the District to other parts of the mega-city (table 9.3). It is now clear that a dual city is emerging: a core, comprising the Federal District, and a periphery made up of the remaining parts of the mega-city. For the most part the core is socially heterogeneous but relatively stable, economically healthy (many residents receive medium to high incomes), and characterized by service employment. The periphery, situated predominantly in the State of Mexico, is occupied by people who have recently migrated there, receive incomes in the low to medium range, and are employed in secondary and tertiary economic activities including the informal economy.

Economic structure

Mexico City is the nation's premier economic centre. Its recent growth is largely the product of a post-war economic system that attempted to develop domestic resources and substitute them for imported goods. A by-product of this system was the concentration of people and production in a few urban centres – especially Mexico City – leaving much of the rest of the country underdeveloped. Now the negative consequences of the system, including excessive concentration, are becoming apparent.

Table 9.2 Rate of growth of urbanization, 1900–1990

Year	Federal District			Conurbated municipalities			ZMVM[a]		
	Urb. pop. (%)	Urb. area (%)	Density (inh./ha)	Urb. pop. (%)	Urb. area (%)	Density (inh./ha)	Urb. pop. (%)	Urb. area (%)	Density (inh./ha)
1900–1910	2.02	3.90	105.82				2.02	3.90	105.82
1910–1920	3.87	1.54	132.71				3.87	1.54	132.71
1920–1930	5.48	6.38	121.85				5.48	6.38	121.85
1930–1940	4.60	3.16	139.99				4.60	3.16	139.99
1940–1950	6.56	6.94	135.05				6.66	7.43	130.33
1950–1960	5.13	4.82	139.16			25.37	5.55	6.94	114.32
1960–1970	3.60	2.86	149.54	22.10	6.62	96.42	5.52	3.79	134.94
1970–1980	0.96	2.51	128.39	10.29	9.65	104.28	3.61	5.03	117.80
1980–1990	0.26	1.33	115.45	2.92	4.78	87.15	1.86	2.97	90.00

Source: Urban Plan of Central Region of the Megalopolis of Mexico, States of Mexico and the Metropolitan Autonomous University, 1993.

a. Metropolitan Zone of the Valley of Mexico.

Table 9.3 Population growth of the Central Region of the megalopolis of Mexico, 1980–1990

State	Total growth	Natural growth	Migration growth
Fed. District	211,246	1,723,585	−1,512,339
Hidalgo	122,279	115,115	7,164
Mexico	2,929,054	1,509,389	1,419,665
Morelos	197,737	193,892	3,845
Puebla	347,484	422,977	−75,493
Tlaxcala	173,156	162,992	10,164
Total	3,980,956	4,127,950	−146,994

Sources: Urban Plan of Central Region of the Megalopolis of Mexico, States of Mexico and the Metropolitan Autonomous University, 1993.

The dynamics of economic growth

Prior to the Second World War, industrialization in Mexico was very limited and was largely restricted to mining and textiles. But after 1940 industrialization took off in step with urbanization. Industries gravitated especially towards Mexico City, attracted by its long-established political and economic importance and its steady economic growth. In 1940, Mexico City accounted for 30.6 per cent of the national gross domestic product (GDP). It dominated some economic sectors more than others (e.g. transportation – 66.6 per cent; industry – 33.7 per cent) (table 9.4). During the next four decades (1940–1980), the MCMZ's share of GDP increased slowly, peaking in 1980 at 38.2 per cent. Over this period, the (standardized) value of the MCMZ's contribution increased from 7 billion pesos in 1940 to 115 billion pesos in 1980 – a sixteenfold increase. Not surprisingly, the city underwent profound social and economic changes.

Trends in different economic sectors showed considerable fluctuations and dramatic contrasts during these 40 years. For example, the MCMZ's share of the national industrial sector rose from 34 per cent (1940) to 43 per cent (1960) and then fell back to 31 per cent (1980). Conversely, the service sector grew continuously, from 41 per cent (1940) to 51 per cent (1980). Finally, the MCMZ transportation sector's share of GDP was cut in half from 67 per cent (1940) to 34 per cent (1980). Data about the changing sectoral shares of the MCMZ's own GDP show that services have generally been most important overall and account for an increasing proportion of the total, whereas industry declined slightly during the period 1940–1980 (Puente, 1987). Reductions in the share of GDP accounted for by industry – in both the country as a whole and the MCMZ – must be viewed against real absolute increases in pro-

Table 9.4 National and Mexico City's gross national product, by economic sector, 1940–1980 (millions of 1950 pesos)

	1940		1950		1960		1970		1980		Rate of growth, 1970–1980	
	National	MCMZ	National	MCMZ	National	MCMZ	National	MCMZ	National	MCMZ	National	MCMZ
Total	22,889	7,010	41,060	12,427	74,215	26,858	151,760	56,731	301,731	115,338	2.0	2.0
Agriculture	5,170	30	9,242	28	13,917	37	17,643	54	25,198	105	1.4	1.9
Industry	6,789	2,286	12,466	3,378	24,603	10,509	52,009	16,086	112,509	34,619	2.1	2.2
Transport	865	576	1,988	1,038	3,638	2,184	4,778	2,775	13,903	4,788	2.9	1.7
Services	10,065	4,118	17,364	7,983	32,057	14,128	77,330	37,816	150,121	75,826	1.9	2.0
MCMZ as % of national												
Total	100.0	30.6	100.0	30.3	100.0	36.2	100.0	37.4	100.0	38.2		
Agriculture	100.0	0.6	100.0	0.3	100.0	0.3	100.0	0.3	100.0	0.4		
Industry	100.0	33.7	100.0	27.1	100.0	42.7	100.0	30.9	100.0	30.8		
Transport	100.0	66.6	100.0	62.2	100.0	60.0	100.0	58.1	100.0	34.4		
Services	100.0	40.9	100.0	46.0	100.0	44.0	100.0	49.0	100.0	50.5		
Economic sector as % of total												
Total	100.0	100.0	100.0	100.0	100.0	100.0	100.0	100.0	100.0	100.0		
Agriculture	22.6	0.4	22.5	0.2	18.7	0.1	11.6	0.1	8.4	0.1		
Industry	29.6	32.6	30.4	27.2	33.2	39.1	34.3	28.4	37.3	30.0		
Transport	3.8	8.2	4.8	8.4	4.9	8.1	3.1	4.9	4.6	4.2		
Services	44.0	58.8	42.3	64.2	43.2	52.7	51.0	66.6	49.8	65.7		

Sources: Mexican National Account System, Gross National Product by Federal States, General Direction of Statistics, SPP, 1982.

duction, which continued throughout the period of record. A declining share of industrial GDP merely indicates that industrial investments were not confined solely to Mexico City; other urban centres also became industrialized.

Although the MCMZ's share of national GDP decreased, absolute numbers of establishments (firms) rose from 27,707 (1960) to 33,049 (1970) and then to 34,549 (1975). Yet assets (capital invested) and numbers of personnel both began to decline by the mid-1970s. What appears to have occurred in Mexico City during the decades between 1940 and 1980 is a proliferation of small-scale establishments using low-paid labour. More industrial establishments were created in the MCMZ, but total assets there declined as investments were relocated to other parts of Mexico. The figures point to a complex transformation of the country's industrial structure. Larger numbers of business enterprises became concentrated in the MCMZ but value-added did not grow proportionately, and investments began to be dispersed more widely throughout Mexico.

Data provided by the most recent economic censuses and estimates made by the National Institute for Geography and Statistics (INEGI) indicate no significant changes in the Mexico City Metropolitan Zone's share of the national GDP for 1985 and 1988. This remains between 37 per cent and 38 per cent. Unfortunately, apart from the Federal District and the State of Mexico, the rest of the mega-city is attracting much less new investment. For residents of the Valley of Mexico, the benefits of shifting (diversifying and deconcentrating) investment towards cities in Puebla, Tlaxcala, Hidalgo, and Morelos (i.e. beyond the Valley of Mexico) would be manifold: (1) industrial and vehicular pollution in the central cities would be reduced; (2) the conversion of highly productive agricultural land and biologically rich ecosystems on the fringes of the present built-up area would be slowed; and (3) there would be a reduction in levels of the catastrophe potential of natural disasters. Mexico City's powerful attraction for economic investment has defeated all policies of deconcentration and dispersal up till now. Some experts view the North American Free Trade Agreement as a structural factor that will make a profound difference by shifting industrial plants from Mexico City to communities on the frontier with the United States.

Environmental hazards and urban vulnerability

Mexico as a whole is affected by a wide range of natural hazards. Geological hazards, particularly earthquakes, subsidence, and volcanoes, are of particular importance, but hydrological risks (e.g. floods and drought) are also significant. Moreover, the city is also at risk from serious air and water pollution and from an increasing number of technological emergencies such as natural gas tank explosions. The potential for social hazards is believed to be increasing, although accurate statistics are not available to substantiate this judgement.

Seismic hazards

Crossed and encircled by numerous geological faults and straddling several tectonic plates, the Mexican Republic is one of the world's most seismically active countries. Five important plates meet in the vicinity of Mexico: Cocos, Pacific, North American, Caribbean, and Rivera. Two seismic phenomena are particularly important: (1) lateral slipping, which takes place at the boundary of the Pacific and North American plates, and (2) subduction of the Cocos plate beneath the North American plate. Both of these phenomena are permanent generators of earthquakes, which have affected Mexico throughout its history. Major continental faults such as the San Andreas cross the entire country, while regional and local faults fracture much of the intervening territory. Of particular importance is the Acambay system of faults near the centre of Mexico and the Ocosingo system of Chipias State in southern Mexico (Sistema Nacional de Proteccion Civil, 1991).

Situated in the centre of the country, Mexico City is affected by many faults that cut across areas of greatest population density such as the historic centre. By themselves, the faults render the city potentially vulnerable to seismic movements, including sudden subsidence. In addition, the city is constructed on what was, until the turn of the century, a system of lakes that covered much of the Valley of Mexico. In other words, Mexico city is founded on saturated mud and clay soils that are particularly unstable. Among other things these soils amplify seismic waves up to 30 times more than do the firmer soils of adjacent higher zones. Movements in the tectonic plates offshore from Michoacan and Guerrero States pose the most serious threat to Mexico City. During the twentieth century, 34 tremors of Richter magnitude 7.0 or greater have occurred, almost all of them centred in this zone. Seismic activity is also manifest in the continental plates, particularly in association with the volcanoes that surround Mexico City. For example, Popocatepetl, whose last major eruption took place in 1920, became active again in December 1995, and Xitle or Teuhtli are also possible earthquake generators (Kovach, 1995, pp. 170–171). Although of low intensity and rarely damaging, these tremors alarm local populations; they are of little importance compared with those of the offshore Pacific plate. On occasions they have reached magnitudes of 7.0 (e.g. Acambay, State of Mexico, 1912).

Mexico's proneness to seismic activity is clearly reflected in a succession of twentieth-century earthquake disasters (table 9.5). All but one had intensities of 7.0 or greater. Several significantly affected Mexico City. In 1912, a magnitude 7.0 earthquake centred in the State of Mexico produced numerous landslides and killed 202 people. In 1932, a magnitude 8.4 event induced strong motion in Mexico City but did little damage. In 1957, a magnitude 7.7 quake, centred on the coast of Guerrero, killed 98 and caused the collapse of buildings in the capital. It also brought about various improvements in earthquake engineering practice (Esteva, 1997). Most recently, there occurred an even more destructive event.

Societal impact of the 1985 earthquake on Mexico City

On 19–20 September 1985, a series of earthquakes – including one of Richter magnitude 8.1 – killed at least 3,050 people, injured 40,750, rendered 80,000 homeless, destroyed 1,970 buildings and damaged 5,700 more. Total material losses were estimated at 1,000 billion pesos (Sistema Nacional de Proteccion Civil, 1991). These are official loss figures and they may understate the death toll. A different authoritative source lists around 4,500 dead but only 14,000 injured (*Atlas de la Ciudad de Mexico*, 1987). Others have accepted much higher estimates (Cevallos, 1995), perhaps as many as 5,000–10,000 dead.

Data on building losses are even more varied. Rivas Vidal and Salinas Amezcua list only 412 buildings totally destroyed, but agree with other sources about the number of buildings that were damaged (5,728). Many damaged buildings eventually had to be demolished; others that ought to have been pulled down for structural safety reasons were repaired at the insistence of their inhabitants. Among the latter was the Nonalco-Tlatelolco housing estate, which comprised 102 separate buildings. When constructed in the early 1960s the estate was intended to be a model of state responses to joint needs for slum clearance, new housing, and improved architectural design. It is significant that this estate was one of the most badly affected by the quake. Critics have speculated that poor-quality construction as a result of corrupt government practices may have been responsible for the débâcle. That explanation takes on added significance in light of other failures of government-sponsored projects. For example, Mexico City lost 30 per cent (3,457 beds) of its installed hospital capacity, all belonging to government social security institutions; most of these were in buildings constructed after 1950. The destruction of the Centro Medico was a particularly egregious loss. Infrastructure damage was heavy, although some systems (e.g. underground rail transport) were unaffected. The city's telephone service was crippled by the quake and contacts with the outside world were severed. It took approximately six weeks before the bulk of telephone communications were reinstated.

Table 9.5 Main catastrophic earthquakes during the twentieth century

Date	Region	Affected population	Magnitude (Richter scale)	Damage
19 Nov. 1912	State of Mexico	Acambay, Timilpan, and *Federal District*	7.0	Landslides; 202 deaths and several injuries
4 Jan. 1920	Puebla, Veracruz	Cosautlan, Teocelo, and Jalapa in Veracruz; Patlanala and Chilchotla in Puebla	6.5	Landslides; 230 deaths and several injuries
3 June, 1932	Jalisco, Colima	Manzanillo, Coyutla, Tecoman, and Colima in Colima; Guadalajara, La Barca, Mascota, and Autlan in Jalisco	8.4	300 deaths and several injuries
26 Jul. 1937	Oaxaca, Veracruz	Maltrata in Veracruz	7.3	34 deaths
15 Apr. 1941	Michoacan, Jalisco	States of Michoacan, Jalisco, and Colima	7.9	Destruction of Colima City's Cathedral; 900 deaths and several injuries
28 Jul. 1957	Guerrero	San Marcos and Chilpancingo in Guerrero, and *Federal District*	7.7	Collapse of several houses and buildings; 98 deaths; and tsunami in Acapulco and Salina Cruz
6 Jul. 1934	Guerrero, Michoacan	Cd. Altamirano, Cutzamala, and Coyuca de Catalan in Guerrero; Tanganhuato and Huetamo in Michoacan	7.2	40 deaths, 140 people injured; heavy material losses
29 Aug. 1973	Oaxaca, Veracruz	Puebla, Veracruz, and Oaxaca	7.3	Collapse of houses and severe damage to buildings; 527 deaths and 4,075 people injured; high material losses
24 Oct. 1980	Oaxaca, Puebla	States of Oaxaca, Guerrero, and Puebla	7.0	300 deaths, 1,000 people injured, 15,000 homeless
19 Sep. 1985	Michoacan, Guerrero	Michoacan, Colima, Guerrero, México, Jalisco, Morelos, and *Federal District*	8.1	3,050 deaths, 40,750 people injured, 80,000 homeless; 1,970 buildings collapsed and 5,700 buildings badly damaged; heavy material losses

Source: National Autonomous University of Mexico, Institute of Geophysics.

Drinking water supplies and drainage networks were also disrupted for long periods. Fractured supply pipes leaked water and 17 major junctions of the drainage system were damaged. Water supplies were not fully restored for a month and a half. Even today it is still possible to find evidence that the aftermath has not ended.

It is sometimes argued that the population of Mexico has historically accepted natural calamities as a part of normal life. Many individuals have exhibited traits of passivity, acceptance, resignation, or stoicism in the face of hazard. Though victims were undoubtedly grieviously distressed and the frequency of such misfortunes was high, the effects were usually limited. This situation changed radically after 20 September 1985. Not only was the earthquake's magnitude unprecedented in Mexico City, it shook local society to such an extent that manifold everyday problems of marginalization, poverty, unemployment, anarchy, and pollution were eclipsed for a considerable period. Both government and civil society were surprised by the results.

Fearful of any situation that might upset the fragile social stability that existed in Mexico City despite acute social polarization, government leaders at first tried to play down the event and proclaimed their own ability to deal with the resulting problems. International offers of aid were initially spurned. But the slowness of rescue operations and the lack of timely information from the government induced residents to take matters into their own hands. In the absence of effective state-sponsored relief, the efforts of informal public groups and individuals were highly valued. Despite a lack of coordination and experience, much was achieved, including the rescue of at least 3,226 people. In addition, the success of public involvement demonstrated the existence of a formerly untapped but vast potential for social initiative, participation, and solidarity among ordinary citizens. This realization was not lost on the government, but it triggered fears that protests might subsequently occur and undermine political stability. As a result, a call was issued for residents to return to normal activities while the government reasserted its authority and control. None the less, it was recognized that informal public action had successfully produced housing and other resources for destitute residents (Connolly, 1993); so the state began to capitalize on the people's initiative, institutionalizing it as a formal government programme. Now the state would provide financial and technical assistance but the people were required to organize themselves and contribute their own labour to carry out requested urban improvement works. In effect this was a belated official recognition of private – and illegal – practices that had long existed in Mexico City, whereby land and buildings are often appropriated by poor squatters and housing is constructed or converted by informal community groups. Whether this type of response has come in time to offset the degeneration of urban life in Mexico City remains to be seen. However, it earned the government much political capital over the following two years, out of which was created the Ministry of Social Development (SEDESOL), previously the Ministry of Ecology and Urban Development (SEDUE).

It is estimated that sufficient elastic energy has built up in the Guerrero section of the Pacific plate boundary that another earthquake of similar magnitude to that of 1985 could occur there soon, with devastating consequences for Mexico City. It is also known that the most seismically vulnerable parts of the capital are located in central residential neighbourhoods such as Cuauhtémoc, Venustiano Carranza, Gustavo A. Madero, and Benito Juárez, which are founded on unstable lacustrine soils. The urgency of implementing preventive measures is obvious. Towards that end it is essential to carry out detailed seismic risk mapping that also takes account of Mexico City's socio-economic heterogeneity. In addition to being useful tools for increasing public awareness of risk, such maps facilitate the setting of priorities for action.

Lack of sensitivity to the social dimensions of earthquakes is reflected in many public safety programmes throughout Mexico City. One example is the earthquake warning system, which relies on distant ground motion sensors and rapid transmission of radio and television signals to targeted vulnerable facilities. Using this system it is possible to provide residents of Mexico City with at least a 1 minute warning of earthquakes occurring in Michoacan and Guerrero states or adjacent offshore waters. That may be sufficient for individuals in homes and small businesses to take protective action, but it is questionable that people in large, densely occupied facilities such as stadiums, major stores, and factories could be evacuated; many might be injured in rushing for exits. The system was suspended shortly after being introduced because of technical malfunctions that triggered delayed or bogus warnings. Since then improvements have been made and several schools have conducted tests that produced hopeful results. However, without broader public education and training the prospects for success of this warning system are still in doubt. Myopia about social factors in disasters begins at the top of disaster agencies in Mexico, including those, such as the National System of Civil Defence, that were established right after the 1985 earthquake. Failure to take account of the composition of populations at risk compromises public policies and hampers the introduction of sophisticated technologies for the prevention of natural risks.

Other environmental hazards of Mexico City

Floods, water shortages, and associated famines have been an important feature of Mexico City's history and at least the first two of these are continuing hazards. The closed, internally drained basin of Mexico is particularly susceptible to disruption of its hydrological system, and human modifications of local ecosystems have been extensive. The floods of 1629 – which submerged the capital for five years – may have killed as many as 30,000, and further severe inundations occurred in 1691. Major drainage works were completed in the late seventeenth century that sufficed for over a hundred years, but additional heavy investment in canals, pumps, and tunnels was required at the end of the nineteenth century (Kandell, 1988, pp. 371–372). Even then, flooding continued as a sporadic troubling hazard (e.g. in 1910).

Environmental pollution has become a grave endemic problem in Mexico City. Almost every day, the permissible limits of atmospheric pollution are exceeded. Elevated levels of tropospheric ozone cause most problems. It is estimated that total pollutant emissions by weight are approximately 5 million metric tons within the Mexico City Metropolitan Zone: 15 per cent from fixed sources (e.g. buildings), 80 per cent from mobile sources (i.e. vehicles), and the rest (5 per cent) from natural sources (Sistema Nacional de Proteccion Civil, 1991). The enclosed basin in which Mexico City sits ensures that locally generated air pollutants are not subject to much dispersal, especially during winter when thermal temperature inversions are common (*New York Times*, 4 February 1996). Anti-pollution programmes have not had any significant effect in reducing these problems, though not for the want of trying. The lead content of petroleum has been considerably reduced, but ozone levels have risen. The Azcapotzalco refinery, which was a primary source of hydrocarbon emissions, has been closed down. An obligatory annual anti-pollution test of road vehicles was introduced and then applied every six months. The "Hoy no circula" programme, which prohibits the use of each vehicle on one day a week, has also been adopted. But beneficial results have been minimal and some measures have produced perverse effects. The "day without driving" programme has largely become a "day off work." Among middle-class residents it has also prompted the purchase of additional cars that can be driven when primary vehicles are barred from use. In addition, the enforcement of pollution controls against highly polluting industries has been extremely lax. Chemical and cement plants have resisted installing anti-pollution equipment, preferring to invest in bribes to enforcement officials.

Nor is the problem solely one of poorly conceived public policy. The diagnosis of pollution causes is inadequate and the means to control pollution are still insufficient. For example, it is difficult to explain why levels of pollution remain high at weekends; this is a time when most industries stop production, a majority of the population rests, and large numbers of residents are absent from the city. It is probable that the contributions of various individual and family behaviours to pollution have been underestimated. These might include cooking with bottled gas and widespread private burning of rubbish.

Fires and explosions constitute other important socially induced technological disasters. Between 1982 and 1984, there was an average per year of about 20,000 fires and explosions throughout Mexico, of which 3,474 took place in the Federal District and 925 in the State of Mexico. Nearly 500 people died throughout the country from these causes, including approximately 100 in the State of Mexico, where there are many industries, urban deprivation is high, and incomes are low (table 9.6). Most of the fires are of domestic origin (68 per cent), ahead of fires in commercial premises (20 per cent), with industrial fires making up the rest (12 per cent). In the Federal District, the frequency of fires varies widely from neighbourhood to neighbourhood. The highly populated and densely industrialized central *delegaciones* of Cuauhtémoc, Madero, and Hidalgo are particularly fire prone (table 9.7). In contrast, the predominantly agricultural, low-density, peripheral *delegaciones* of Tláhuac, Milpa Alta, and Xochimilco experience fewer than 5 per cent of all fires despite the fact that these are places where slash-and-burn cultivation is practised and both the burning of rubbish and the ignition of celebratory bonfires are common. As urban development spreads into these areas, the possibility is growing that loosely supervised rural fires might spread to built-up areas, especially during the dry months of February, March, and April.

Catastrophic industrial explosions have occurred in Mexico City and are a continuing threat. For example, on 19 November 1984, 334 people were killed when a natural gas storage facility exploded in the north-eastern neighbourhood of San Juanico. Perhaps more indicative of the complex nature of urban industrial hazards is the example of Guadalajara's sewer system explosion in 1992. This was fuelled by leakage of gasoline from a refinery owned by the national petroleum company of Mexico (Pemex). An extensive residential zone blew up, causing widespread damage and loss of lives. The negligence of the local authorities was patently obvious. Days before, inhabitants of the affected neighbourhood had reported the leak, but the municipal authorities had not responded. Other contributory factors included lax and unobserved public safety regulations, lack of maintenance and obsolescence of the sewer system, and absence of coherent urban planning. These are the kinds of problem that are all too common in Mexico's large urban centres, especially Mexico City, and which make living there an experience of chaos that is permanently on the verge of catastrophe.

In short, Mexico City is a place of contradictions. Far from being a city that has solved problems of environmental risk and hazard, it has become the setting where a multitude of such problems converge (Puente, 1988).

Table 9.6 National urban fires and explosions, by federal state, 1982–1984 (annual means)

Federal state	No. of fires and explosions (annual mean)	Domestic	Commercial	Industrial	No. of people killed
Aguascalientes	173	111	48	14	2
Baja California	1,539	1,023	318	198	19
Baja California Sur	171	129	35	7	1
Campeche	70	39	18	13	–
Coahuila	1,082	736	256	90	8
Colima	134	106	21	7	2
Chiapas	36	20	12	4	1
Chihuahua	1,845	1,317	376	152	21
Distrito Federal	3,474	2,471	635	368	15
Durango	252	194	31	27	7
Guanajuato	1,103	824	130	149	28
Guerrero	167	97	63	7	1
Hidalgo	217	134	73	10	93
Jalisco	520	313	142	65	6
México	925	510	195	220	104
Michoacán	483	370	89	24	7
Morelos	212	131	59	22	7
Nayarit	163	124	28	11	8
Nuevo León	1,635	911	331	393	15
Oaxaca	184	158	24	2	6
Puebla	367	208	95	64	13
Querétaro	288	186	78	24	7
Quintana Roo	213	137	27	49	–
San Luis Potosí	369	264	79	26	4
Sinaloa	575	327	191	57	13
Sonora	1,787	1,212	367	208	62
Tabasco	206	123	67	16	3
Tamaulipas	971	738	188	45	10

Tlaxcala	–	–	–	–	–
Veracruz	620	445	139	36	21
Yucatán	205	144	44	17	–
Zacatecas	40	33	4	3	–
Total	20,026	13,535	4,163	2,328	484

Source: Secretary of Budget and Planning, National Institute of Statistics, Geography and Information.

Table 9.7 Fires in Mexico City (Federal District), by *delegacion*, 1974, 1979, 1984

Delegacion	1974	1979	1984	Mean	%
Gustavo A. Madero	212	311	263	262	11
Venustiano Carranza	182	266	225	224	9
Cuauhtémoc	424	621	525	523	21
Iztacalco	99	146	123	116	5
Benito Juárez	166	243	205	205	8
Iztapalapa	133	194	164	164	7
Coyoacán	110	160	136	135	6
Tláhuac	48	70	59	52	2
Milpa Alta	5	8	7	7	0
Xochimilco	35	51	43	43	2
Tlalpan	100	146	124	123	5
Magdalena Contreras	22	33	28	28	1
Alvaro Obregón	147	215	182	181	7
Cuajimalpa	12	17	14	14	0
Miguel Hidalgo	209	306	258	258	11
Azcapotzalco	99	145	123	122	6
Total	2,003	2,932	2,479	2,457	100

Source: Based on Department of Protection and Transport, September 1984.

In order to address the hazards it is important to understand the dominant social actors who produce the city that generates them. These are
individuals and institutions that affect the city's quality of life as well as its
vulnerability to natural and socially induced risks.

Assessing the vulnerability of Mexico City

Mega-cities of developing countries are vulnerable to environmental
hazards for a complex mixture of reasons (Eibenschutz and Puente,
1991). But the crux of the problem is an inability to cope with very rapid
population growth and extreme socio-economic polarization between rich
and poor neighbourhoods. In Mexico City, inner neighbourhoods are compact, multistoreyed, well served and well equipped, and oriented towards
consumption. Outer neighbourhoods are sprawling low-rise places, deficient in housing, infrastructure, and services, and oriented to production.
Much of the outer city is occupied illegally by squatters; settlements
spread indiscriminately over land of high agricultural productivity and
low seismic resistance. It is probable the occupants of these informal settlements are also undergoing profound social transformations, but little is
known about the urban acculturation of the new arrivals. However, it is

clear that the informally settled areas are disproportionately poorly serviced and vulnerable to disaster.

A spatially distinctive hierarchy of vulnerabilities

In principle, cities are vulnerable to disasters in four distinct but inter-acting ways. First, they are internally vulnerable because of weaknesses of social and material infrastructures (e.g. people who lack the resources to improve protection against disaster; centralized utility networks with-out adequate backup systems). Secondly, cities are more vulnerable if they are dependent on rural hinterlands that are themselves vulnerable (e.g. in-migrant refugees from rural famines and civil wars impose heavy burdens). Thirdly, cities that are lower in the national urban hierarchy may be more vulnerable than those near the top of the system because their claims on public and private resources are less effective. Fourthly, and for similar reasons, cities that lie at the centre of the global eco-nomic system (e.g. the polycentre) are less vulnerable than those on the periphery. Tokyo and Los Angeles are less vulnerable than Calcutta and Mexico City. For a given city, one type of vulnerability may be of overwhelming importance or there may be several lesser vulnerabilities acting together synergistically. Increasing interdependence among dif-ferent levels of the global urban system is likely to confer benefits on the entire system by reducing the dependence of any single city – though whether this might lead to the build-up of a potentially catastrophic system-wide failure is not clear. How shall the multiple vulnerabilities of mega-cities be identified and assessed? The next section takes up this question.

Factors and gradients of vulnerability: A matrix approach

The following methodology for assessing urban vulnerability rests on two premises: (1) the material conditions of a city are good indicators of vulnerability; (2) the main components of vulnerability can be mapped at the scale of urban neighbourhoods. It only remains to create a matrix that displays the appropriate indicators (factors) on one axis and the areal units of analysis on the other. For the purposes of this chapter, five clusters of vulnerability factors are included. These are:

- natural factors (e.g. geomorphic, climatic, biotic, edaphic)
- material factors (e.g. infrastructure services, equipment, housing, economic base, spatial organization)
- environmental factors (e.g. pollution levels, biotic variety, amenity lands)

- socio-economic factors (e.g. household composition, mortality, life expectancy, morbidity, occupation, income, nutrition, social organization)
- psychological factors (e.g. perception of environment, tension, anxiety)

Natural factors

The contribution of natural factors towards urban vulnerability varies widely from city to city. In Mexico City, seismic factors and landforms are especially important. Many of the landforms have been subject to extensive human manipulation. For example, along the shores of the original lakes that dotted the valley floor, artificial floating "garden" platforms (*chinampas*) once provided food for local residents. Today, these water bodies have been drained and desiccated, the hydrological cycle has been altered, forest cover has been reduced from 54 per cent to 14 per cent, and the floating agricultural lands have been taken over by urban uses. The seismic vulnerability of dried-out lake beds is particularly high, but the city's soils vary in complex ways and there is a trend toward greater vulnerability almost everywhere.

Material factors

Deficiencies in the material fabric of mega-cities contribute to disaster-vulnerability. Buildings and infrastructure are strongly affected by prevailing technologies. Some technologies can be transferred successfully between countries but others are less versatile. When a technology that has been specifically developed for one country – or one city – is transferred to another, perverse effects may be generated. These include: inappropriate fit with local ecosystems and environments; high costs of construction and operation; breakdowns, deterioration, and obsolescence; and the unanticipated consequences of replacing existing traditional technologies that depend on human labour with technologies that are predominantly non-human. Poor-quality urban techologies are a widespread problem in developing countries. Inadequate construction and maintenance can render even the best-designed technology inoperable. The wastage of water in Mexico City because of leaks in supply pipes is well known. Up to 45 per cent of the initial supply is believed to be lost – a staggering proportion in view of the fact that much water must be brought to the city from remote sources outside the Valley of Mexico and many residents receive no formal water service.

Material contributions to vulnerability change in response to the process of urbanization. For example, growing cities often swallow up potentially hazardous industries that were originally outside the built-up zone. Once surrounded by housing, these sites face an increased proba-

bility of disaster. A sound urban policy can reduce such risks by chan-
neling new development onto suitable land and by promoting a coherent
spatial arrangement of residential and industrial uses.

Environmental factors

Environmental pollution is an increasingly serious problem in third world
cities. In the pre-urban past, significant pollution was often non-existent
or inconsequential. Later, when industrial development became viewed
as a corollary of modernization, it was regarded as a price that society
must pay to obtain higher incomes. It is just recently that some countries
have recognized pollution as a public health problem to which cities are
highly sensitive and vulnerable.

Pollution arises from both fixed and mobile sources. Industrial facilities
in the mega-cities of developing countries have rarely been subject to
policies of pollution control. Equally important is pollution generated
by urban transportation systems, especially those that depend on motor
vehicles. In recent years, local authorities have been obliged to put up
with crawling traffic, many frequent traffic jams, and other forms of
vehicular paralysis. The supposed advantages of flexibility and speed that
were associated with motor vehicles are rapidly disappearing. None the
less, these cities must live with the permanent costs of neighbourhood
social disruption and increased pedestrian hazards that have followed
in the wake of motorization. Similar problems of overuse and under-
management have also affected water resources. Lack of treatment
facilities has led to the contamination of streams where wastes are
deposited and of the associated aquifers.

Socio-economic factors

Mega-cities of developing countries are typically characterized by spatial
patterns of great social inequality. These patterns are affected by the
decisions of three different groups: real estate agents and urban developers; the state; and the poor or informal sector. Today the last of these
three is increasingly prevalent in the big cities of developing countries.
The spatial differentiation of different social classes occurs in several
stages (Puente, 1988). These stages are most noticeable for the informal
sector because of the slowness with which the conditions of the poor change.

The vulnerability of a specific residential area in an informal settlement
varies according to the stage of urbanization. Initially, the less the degree
of development, the greater the vulnerability. At the outset, infrastructure is almost non-existent and is of very poor quality. Housing is equally
poor and, as it is extended or upgraded, vulnerability may remain high

because the technologies available to the users are unsophisticated and there is little compliance with building codes or other standards. These disadvantages may be offset by the improved access to information that comes with better education, thereby increasing the likelihood that residents will join in timely evacuations when threatened by disaster. This produces a paradox. Efforts to enlist the poor in government-sponsored self-help schemes that are intended to reduce vulnerability depend on generating a minimum level of social cohesion and interaction among the target groups. But self-reliant communities are often seen as threatening by the state and may attract government-inspired attempts to undermine or manipulate them.

Clearly, the reinforcement of local social organizations could be an excellent instrument to minimize vulnerability. Not only would such institutions be useful as vehicles for publicizing and facilitating anti-disaster measures, they would help to avoid conflicts about relocation of people from high-risk settlements. Today such measures are almost non-existent, notwithstanding the risks to which mega-cities – including Mexico City – are exposed. Of course, it must also be remembered that the participation and organization of the poor residents of informal settlements are not – by themselves – sufficient to reduce the disaster-vulnerability of third world mega-cities. There also must be a clear consensus about the goals of urban development. Otherwise, duplications and contradictions will be generated. The state has a leading responsibility to articulate and foster these goals, which until now it has barely assumed.

Psychological factors

Psychological factors that affect vulnerability are difficult to measure but none the less important. Such factors are the basis of individual and social anomie and pathological behaviour. They also affect potentials for increased social interaction, cohesion, and organization. Migrant populations are slow to adopt urban lifestyles in Mexico City. The migration process generates profound geographical and cultural imbalances that impair the satisfaction of basic human needs for empathy and identity. Uprooted from places of origin, faced with behaviour patterns different from their own, and surrounded by an indifferent and even hostile environment, newcomers have developed distinctive behaviours that may protect the psyche but do not necessarily reduce vulnerability to other hazards. Guided by primary instincts for survival, migrants react against their new surroundings, often by means of aggressive confrontations that not infrequently lead to criminal or anti-social conduct. Tension, anxiety, and aggression are psycho-social traits that undermine the achievement of identity, social well-being, and security. They counteract integration, cooperation, and social organization.

The preceding are some – but not necessarily all – of the factors that affect vulnerability in the city. Together they determine the spatial pattern of vulnerability, and occasionally one of them largely dominates it. By assigning a value to each factor it is possible to create a comprehensive index, which is here referred to as the Compound Index of Urban Vulnerability. This index is derived by dividing the sum of the (weighted) values for each factor by the number of factors. Separately, the relative importance of each factor can be shown for different areas of enumeration; positive scores (i.e. high vulnerability) for some factors might be offset by negative scores (i.e. low vulnerability) in others. For example, poor-quality housing might coexist with relatively high levels of education.

Application of the matrix to Mexico City

The purpose of this matrix is to identify Mexico City's most vulnerable areas, so that priorities for hazard reduction can be established. It has not yet been possible to find appropriate data for all matrix categories. Therefore the vulnerability indices that are derived here for Mexico City are preliminary and could be improved with better data and finer analytic procedures. Nevertheless, this instrument has great potential utility in the formulation of effective policies and decisions about urban hazard reduction. The measures of vulnerability that were included in the Mexico City Vulnerability Index were:

1. *Socio-economic factors:*
   - Size and density of urban concentration
   - Material conditions of housing
   - Density and type of housing
   - Level of education
   - Employment (sectoral occupation of the population)
   - Income level
   - Economic activity (industrial density, especially of dangerous industries such as petroleum and chemicals).
2. *Regional infrastructure:*
   - Electricity network and power stations
   - Gas network
   - Oil network
   - Water supply network.
3. *Urban spatial structure:*
   - Level and quality of the physical urban services: sewerage, electricity, and road and transport system.

4. Natural factors:
   – Soft soils with high seismic amplification
   – Seismic zones with high amplification
   – Seismic zones with medium amplification
   – Seismic zones with medium- to low-range amplification
   – Landslides
   – Mining zones
   – Mudslide zones
   – Flooding zones with medium and low seismic amplification.

For analytical purposes, the study area has been divided into two broad zones: (1) the Federal District; and (2) the metropolitan zone, which lies

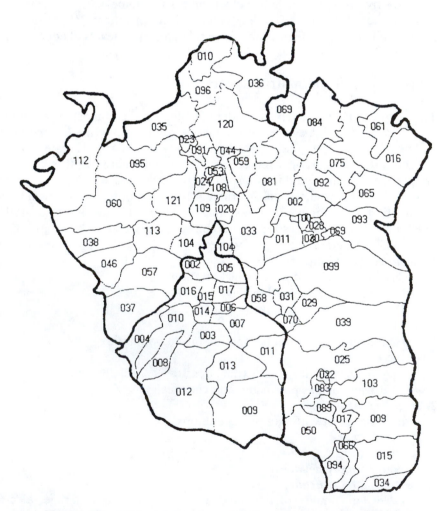

Fig. 9.2 The *delegaciones* and *municipios* of the mega-city of Mexico

mainly within the federal state of Mexico. This division ignores and cuts across the activities that link both parts of the mega-city, but it recognizes the basic distinction between a fast-growing zone (State of Mexico) and a slow-growing one (Federal District). The spatial units of analysis used in the matrix are those employed in Mexican population censuses: *delegaciones* in Mexico City and *municipios* in the MCMZ (see fig. 9.2). They

KEY:

Federal District *delegaciones*:

002	Azcapotzalco	010	Alvaro Obregón
003	Coyoacan	011	Tláhuac
004	Cuajimalpa de Morelos	012	Tlalpan
005	Gustavo A. Madero	013	Xochimilco
006	Iztacalco	014	Benito Juárez
007	Iztapalapa	015	Cuauhtémoc
008	Magdalena Contreras, La	016	Miguel Hidalgo
009	Milpa Alta	017	Venustiano Carranza

State of Mexico *municipios*:

002	Acolman	058	Netzahualcoytl
009	Amecameca	059	Nextlalpan
010	Apaxco	060	Nicolas Romero
011	Atenco	061	Nopaltepec
015	Atlautla	065	Otumba
016	Axapusco	066	Otzoloapan
017	Ayapango	069	Papalotla
020	Coacalco	070	Paz, La
022	Cocotitlan	075	San Martin de las Piramides
023	Coyotepec	081	Tecamac
024	Cuautitlan	083	Temamatla
025	Chalco	084	Temascalapa
028	Chiautla	089	Tenango del Aire
029	Chicoloapan	091	Teoloyucan
030	Chiconcuac	092	Teotihuacan
031	Chimalhuacan	093	Tepetlaoxtoc
033	Ecatepec	094	Tepetlixpa
034	Ecatzingo	095	Tepotzotlan
035	Huehuetoca	096	Tequixquiac
036	Hueypoxtla	099	Texcoco
037	Huixquilucan	100	Tezoyuca
038	Isidro Fabela	103	Tlalmanalco
039	Ixtapaluca	104	Tlalnepantla
044	Jaltenco	108	Tultepec
046	Jilotzingo	109	Tultitlan
050	Juchitepec	112	Villa del Carbon
053	Melchor Ocampo	113	Villa Guerrero
057	Naucalpan de Juarez	120	Zumpango
		121	Cuautitlan-Izcalli

State of Hidalgo *municipio*:

069	Tizayuca

vary in size and economic composition: some are already highly urbanized, industrialized, and thoroughly integrated into the city; others have small scattered populations, and are quasi-rural but physically connected to the city.

It is assumed that vulnerability is partly a function of population size, and this is built into the socio-economic component of the matrix by means of weighted values. There are three units of over 1 million people, but most possess fewer than 15,000 inhabitants. Building materials, housing types (detached, high-rise, etc.), and density form a second key component of vulnerability. Higher education is a third factor because it correlates with improved capacities for social interaction and the management of problems. Income levels are included because housing improvements are likely to be a function of economic surpluses, though recent data call this assumption into question (Puente, 1988). Occupation types and economic sector (transportation, industry, etc.) are included because they reflect the potential for human-made disasters – especially those involving the chemical and petroleum industries. Likewise, infrastructure measures capture risks associated with proximity to oil and gas pipelines and other hazardous technologies. Urban spatial structure is a surrogate for the quantity and quality of services that are important factors of vulnerability. Of course, natural factors are of the utmost importance. All single factors were weighted positively or negatively so that – in combination with other factors – they might intensify or reduce composite vulnerability; no single factor is pre-eminent. Composite vulnerability is therefore based on both absolute and relative factor values. This is an improvement over vulnerability analyses that do not take account of interaction among individual factors. For example, in this analysis, material adjustments to housing can counterbalance soil vulnerability.

Results

The general pattern of socio-economic vulnerability is shown in table 9.8 (column 1). The 16 Federal District *delegaciones* exhibit varied vulnerability: two high, four high–medium, eight low–medium, and two low. One of the high-vulnerability zones is part of the old city (the *delegacion* of Gustavo A. Madero). The other (Iztapalapa) has become heavily populated only recently, but is now fully urbanized. Three additional old-city neighbourhoods are ranked high–medium vulnerable and only one possesses low vulnerability (Benito Juárez). The remaining low-vulnerability area is semi-rural Milpa Alta.

A similarly heterogeneous pattern shows up in the 58 *municipios* of the metropolitan zone: 5 high vulnerable, 11 high–medium, 11 medium, 16

Table 9.8 General compound index of vulnerability of Mexico City

Delegacion	Socio-economic compound index	Urban spatial structure compound index	Regional infrastructure compound index	Natural (geographical, topographical, and geotechnical) compound index	General compound index
Alvaro Obregón	4.0	4.0	4.0	5.0	4.3
Azcapotzalco	4.0	4.0	4.0	5.0	4.3
Benito Juárez	1.0	1.0	1.0	5.0	2.0
Coyoacán	2.0	2.0	2.0	5.0	2.8
Cuajimalpa de Morelos	2.0	2.0	2.0	4.0	2.5
Cuauhtémoc	4.0	4.0	4.0	5.0	4.3
Gustavo A. Madero	5.0	5.0	5.0	5.0	5.0
Iztacalco	2.0	2.0	2.0	5.0	2.8
Iztapalapa	5.0	5.0	5.0	4.0	4.8
Magdalena Contreras, la	2.0	2.0	2.0	2.0	2.0
Miguel Hidalgo	2.0	2.0	2.0	5.0	2.8
Milpa Alta	1.0	1.0	1.0	1.0	1.0
Tláhuac	2.0	2.0	2.0	4.0	2.5
Tlalpan	2.0	2.0	2.0	1.0	1.8
Venustiano Carranza	4.0	4.0	4.0	5.0	4.3
Xochimilco	2.0	2.0	2.0	4.0	2.5
Mean	2.8	2.8	2.8	4.1	3.1

Sources: General Housing and Population Census, 1990; Economic Census of the State of Mexico, 1989; Annual Reports of the Government of the State of Mexico; Risk and Vulnerability Regional Study of the Valley of Mexico and its Metropolitan Zone, 1992.

Weighting: 5 = high; 4 = high–medium; 3 = medium; 2 = low–medium; 1 = low.

Table 9.9 General compound index of vulnerability, by degree of vulnerability (Mexico City)

Delegacion	General compound index
High vulnerability	
Gustavo A. Madero	5.0
Iztapalapa	4.8
Venustiano Carranza	4.3
Cuauhtémoc	4.3
Azcapotzalco	4.3
Alvaro Obregón	4.3
Low–medium vulnerability	
Miguel Hidalgo	2.8
Iztacalco	2.8
Coyoacán	2.8
Xochimilco	2.5
Tláhuac	2.5
Cuajimalpa de Morelos	2.5
Low vulnerability	
Magdalena Contreras, La	2.0
Benito Juárez	2.0
Tlalpan	1.8
Milpa Alta	1.0
Mean	3.1

Sources: General Housing and Population Census, 1990; Economic Census of the State of Mexico, 1989; Annual Reports of the Government of the State of Mexico; Risk and Vulnerability Regional Study of Valley of Mexico and its Metropolitan Zone, 1992.

low–medium and 15 low (table 9.10, column 1). The high-vulnerable *municipios* in the State of Mexico are located close to the Federal District's political boundaries. Industries, especially chemicals, are concentrated in three of them (Tlanepantla, Naucalpan, Ecatepec). Another unit (Netzahualcoytl) is a densely populated low-income residential area – a city within the city.

Within the Federal District, infrastructure has been weighted similarly to socio-economic factors. The District is a long-settled and maturely developed area whose infrastructure reflects the social make-up of its population (see table 9.8, column 3). The same cannot be said of the MCMZ. There, infrastructures are still being developed as informal settlements give way to more formal ones. As shown in table 9.10 (column 3), the infrastructure of 7 *municipios* is highly vulnerable, 18 are high–medium, 17 are low–medium, and 16 are low.

Table 9.10 General compound index of vulnerabilty of MCMZ

Municipio	Socio-economic compound index	Urban spatial structure compound index[a]	Regional infrastructure compound index	Natural (geological, topographical, and geotechnical) compound index	General compound index
Acolman	2.0	3.0*	5.0	5.0	3.8
Amecameca	4.0	4.0	4.0	5.0	4.3
Apaxco	1.0	3.0*	4.0	1.0	2.3
Atenco	1.0	3.0*	4.0	5.0	3.3
Atizapan de Zaragoza	4.0	2.0	1.0	5.0	3.0
Atlautla	2.0	3.0*	2.0	2.0	2.3
Axapusco	3.0	3.0*	5.0	1.0	3.0
Ayapango	1.0	1.0*	4.0	1.0	1.8
Chalco	4.0	5.0	1.0	5.0	3.8
Chiautla	1.0	1.0*	1.0	5.0	2.0
Chicoloapan	3.0	2.0	1.0	2.0	2.0
Chiconcuac	1.0	1.0*	2.0	1.0	1.3
Chimalhuacan	4.0	5.0	2.0	5.0	4.0
Coacalco	4.0	2.0	5.0	5.0	4.0
Cocotitlan	1.0	1.0*	1.0	5.0	2.0
Coyotepec	2.0	2.0*	2.0	4.0	2.5
Cuautitlan Izcalli	4.0	1.0	2.0	5.0	3.0
Cuautitlan	4.0	2.0	2.0	5.0	3.3
Ecatepec	5.0	4.0	5.0	5.0	4.8
Ecatzingo	2.0	1.0*	4.0	1.0	2.0
Huehuetoca	2.0	3.0*	4.0	1.0	2.5
Hueypoxtla	1.0	3.0*	1.0	1.0	1.5
Huixquilucan	3.0	1.0	2.0	5.0	2.8
Isidro Fabela	1.0	1.0*	1.0	2.0	1.3

Table 9.10 (*continued*)

Municipio	Socio-economic compound index	Urban spatial structure compound index[a]	Regional infrastructure compound index	Natural (geological, topographical, and geotechnical) compound index	General compound index
Ixtapaluca	4.0	4.0	4.0	1.0	3.3
Jaltenco	2.0	3.0*	1.0	5.0	2.8
Jilotzingo	1.0	1.0*	1.0	2.0	1.3
Juchitepec	1.0	1.0*	4.0	1.0	1.8
Melchor Ocampo	2.0	3.0*	2.0	5.0	3.0
Naucalpan	5.0	1.0	1.0	5.0	3.0
Netzahualcoytl	5.0	2.0	2.0	5.0	3.5
Nextlalpan	1.0	1.0*	1.0	5.0	2.0
Nicolas Romero	4.0	4.0	1.0	4.0	3.3
Nopaltepec	2.0	1.0*	4.0	1.0	2.0
Otumba	3.0	3.0*	4.0	1.0	2.8
Ozumba	2.0	3.0*	1.0	1.0	1.8
Papalotla	2.0	1.0*	4.0	1.0	2.0
Paz, La	4.0	2.0	1.0	4.0	2.8
San Martin de las Piramides	3.0	1.0*	2.0	5.0	2.8
Tecamac	3.0	4.0	4.0	4.0	3.8
Temamatla	1.0	1.0*	2.0	5.0	2.3
Temascalapa	2.0	3.0*	1.0	1.0	1.8
Tenango del Aire	2.0	1.0*	4.0	1.0	2.0
Teoloyucan	3.0	3.0*	2.0	5.0	3.3
Teotihuacan	2.0	3.0*	5.0	2.0	3.0
Tepetlaoxtoc	2.0	3.0*	5.0	1.0	2.8
Tepetlixpa	1.0	1.0*	4.0	1.0	1.8
Tepotzotlan	3.0	3.0*	1.0	4.0	2.8

326

Tequixquiac	1.0	3.0*	2.0	1.0	1.8
Texcoco	4.0	2.0	2.0	5.0	3.3
Tezoyuca	1.0	1.0*	2.0	5.0	2.3
Tizayuca, HIDALGO	2.0	3.0*	2.0	5.0	3.0
Tlalmanalco	3.0	3.0*	4.0	2.0	3.0
Tlalnepantla	5.0	4.0	5.0	5.0	4.8
Tultepec	3.0	3.0*	2.0	5.0	3.3
Tultitlan	5.0	4.0	4.0	5.0	4.5
Villa del Carbon	2.0	3.0*	4.0	5.0	3.5
Zumpango	3.0	4.0	4.0	5.0	4.0
Mean	2.6	2.4*	2.7	3.4	2.8

Sources: General Housing and Population Census, 1990; Economic Census of the State of Mexico, 1989; Annual Reports of the Government of the State of Mexico; Risk and Vulnerability Regional Study of Valley of Mexico and its Metropolitan Zone, 1992.
Weighting: 5 = high; 4 = high–medium; 3 = medium; 2 = low–medium; 1 = low.
a. Figures with an asterisk are estimates based on limited information.

Urban structure illustrates a contrasting pattern (tables 9.8 and 9.10, column 2). Chalco – which was among the least vulnerable units according to measures of infrastructure development – now has the highest level of vulnerability (table 9.10, column 2). Of 20 *municipios* that are sufficiently urbanized to make an assessment, only 2 have a high level of vulnerability; 8 are high–medium, 1 medium, 6 low–medium, and 3 low. Ecatepec and Tlalnepantla are the *municipios* that are now most developed but they still show high–medium levels of vulnerability because the process of urbanization is far from complete. Naucalpan, on the other hand, is characterized by low vulnerability on this factor. Surprisingly, the recently created low-income residential area of Netzahualcoytl exhibits only medium–low vulnerability.

The natural hazard dimensions of the enumeration units comprise the final assessment factor. Within the Federal District, vulnerability is relatively homogeneous: 9 out of the 16 spatial units have high vulnerability, 4 high–medium, 1 low–medium, and only 2 low (table 9.8, column 4). A combination of soft subsoils and seismic risks accounts for the elevated vulnerability. Central *delegaciones* are particularly at hazard. Furthermore, these units are densely populated and heavily built up. The two *delegaciones* with low vulnerability, Tlalpan and Milpa Alta, have only recently been incorporated into the urban area. They retain many fields and open spaces. The overall level of vulnerability in the MCMZ area is similar to that of the city but for different reasons. Most of the *municipios* (29) are highly vulnerable, though 18 possess low vulnerability (table 9.10, column 4).

Compound vulnerability scores and spatial patterns have been constructed by combining the separate indices discussed above (tables 9.8 and 9.10, column 5). Six *delegaciones* are highly vulnerable, all located in the central historic city. Gustavo A. Madero *delegacion* tops the list. If this preliminary finding is borne out by more detailed studies, such places deserve to become the focus of effective vulnerability-reduction programmes. However, these are very large *delegaciones* that exhibit considerable internal heterogeneity, and finer-grained assessments are probably warranted. In contrast, only one of the central *delegaciones* in the Federal District has a low degree of vulnerability (Benito Juárez). Other low-vulnerability units (Magdalena Contreras, Tlalpan, and Milpa Alta) are all located on the southern fringe of the city, on seismically stable soils (tables 9.8 and 9.9).

There is a wider range of compound vulnerability in the MCMZ (tables 9.10 and 9.11 and fig. 9.3). Ten *municipios* are highly vulnerable, 17 high–medium, 7 medium, 14 low–medium, and 10 low. The highly vulnerable group includes places that are adjacent to the city (Tlalnepantla, Ecatepec), as well as others that are not fully urbanized and contiguous to the

Table 9.11 General compound index of vulnerability, by degree of vulnerability (MCMZ)

Municipio	General compound index
High vulnerability	
Tlalnepantla	4.8
Ecatepec	4.8
Tultitlan	4.5
Amecameca	4.3
Zumpango	4.0
Coacalco	4.0
Chimalhuacan	4.0
Tecamac	3.8
Chalco	3.8
Acolman	3.8
High–medium vulnerability	
Villa del Carbon	3.5
Netzahualcoytl	3.5
Tultepec	3.3
Texcoco	3.3
Teoloyucan	3.3
Nicolas Romero	3.3
Ixtapaluca	3.3
Cuautitlan	3.3
Atenco	3.3
Tlalmanalco	3.0
Tizayuca, HIDALGO	3.0
Teotihuacan	3.0
Naucalpan	3.0
Melchor Ocampo	3.0
Cuautiltlan Izcalli	3.0
Axapusco	3.0
Atizapan de Zaragoza	3.0
Medium vulnerability	
Tepotzotlan	2.8
Tepetlaoxtoc	2.8
San Martin de las Piramides	2.8
Paz, La	2.8
Otumba	2.8
Jaltenco	2.8
Huixquilucan	2.8
Low–medium vulnerability	
Huehuetoca	2.5
Coyotepec	2.5
Tezoyuca	2.3
Temamatla	2.3
Atlautla	2.3
Apaxco	2.3
Tenango del Aire	2.0

Table 9.11 (*continued*)

Municipio	General compound index
Papalotla	2.0
Nopaltepec	2.0
Nextlalpan	2.0
Ecatzingo	2.0
Cocotitlan	2.0
Chicoloapan	2.0
Chiautla	2.0
Low vulnerability	
Tequixquiac	1.8
Tepetlixpa	1.8
Temascalapa	1.8
Ozumba	1.8
Juchitepec	1.8
Ayapango	1.8
Hueypoxtla	1.5
Jilotzingo	1.3
Isidro Fabela	1.3
Chiconcuac	1.3
Mean	2.8

Sources: General Housing and Population Census, 1990; Economic Census of the State of Mexico, 1989; Annual Reports of the Government of the State of Mexico; Risk and Vulnerability Regional Study of Valley of Mexico and its Metropolitan Zone, 1992.

city. Chalco *municipio* is included in this category. Other municipalities adjacent to the city, Netzahualcoytl and Naucalpan, which might be expected to have the same ranking, have only a high–medium vulnerability. All *municipios* in the low-vulnerability category are loosely attached to the mega-city and are only tenuously integrated into its functions. They contain much agricultural land and little industry.

How do the results of this vulnerability analysis differ from more conventional ones that rely solely on natural hazard factors? If only natural factors are used, more than half of the MCMZ's area is classified as highly vulnerable. When social, economic, and other factors are added, the high-vulnerability area is reduced to about one-third of the mega-city and lands that were formerly designated as high vulnerable are reclassified as of low vulnerability. The high vulnerability of industrial zones is reinforced, but major changes occur in northern and central parts of the MCMZ. Conversely, the south of the city is revealed as an area of low vulnerability. However, there is also no obvious correlation between disadvantaged socio-economic conditions and high vulnerability.

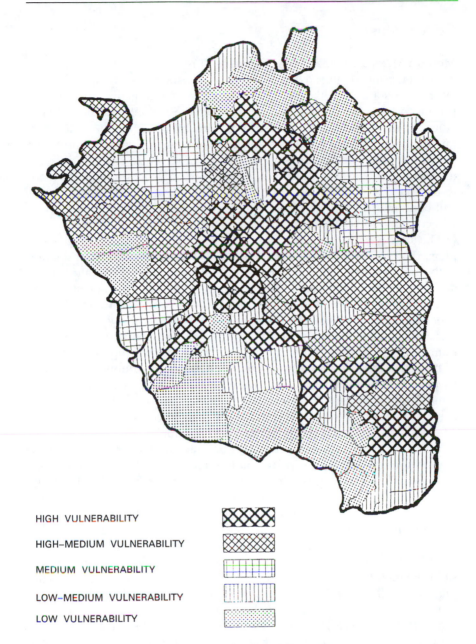

HIGH VULNERABILITY

HIGH–MEDIUM VULNERABILITY

MEDIUM VULNERABILITY

LOW–MEDIUM VULNERABILITY

LOW VULNERABILITY

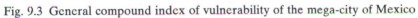

Fig. 9.3 General compound index of vulnerability of the mega-city of Mexico

Conclusions

Modern Mexico City has been created by three main groups that interact in complex and highly dynamic ways: real estate investors and developers, the state, and the poor. The process of interaction has, in fact, produced two cities. These can be labelled the formal and the informal cities, or the solvent and the insolvent, but, by whatever name they are known, each of them is subject to a different type of vulnerability. To a greater or lesser degree the same is true of other mega-cities in developing countries. Unless this dichotomy is understood and attempts are made to eliminate it, polarization will not diminish, growing numbers of people will continue to be vulnerable to worsening hazards, and the mega-cities will continue to be dysfunctional places.

The state has the greatest responsibility to develop urban hazard action plans that are, at the same time, internally coherent and efficient, externally adjusted to changing global economic and political arrangements, and responsive to the needs of diverse public and private interest groups. But states have often resisted adopting comprehensive urban planning procedures, thereby helping to increase the vulnerability of mega-cities. Herein lies a true paradox; at a time when governmental intervention in human affairs is often considered to be excessive, the empirical evidence of hazard in third world mega-cities demonstrates that public intervention to protect life, livelihood, and property has actually been insufficient, doubtful, erratic, and always late in coming. The alternative – market-driven forces – has been dominant for the past 40 years and has produced cities that are anarchic, dysfunctional, and unequal.

Whatever planning exists is largely confined to the formal city, yet the city's vulnerability is principally conditioned by the laws of supply and demand. These have generally worked to produce lower levels of vulnerability. In contrast, the informal city is illegal and therefore unplanned. It also typically possesses higher levels of vulnerability. But it is precisely the informal city that requires strict planning. We cannot permit human settlements to be created in areas that are vulnerable to geomorphic hazards. Such actions are socially unjust and fiscally irresponsible. Human life and well-being are directly at risk and sooner or later the state will be forced to bear an increased financial burden for the infrastructure that will eventually be required to offset these risks.

States must give high priority to urban planning and must treat it as a coordinating function that bridges all other sectors of society, not – as is so often the case in developing countries – as a separate sector. The task of planning is becoming more difficult because mega-cities are fragmenting into different political-administrative entities whose leaders frequently lose sight of the needs of the whole community. Nevertheless, the essential unity of the mega-city is indisputable, although sometimes it is camouflaged by substantive complexity. In order to reduce urban hazards, mega-city governments require, at a minimum, the capacity to coordinate decisions about land use and urban functions within their boundaries. This implies the formulation of strategies of metropolitan development and the establishment of permanent processes of mega-city planning and management.

This chapter has emphasized social and legal issues because it is necessary to come to grips with the social forces that underlie vulnerability to hazards. Nature is sometimes susceptible to the actions of humans, who can generate catastrophes by destabilizing some natural processes. But we must also remember that other natural processes are not necessarily subordinate to social ones. The social and natural causes of natural disasters should both be accommodated in the urban planning process. What is needed is a new axiology or a new ethic to guide the management of mega-city hazards. Such an ethic might take its cue from the concept of sustainable development – a merging of economic and environmental planning in pursuit of long-term sustainability. In the pursuit of sustainable development, the mega-city may make itself less artificial; indeed, it may come to acquire its own nature.

Notes

1. The name "City of Mexico," in the strict sense, refers exclusively to the old colonial city, which today forms the commercial centre of the metropolis.
2. Other metropolitan zones such as Guadalajara and Monterrey are also growing fast (5.6 per cent and 5.83 per cent during 1960–1970, and 1980 populations of 2.3 and 2.0 million, respectively). Despite this, Mexico City still maintains its unquestionable primacy.

REFERENCES

Atlas de la Cuidad de Mexico. 1987. Departamento del Distrito Federal and El Colegio de Mexico.

Cevallos, Diego. 1995. "Mexico: 10 years later earthquake continues to awaken opposition." Inter Press Service, 19 September.

Connolly, Priscilla. 1993. "The go-between: CENVI, a Habitat NGO in Mexico City." *Environment and urbanization* 5(1): 68–90.

Eibenschutz, Roberto and Sergio Puente. 1991. "Environment degradation, vulnerability and metropolitan development in developing countries: A conceptual approach." Report to the World Bank.

Esteva, Luis. 1997. "An overview of seismic hazard, seismic risk and earthquake engineering in Mexico City." Paper delivered at the First International Earthquakes and Megacities Workshop, 1–4 September, Seeheim, Germany.

Ezcurra, Exequiel and Marisa Mazari-Hiriart. 1996. "Are megacities viable? A cautionary tale from Mexico City." *Environment* 38(1): 6–15, 26–35.

Ibarra, V., S. Puente, and F. Saavedra. 1988. *La ciudad y el medio ambiente en America Latina: Seis estudios de caso.* Mexico: El Colegio de Mexico.

Ibarra, V., S. Puente, and M. Schteingart. 1984. "La ciudad y el medio ambiente." *Revista de Economia y Demografia*, no. 57 (El Colegio de Mexico).

Kandell, Jonathan. 1988. *La Capital: The biography of Mexico City.* New York: Henry Holt.

Kovach, Robert L. 1995. *Earth's fury: An introduction to natural hazards and disasters.* Englewood Cliffs, NJ: Prentice-Hall.

Linden, Eugene. 1996. "The exploding cities of the developing world." *Foreign Affairs* 75(1): 52–65.

Puente, Sergio. 1987. "Estructura industrial y participacion de la zona metropolitana de la ciudad de Mexico en el producto interno bruto." In *Atlas de la Ciudad de Mexico.* Departamento del Distrito Federal and El Colegio de Mexico.

——— 1988. "Calidad material de vida en la Zona Metropolitana de la Ciudad de Mexico. Hacia un enfoque totalizante." In Sergio Puente and Jorge Legorreta (eds.), *Calidad de vida y medio ambiente.* Mexico: D.D.F. and Plaza y Valdez.

Roberts, Bryant. 1980. *Ciudades de campesinos: La economia politica de la urbanizacion.* Mexico: Siglo XXI.

Sistema Nacional de Proteccion Civil. 1991. *Atlas nacional de riesgos.* Mexico: Secretaria de Gobernacion.

Vargas, Carlos Tejeda. 1995. "A plan for all reasons: Future vision from Mexico City." *Business Mexico*, April.

10

Natural hazards of the San Francisco Bay mega-city: Trial by earthquake, wind, and fire

Rutherford H. Platt

Editor's introduction

To outsiders and residents alike, the earthquake and fire of 1906 seem to be a permanent part of the image of San Francisco and the San Francisco Bay Area. Moreover, research has shown that the 1906 event is one of the most frequently employed yardsticks by which all earthquakes – and indeed all urban natural disasters – are evaluated throughout the United States. But it is not an entirely accurate symbol of metropolitan hazard. For one thing, geologists and historians point out that larger earthquakes and fires also affected an (admittedly smaller) San Francisco in the previous century. For another, the local record of natural disasters includes many different kinds of event from slope failures and flash-floods to urban wildfires, all of which have added contrasting strands to the region's hazard experience. Finally, preparedness and mitigation initiatives in the decades since 1906 have upgraded the Bay Area's hazard-resistance capabilities. Though there is still far to go before San Francisco – and more especially the East Bay communities – can be declared models of sound hazard management, there are also grounds for believing that they are not all teetering on the brink of an apocalyse.

Introduction

The mega-city that enfolds San Francisco Bay holds a special status in North America. Its core community is considered by many to be the most beautiful and vibrant urban place on the continent. The city of San Francisco is justly renowned for spectacular views of ocean cliffs and sheltered anchorages, a temperate marine climate adorned with rolling fogs, the sublime Golden Gate Bridge, captivating public parks, striking architecture, cable cars, restaurants, theatres, opera, symphony, art museums, a flourishing Chinatown, and a peerless night-life. It is one of the world's truly cosmopolitan cities.

Several other parts of the Bay Area are celebrated in their own right. Marin County is the quintessence of American suburban affluence; the Sonoma and Napa valleys comprise California's premier wine-growing region; Berkeley and Palo Alto are seats of distinguished universities; and Silicon Valley contains the global nexus of post-industrial technology. These superlative characteristics notwithstanding, the Bay Area mega-city (fig. 10.1) is a place of great physical and human hazard. In 1906, San Francisco was substantially destroyed by an 8.2 Richter magnitude earthquake and subsequent fires that spread from ruptured chimneys and exploding gas mains. On 17 October 1989, the 7.1 magnitude Loma Prieta earthquake, centred 100 km south of San Francisco, reminded residents that another "Big One" may not be far in the future. In the words of a British Broadcasting Corporation film, this is "The City that Waits to Die." Meanwhile, the Bay Area's people have been dying in a different and more literal sense. San Francisco was where the AIDS epidemic first came to public attention in the United States as it ravaged the city's prominent homosexual community during the 1980s. Exuberance and foreboding are opposite sides of life in this mega-city.

The Bay Area urban region is officially designated as the "San Francisco–Oakland–San Jose CMSA" (Consolidated Metropolitan Statistical Area).[1] With a population of 6.3 million, it is now the fourth-largest conurbation in the United States after New York, Los Angeles, and Chicago. The CMSA extends approximately 200 km from Santa Cruz in the south to Santa Rosa in the north, and about 65 km from the Pacific Ocean to the city of Livermore, east of San Francisco Bay. It contains nine counties and several hundred municipalities in a belt of territory that roughly encircles the Bay and its northerly extensions, San Pablo Bay and Suisun Bay. The city of San Francisco (724,000) contains just 11.6 per cent of the total population.

Like San Francisco itself, the Bay Area is a region of widespread physical and human-caused hazards. Foremost among these is the risk of earthquakes along a network of north–south-trending faults that under-

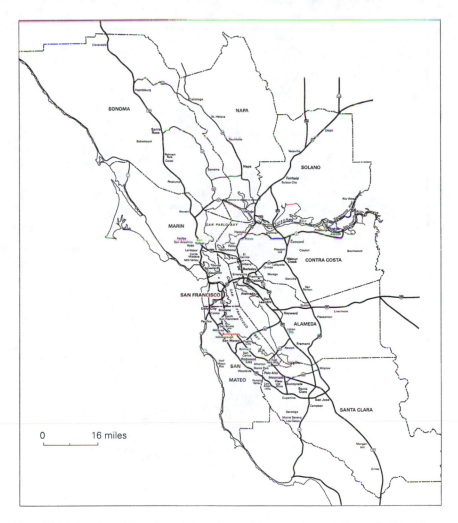

Fig. 10.1 The San Francisco Bay Area (Source: Association of Bay Area Governments)

lies the mega-city and stretches far beyond (fig. 10.2). More localized physical risks include landslides, flooding, soil liquefaction, and wildfires. Compounding the effects of natural disasters is a significant potential for failures of lifelines, including transportation, communications, water and sewer pipes, dams, energy systems, food, and medical care delivery. And, like other mega-cities, this one also suffers daily human-created hazards of traffic congestion, air and water pollution, chemical and toxic wastes, and urban crime. But these latter conditions lie beyond the scope of this

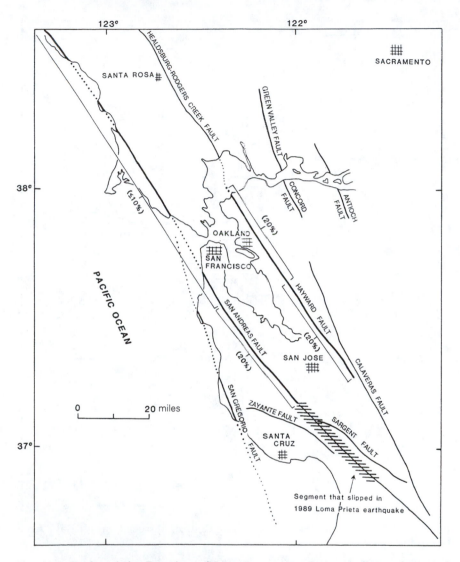

Fig. 10.2 Faults of the Bay Area (Note: Percentage figures indicate chance of occurrence of an earthquake in the 30 years from 1988. Source: Plafker and Galloway, 1989)

chapter. Although the term "Bay Area" is synonymous in American culture with natural calamity, the region is earning a reputation for leadership in the development of new approaches to disaster preparedness, response, and recovery. As recounted below, this mega-city can mobilize massive resources – fiscal, technical, and intellectual – in the ongoing process of fitting its built environment to the pervasive natural limitations of its site.

Physical setting

Unlike the Atlantic coast, the Pacific littoral offers few safe natural harbours. Between Puget Sound to the north and San Diego Bay to the south, there is only one major opening through the treacherous sea cliffs and coastal barriers: the Golden Gate. This fabled strait leads from the open ocean into the vast, sheltered recesses of San Francisco Bay, which is – in effect – an inland sea extending about 100 km from north to south. The Sacramento River, which drains into the Bay, historically afforded access by riverboat to the state's interior central valley. San Francisco thus possessed three physical advantages that ensured its rapid growth: (1) a sheltered deepwater port; (2) easy water access to cross-bay satellite communities (e.g. Berkeley, Oakland, Alameda); and (3) river access to a vast agricultural and mining hinterland that extended eastward to the Sierra Nevada mountains.

The site of San Francisco has always appeared highly favourable to urban settlement. In the words of one of the city's early enthusiasts:

[It] seems topographically marked for greatness, rising on a series of hills, with a great harbor on one side, a great ocean on the other, ... mighty waters ever passing by to the outlet of the widespread river system of the country. (Bancroft, 1888; quoted in Vance, 1964, pp. 4–5)

Lying at the northern tip of a broad peninsula, it is accessible from the sea or overland from a region of fertile valleys and low hills to the south. This site was originally bordered by bay-side marshes that were easily drained and converted to buildable land. Hills on the peninsula (e.g. Nob Hill, Telegraph Hill, Russian Hill) afforded fresh breezes and superb views for wealthy inhabitants. The imposition of a typical American grid street system on irregular and occasionally steep slopes fostered San Francisco's most famous means of transportation, the cable car, and contributed to the cardiovascular well-being of local residents!

In the midst of these favourable attributes lies the mega-city's nemesis, the San Andreas Fault. It marks a major boundary between the Pacific tectonic plate and the North American plate. Skirting the mainland north of the Bay Area, this fault passes San Francisco just offshore and crosses the coast a few kilometres south of the present city limits. It continues inland, bisecting the base of the peninsula and extending south-east towards southern California (fig. 10.2).

The eastern side of the bay ("East Bay") was originally fringed by a marshy shoreline and backed by steep topography (e.g. San Pablo Hills, San Leandro Hills). This shore has been extensively reclaimed by filling wetlands, and subsequent urban development has occurred on unstable soils that are subject to mobilization during earthquakes. Another fault (the Hayward Fault) marks the junction between bay-side flatlands and hills a few kilometres inland. This branch of the San Andreas system is "an offspring even more dangerous than its treacherous parent" (Yanev, 1991, p. 36). Most predictions of future Bay Area earthquakes focus on the Hayward Fault (see below).

Originally cloaked in grasslands with pockets of oak, the East Bay Hills blocked expansion of settlement beyond the bay shore during the nineteenth century. With the advent of automobiles and improved building technologies, the steep west-facing slopes of the hills, which have spectacular views of the bay and distant San Francisco, became élite residential districts beginning in the 1920s. There the proximity of structures, ornamental vegetation, and residual grasslands has always posed a chronic risk of wildfire at times of drought. Much of this area burned in the East Bay Hills "firestorm" of 20 October 1991.

Settlement and growth

European settlement first reached the Golden Gate in 1776 when the Mission of San Francisco de Asis was founded by Spanish Franciscan monks and military authorities. This mission and its associated *presidio* formed the nucleus of the future city. The United States gained sovereignty over California from Mexico in 1846. Until then, American contact with this area was largely via New England-based ships trading along the coast in competition with Russian vessels. An early (1835) American visitor, Richard Henry Dana, described the locality without making reference to its yet-unrecognized lurking hazards:

A few days after our arrival [in San Fransicso Bay], the rainy season set in, and for three weeks it rained almost every hour, without cessation. This was bad for our trade.... The mission of San Francisco, near the anchorage, has no trade at all, but those of San Jose, Santa Clara, and others, situated on large creeks or rivers which run into the bay, and distant between fifteen and forty miles from the anchorage, do a greater business in hides than any in California. Large boats, manned by Indians, and capable of carrying nearly a thousand hides apiece, are attached to the missions, and sent down to the vessels with hides, to bring away goods in return. (Dana, 1964, pp. 221–222)

A few years later, San Francisco would be transformed, first by political annexation to the United States and second by the discovery of gold in the Sierra Nevada mountains. The promise of riches attracted thousands of "49ers" from Back East and abroad to pan for gold in the streams and to dig mines on newly established claims issued by federal land offices. San Francisco became the commercial centre for the region by serving the needs of would-be miners in transit and those who returned with sudden wealth. Banks, warehouses, retail stores, and hotels quickly appeared along wide, new streets leading from the wharves straight up the hills and across the peninsula. Agriculture and industry also contributed to the local urban economy. The city's West Coast economic primacy was assured when rail lines arrived from the East in the 1860s, terminating in Oakland.

By the 1850s, San Francisco was beginning to emerge as a city of wealth, philanthropy, gracious architecture, and culture. Separated from the East Coast by a week of rail travel, and from Europe by at least a month at sea, it took on the sobriquet "Paris of the West Coast." In 1866, the New York landscape architect Frederick Law Olmsted, Sr. was invited to propose a plan for what would eventually become Golden Gate Park, the San Francisco counterpart to his celebrated design for New York's Central Park. By 1875, the Palace Hotel, financed with Comstock Lode mining profits, opened as "the finest hotel in the world" (until its destruction in the 1906 fire). Thirty years later (1905), the Association for the Improvement and Adornment of San Francisco received a "City Beautiful" plan which it had commissioned from the great Chicago architect Daniel H. Burnham. This plan was adopted a month after the earthquake, but much of it was never accomplished. Although the Bay Area was tied commercially and culturally to the Atlantic world, it was also linked by sea to distant ports of the Pacific Rim. It was there that contemporary boosters envisaged further economic gains:

The great triangle of the Pacific is destined to have its lines drawn between Hong Kong, Sydney and San Francisco. Of these three ports, Hong Kong will have China behind it, Sydney, Europe, and San Francisco, America; and with America for a backing, San Francisco can challenge the world in the strife for commercial supremacy. (Keeler, 1903, p. 94)

San Francisco remained the financial and cultural capital of the region, while the East Bay communities of Oakland, Berkeley, and their neighbours flourished both as bedroom suburbs to San Francisco and as separate cities (table 10.1). Oakland later took on the role of major ocean port and military naval facility. Berkeley, home to the University of California's leading campus, became – and still remains – an icon of counter-culture and social protest.

Table 10.1 Populations of San Francisco and East Bay cities, selected years 1870–1990

Year	San Francisco	Oakland	Alameda	Berkeley
1870	149,000	10,500	1,557	–
1900	343,000	67,000	16,000	13,000
1910	416,912	150,000	23,000	40,434
1920	506,676	216,000	29,000	56,000
1950	775,000	384,000	64,000	114,900
1990	724,000	372,000	n.a.	n.a.

Sources: Vance (1964) and Bureau of the Census, *Statistical Abstract of the United States*, Washington, D.C., 1992.

The 1906 earthquake and its aftermath

By the turn of the century, San Francisco had a population of one-third of a million. Many other cities had experienced catastrophic fires, but no city in North America had been consumed by earthquake-related fires. Like the sinking of the ocean liner *Titanic* in 1912, the San Francisco earthquake and fire of 1906 shattered conventional assumptions about the ability of "modern technology" to overcome natural perils. And just as the loss of the *Titanic* led to tighter safety requirements for ocean vessels and an ice patrol in the North Atlantic, the San Francisco earthquake stimulated actions there and elsewhere to mitigate future disasters.

The earthquake struck at 5:12 a.m. on Sunday, 18 April, as a tremor that lasted 40 seconds. A major aftershock took place 13 minutes later, followed by many others of lesser magnitude. Because it occurred at an early hour, deaths were relatively few; falling masonry would have been lethal to daytime crowds, especially in the downtown business district of San Francisco. On the other hand, the duration of shaking proved too much for the city's hastily built housing stock and many victims were trapped under debris. Some structures collapsed, others leaned, and thousands of masonry chimneys fell onto roofs, in some cases starting fires. Fissures opened in pavements miles from the San Andreas Fault and hundreds of church bells pealed as the earthquake jolted terrified inhabitants awake (Bronson, 1986). Nearby cities such as San Jose and Palo Alto were badly damaged, but Oakland and Berkeley were less affected and were able to render vital assistance to San Francisco. The earthquake also struck many other communities along a 320 km corridor between Salinas and Fort Bragg.

Post-earthquake fires took some time to get under way. Meanwhile, residents fled to the streets, unaware that most of the damage was yet to come. Unlike other great urban fires (e.g. London in 1666, Chicago in 1871, Oakland in 1991), wind was not a major factor; nor was the region experiencing a drought. However, near total failure of the water supply contributed directly to the disaster. Before the quake it was already known that San Francisco's water system was inadequate for fighting a major fire (Bronson, 1986). The earthquake promptly disabled the system by rupturing water mains and cracking storage tanks. As gas lines exploded and cooking fires set homes ablaze, the fire department could not extinguish even modest fires. Water from the bay was pumped onto some nearby structures – an event that was to be repeated in the Mission District of San Francisco after the Loma Prieta quake 83 years later. Some use was made of dynamite to clear fire breaks. But, by and large, firemen looked on helplessly as the city burned.

The San Francisco earthquake and fire of 1906 are widely considered America's worst urban natural disaster and one of the world's great urban conflagrations. Flames burned over 1,146 hectares or 490 blocks – an area 50 per cent larger than that devastated by the Chicago fire of 1871 and six times bigger than the district that was burned by the "Great Fire" of London in 1666 (see Platt, 1991, and chap. 7 in this volume). Approximately 500 people died as a direct result. Three-fifths of the city's inhabitants lost their homes and the entire business district was destroyed (Bronson, 1986). The 1906 catastrophe illustrated many characteristics of a contemporary mega-city natural disaster. Multiple interrelated hazards were involved (i.e. earthquake and fire). Lifelines (water, communications, transportation) failed and caused secondary impacts. Widespread structural damage illuminated the inadequacy of existing building standards and the prevalent use of wood for smaller structures. Much of the working-class population lost homes and jobs. Finally, aid from external sources assisted the immediate response and to a lesser extent enabled the longer-term recovery.

Despite many similarities with modern disasters, there was one striking contrast. Missing from the San Francisco catastrophe was any significant economic or material assistance from the national (federal) government. The US Army helped to keep order and allocated supplies, but otherwise most of the aid came from private or quasi-public sources. The Red Cross was entrusted with overall coordination of the relief effort. Financial aid totalling US$9 million (1906 dollars) was provided by voluntary contributions from individuals and other cities (Bronson, 1986). International assistance totalling US$474,000 was also contributed from 14 foreign nations. But the response was largely a spontaneous, "grass-roots" reaction to a widely publicized disaster. Significant federal participation in disaster assistance and recovery did not begin in the United States until the 1950s.

The foremost public action in response to this disaster was a dam, reservoir, and aqueduct system that conveyed water from the Hetch Hetchy valley in the Sierra Nevada over a distance of 190 miles to the city. San

Francisco's proposals for the Hetch Hetchy dam precipitated a precedent-setting 10-year-long controversy between advocates of wilderness preservation, headed by John Muir (founder of the Sierra Club), and proponents of the "wise use of natural resources" represented by Gifford Pinchot, Director of the US Forest Service and adviser to President Theodore Roosevelt. The dam was finally approved and construction began in 1913 during the administration of President Woodrow Wilson (Nash, 1982). The Hetch Hetchy dam and reservoir now provide San Francisco with high-quality drinking water, but the system's reliability in the event of another major earthquake remains in doubt. Pipes from the reservoir cross the Hayward Fault and could rupture in an earthquake, once again leaving San Francisco without its primary water supply.[2]

San Francisco recovered rapidly from the 1906 catastrophe chiefly because the downtown business district was well insured. Total insurance payments amounted to US$5.44 billion (1992 dollars), by far the largest urban fire insurance loss in American history.[3] Reconstruction of the central city began immediately. By 1910, 417,000 people were in residence, an increase of 22 per cent since 1900 (table 10.1). Housing was constructed rapidly both to replace what was lost and to accommodate newcomers. However, like London after 1666, leaders of San Francisco declined to alter the city's basic pattern of streets and land uses during the rebuilding process. They even ignored Daniel H. Burnham's "City Beautiful" plan for the redesign of the city:

Few cities ever found themselves demolished, with a ready-made plan for a new and grander city already drawn up, awaiting implementation, and with money pouring in to help realize the plan. San Francisco chose to ignore its Burnham Plan, and decided instead to build at a rate and manner which made the city not only less beautiful than was possible, but more dangerous. The rubble of the 1906 disaster was pushed into the Bay; buildings were built on it. Those buildings will be among the most vulnerable when the next earthquake comes. (Thomas and Witts, 1971, p. 274)

This statement was prophetic: the city's Marina District, built on 1906 rubble, was badly damaged in 1989.

The Loma Prieta earthquake of 1989

The Bay Area was relatively unscathed by natural disasters for 83 years after 1906. During this period, the relative dominance of San Francisco diminished. Urban development flourished elsewhere in the mega-city, greatly facilitated by the building of the Golden Gate and Oakland Bay bridges during the 1930s. Decentralization was further encouraged after

Table 10.2 Population change in San Francisco Bay Area, 1970–1990 ('000)

	1970	1980	1990	% change 1980–90
San Francisco–Oakland–San Jose CMSA	4,754	5,368	6,253	16.5
San Francisco PMSA	1,482	1,489	1,604	7.7
Oakland PMSA	1,628	1,762	2,083	18.2
San Jose PMSA	1,085	1,296	1,498	15.6
Santa Cruz PMSA	124	188	230	22.3
Santa Rosa PMSA	205	300	388	29.3
Vallejo PMSA	251	334	451	35.0
City of San Francisco	716	679	724	6.6
City of Oakland	362	339	372	9.7

Source: Bureau of the Census, *Statistical Abstract of the United States*, Washington, D.C., 1991, table 36.
CMSA = Consolidated Metropolitan Statistical Area
PMSA = Primary Metropolitan Statistical Area (PMSAs are components of a CSMA)

Table 10.3 Probability of one or more large earthquakes on the San Andreas Fault System, as estimated in 1988

	Expected magnitude	Next			
		5 years	10 years	20 years	30 years
San Francisco Bay Area	7.0	0.1	0.2	0.3	0.5
Southern San Andreas	7.5–8.0	0.1	0.2	0.4	0.6
San Jacinto Fault	6.5–7.0	0.1	0.2	0.3	0.5

Source: US Geological Survey Working Group (1988).

the Second World War by federal guarantees for new home construction in outlying areas, by new connecting freeways, and later by the Bay Area Rapid Transit (BART) system, which links East Bay cities together and includes San Francisco. The combined population of the East Bay (Oakland PMSA) now exceeds that of San Francisco and is growing faster (table 10.2).

During the long period of deceptive quiet that followed the 1906 earthquake, another major earthquake on the San Andreas Fault was awaited by scientists. In 1988, a US Geological Survey Working Group on California Earthquake Probabilities (1988) estimated that a magnitude 7.0 earthquake would occur on the northern San Andreas (Bay Area) with a probability of 10 per cent within five years and 50 per cent within 30 years (table 10.3).

Concurrently, the US Geological Survey and the California Office of Emergency Services (OES) were closely monitoring seismic activity in the region. Two small earthquakes in June 1988 and August 1989 persuaded OES to advise local governments about an impending larger earthquake. This represented "the most significant use of earthquake forecasting for public policy purposes since the 1975 Haicheng earthquake was predicted in the People's Republic of China" (FEMA, 1990, p. 15). These forecasts were soon fulfilled: the San Andreas Fault released a 7.1 magnitude earthquake at 5:04 p.m. (local time) on 17 October 1989. The epicentre was located about 96 km south-south-east of San Francisco, near the community of Loma Prieta. The fault ruptured over a distance of 40 km at a depth of 19 km below the surface; no surface faulting appeared (FEMA, 1991; Bolin, 1993). Seismic shaking lasted 15 seconds, followed by numerous aftershocks. The event was felt over an area of about 1 million km^2 extending southward as far as Los Angeles and northward to the Oregon border (fig. 10.3). It caused 62 known deaths and 3,757 injuries, and left 12,000 people homeless. Property damage of over US$6 billion was sustained and there were severe disruptions of public transportation, utilities, and communications (Plafker and Galloway, 1989).

This earthquake coincided with the beginning of the third game of a baseball World Series that ironically matched two teams from the Bay Area: the San Francisco Giants and the Oakland Athletics. Sports commentators quickly shifted from reporting game preparations to the unfolding spectacle of disaster. It was an unparalleled opportunity to raise public awareness of seismic hazard. Warnings of a greater disaster to come were trumpeted by the news media around the world, as on the cover of *Newsweek* (30 October 1989): "Bracing for the Big One: The Lessons of the San Francisco Earthquake."

Infrastructure damage

French (1990) has identified four consequences of infrastructure damage associated with the Loma Prieta earthquake:
1. direct physical and economic damage to the systems;
2. diminished ability to carry out emergency response activities;
3. inconvenience due to temporary service interruption;
4. longer-term economic losses due to limits on recovery.

Although the quake's epicentre was in a rural upland well south of the mega-city, it caused dramatic and costly damage to infrastructure and older private buildings on both sides of San Francisco Bay. The most deadly outcome involved the collapse of a double-deck commuter freeway that was constructed on bay mud in Oakland. A 2.4 km section of

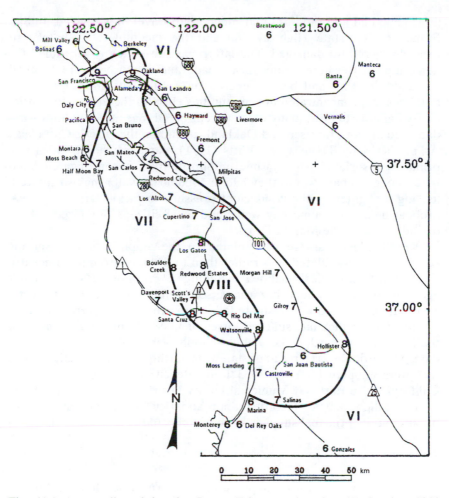

Fig. 10.3 Areas affected by the Loma Prieta earthquake, 17 October 1989 (Source: Plafker and Galloway, 1989)

this road failed, trapping hundreds of vehicles and accounting for 41 of the deaths attributed to the earthquake. For several days, national television broadcasts of rescue efforts transfixed audiences throughout the country. The last living victim was removed from his car after 90 hours (*Newsweek*, 30 October 1989, p. 32).

The only direct road link between San Francisco and the East Bay was severed when a 15 m section of the Bay Bridge collapsed. Its closure for repairs forced commuters to make long detours around the bay, or to rely on the underground BART rapid transit system, which was not damaged. The cross-bay ferry service, which had previously been discontinued after new road and rail links had speeded up travel, was revived. In San Francisco, Interstate Highway 280 and the incomplete Embarcadero Freeway were also damaged. The latter was subsequently scheduled for demolition. Elsewhere, many local roads and highways were blocked by landslides that had been triggered by ground shaking. Damage to buildings was concentrated in communities close to the epicentre, particularly Santa Cruz. But many older unreinforced masonry buildings were damaged in San Francisco and Oakland (including the latter's City Hall). As predicted by Thomas and Witts (1971), damage in both cities was primarily associated with seismic mobilization of unconsolidated filled lands along the bay. Just as they had in 1906, local water mains ruptured, leaving firefighters without sufficient pressure to tackle large blazes. A fireboat helped to pump bay water and local citizens organized bucket brigades to save the Marina District.

Overall, much damage was related to the amplification of seismic energy in unconsolidated soils rather than to distance from the epicentre (FEMA, 1990). Critical facilities, including several hospitals, were damaged in locations where hazardous soil conditions could have been anticipated. Stanford University in Palo Alto, which was severely affected by the 1906 earthquake, suffered losses of US$160 million as a result of the Loma Prieta earthquake. Public schools were generally not severely damaged, probably as a result of statewide earthquake construction codes that were adopted after the 1933 Long Beach earthquake in southern California. The East Bay Municipal Utility District, which supplies water to 1.5 million people, recorded approximately 200 local water mains breaks, but its principal supplies were not affected (FEMA, 1990).

Emergency response

Preparation for a major earthquake in northern or southern California has been under way for many years. A large earthquake in either the Bay Area or Greater Los Angeles has the potential to become a national, or even an international, catastrophe (e.g. through its impact on the global financial and insurance industries). Planning involves not only local, county, and regional authorities but also the State of California, the Federal Emergency Management Agency (FEMA), and other federal agencies. Only three months before Loma Prieta, a training exercise ("Response '89") tested response capabilities in the event of a hypothetical 7.5 magnitude earthquake on the Hayward Fault. After the real earthquake, FEMA reported that emergency responders – both governmental and non-governmental – functioned smoothly but resources were stretched to the limit: "The entire system of emergency responders had reached a total saturation point, and if the disaster had been of any larger proportions, the total system would have broken down" (FEMA, 1991, p. 40).

Emergency response was hindered by overload of telephone circuits, blockage of highways, rupture of gas lines, loss of water pressure, and interruption of electrical power supplies. But several factors helped to reduce potential losses: the epicentre was relatively remote from most of the Bay Area; schools were not in session; it was a windless day. Commuter traffic was also less than normal for 5 p.m. because many people were indoors watching the World Series. Key personnel were still at their workplaces and available to staff emergency operations centres within minutes after the earthquake.

Property damage and individual mitigation

The earthquake damaged more than 22,000 residential structures, 1,567 commercial buildings, and 137 public buildings (FEMA, 1991, p. 21). Local building inspectors marked 3,957 structures with "red tags" to indicate that they were uninhabitable. Most of the damaged buildings had been constructed before the state's Uniform Building Code was updated in 1973 (and revised further in 1988) to reflect seismic risks. Modern office towers in downtown San Francisco swayed as they were designed to. The preventive removal of projecting masonry and ornamentation from most older buildings also helped to avoid showers of rubble on people in the streets below (Plafker and Galloway, 1989), although 12 people died from this cause. Unreinforced masonry construction had long been prohibited for new buildings, thus reducing the risk of total building collapse as occurred during recent earthquakes in Mexico City, Armenia, and elsewhere. Damage to neighbourhoods, homes, and personal property was widespread throughout southern and central parts of the Bay Area. A summary of sample damage incurred by residents of San Francisco and Santa Cruz counties is provided in table 10.4 (O'Brien and Mileti, 1993). Approximately two-thirds of San Francisco respondents and 83 per cent of those in Santa Cruz County reported that their homes suffered slight to moderate damage. Damage in "the neighbourhood" was reported by 85.3 per cent of those contacted in San Francisco and by 98 per cent in Santa Cruz.

The geographic spread and scale of damage are all the more noteworthy in light of the much-documented failure of Californians to purchase earthquake insurance and their reluctance personally to engage in most other types of hazard mitigation. In 1990, the California Legislature established a mandatory earthquake insurance programme that provided up to US$15,000 in benefits; this was funded out of a surcharge on household insurance policies. The propensity to buy earthquake insur-

Table 10.4 Types and distribution of damage due to Loma Prieta earthquake reported by residents surveyed in San Francisco and Santa Cruz counties

	San Francisco County		Santa Cruz County	
	%	No.	%	No.
Neighbourhood damage				
None	14.7	108	2.0	18
Slight	48.1	353	13.0	119
Moderate	20.7	152	50.0	459
Severe	16.5	121	35.1	322
Residence damage				
None	37.7	27	13.4	123
Slight	55.2	405	51.1	469
Moderate	6.4	47	31.7	291
Severe	0.7	5	3.8	35
Household items				
None	46.9	344	4.7	43
Slight	43.1	316	34.9	320
Moderate	9.5	70	50.8	468
Severe	0.5	4	9.7	89

Source: Modified from O'Brien and Mileti (1993), table 3.

ance appears to have been unaffected by the Loma Prieta experience, for Palm and Hodgson (1992) report that there was no substantial increase in earthquake insurance purchases thereafter.[4] Moreover, those who bought insurance tended to be located further from Loma Prieta (table 10.5).

Even in Santa Clara County, whose western boundary virtually coincides with the segment of the San Andreas Fault that ruptured in 1989, the prevalence of earthquake insurance reported by survey respondents was just 51 per cent in 1990. Only 29 per cent of respondents in Contra Costa County, which experienced widespread effects of Loma Prieta, carried earthquake insurance. On the other hand, most victims would not have benefited greatly from such insurance because only 3.8 per cent of residence damage reported for Santa Cruz County (adjoining Santa Clara County) was rated as "severe." Most of the costs of "slight" and "moderate" damage would probably not have exceeded the 10 per cent deductible and would thus have been borne by owners. Palm and Hodgson (1992) also found a comparable reluctance to undertake other personal mitigation activities involving any monetary cost (table 10.5).

About 12,000 people were made temporarily homeless by the earthquake, many of them poor. As in other disasters, the Red Cross moved quickly to set up shelters and emergency feeding stations. About one week after the event, 44 shelters were in operation. These housed approximately 2,500 persons per night, with the remaining 9,500 either

Table 10.5 Earthquake mitigation actions taken before (1989) and after (1990) Loma Prieta by residents surveyed in four California counties

Survey counties	% with earthquake insurance before Loma Prieta	% increase in earthquake insurance after Loma Prieta	% incurring expense for other earthquake mitigation actions	
			1989	1990
Santa Clara	40	+11	9.0	31.0
Contra Costa	22	+7	7.8	10.4
Los Angeles	40	+6	5.0	3.3
San Bernardino	34	+1	5.0	6.0

Source: Modified from Palm and Hodgson (1992), fig. 6 and table 13.

sharing accommodation with friends and relatives or living in tents, vehicles, and other improvised quarters (Bolin and Stanford, 1993).

Federal emergency response and mitigation

Unlike in 1906, the US national government now regularly provides massive assistance in the wake of natural disasters. This is triggered by a "major disaster declaration" issued by the President and affecting speci-fied counties. The lead federal agency – FEMA – had been widely accused of reacting slowly to hurricane Hugo, a major hurricane that spread devastation across a vast area, including the Virgin Islands, Puerto Rico, and North and South Carolina, two months before the Loma Prieta earthquake. In that case, 113 counties, most of them rural and isolated, were declared eligible to receive federal aid. By contrast, the Loma Prieta disaster prompted a Presidential declaration for only 12 counties in a compact zone within or adjacent to the Bay Area. In the glare of national media coverage, FEMA rushed to open 18 Disaster Assistance Centers, including five that were mobile. By the end of January 1990, 77,654 indi-viduals, families, and businesses had registered at these centres for fed-eral assistance, including about 40,000 applicants for temporary housing and 31,255 applicants for individual and family grants of US$11,000 per household. The Small Business Administration issued 34,976 home and personal property loans and 14,773 business loans. Federal assistance by then had amounted to about US$202 million (FEMA, 1990, p. 27). As of August 1993, total federal costs for the Loma Prieta disaster were pro-jected to reach US$689 million (FEMA, unpublished data).

The victim population, like that of California in general, was highly diverse. Special problems in the administration of disaster assistance, particularly temporary housing, appeared in some localities. According to Bolin and Stanford (1993, p. B48), sheltering problems in Watsonville and elsewhere reflected longstanding ethnic and class antagonisms: "The earthquake and the subsequent housing crisis highlighted the preexisting deficiencies and inequities in Watsonville housing and provided the circumstance in which a new political agenda could be formulated, with housing as its focus."

One dispute involved demands by Mexican-Americans that mobile homes provided as emergency housing remain indefinitely because they were more habitable than the substandard housing that was damaged by the earthquake. It is an irony of US disaster policy that the federal government is empowered to address social needs in an emergency recovery context (albeit temporarily) that are often overlooked in the absence of a disaster. The post-disaster problems of California's poor foreign-born population may increase if the provisions of California

Proposition 187 (passed by voters in November 1994) are observed; these may prohibit certain kinds of disaster relief to illegal aliens.

The Federal Disaster Relief and Emergency Assistance Act (Stafford Act) requires that:

the State or local government [receiving disaster assistance] shall agree that the natural hazards in the areas in which the proceeds of the grants or loans are to be used shall be evaluated and appropriate actions shall be taken to mitigate such hazards, including safe land-use and construction practices. (Sec. 409)

Accordingly, FEMA and the California Office of Emergency Services convened a State/Federal Hazard Mitigation Survey Team two weeks after the earthquake. The team involved 60 participants, including 32 federal, state, regional, and private utility agencies with earthquake-preparedness responsibilities. Their ensuing report (FEMA, 1990) offered 57 recommendations under six headings: (1) hazard identification and monitoring; (2) land-use planning and regulation; (3) repair and reconstruction; (4) response planning; (5) insurance; and (6) funding.

The next Bay Area earthquake

Once the dust settled, it was clear that the Loma Prieta earthquake was not the "Big One" that has been long expected in the Bay Area. A new working group was assembled by the US Geological Survey to assess the risk of further large earthquakes in the region. This sought to: (1) reassess earlier seismic data in light of Loma Prieta; (2) evaluate the effects of Loma Prieta on other faults in the Bay Area; and (3) incorporate new data on slip rate and recurrence interval on the Hayward and Rodgers Creek faults. The report of this group estimated *a 67 per cent probability of one or more large earthquakes in the Bay Area within the next 30 years*. This is a substantial increase over the 50 per cent probability estimated by the 1988 Working Group (US Geological Survey Working Group, 1990).

The Association of Bay Area Governments (ABAG, 1992) has compiled estimates of housing units likely to be rendered uninhabitable by a 7.0 magnitude quake (table 10.6). These figures reflect only initial damage inflicted by ground shaking and do not include possible further damage owing to fires, landslides, and other causes. Nevertheless, for almost all of the scenarios that were projected, the numbers of dwellings rendered uninhabitable are greater than those affected by the Loma Prieta quake. For one scenario (i.e. an earthquake that affects both north and south portions of the Hayward Fault) up to 29 times as many houses might be involved. The costs to insurers, government, and home-owners

Table 10.6 Estimated total dwelling units rendered uninhabitable by selected earthquake scenarios

Fault involved	No. of uninhabitable units
San Andreas – Peninsula	8,154
Hayward – north and south	57,045
Hayward – north segment only	39,789
Hayward – south segment only	27,777
Healdsburg – Rodgers Creek	10,516
Loma Prieta (as per model)	3,323
Loma Prieta 1989 (actual)	3,957

Source: Adapted from ABAG (1992), table 1.

would all be vastly greater than those incurred in 1989. Post-disaster problems of emergency response, sheltering, and rehousing are likely to be overwhelming.

Ominously, the next major Bay Area earthquake is most likely to occur on the Hayward Fault in the heavily urbanized East Bay. The Working Group estimated a risk of nearly 30 per cent that an earthquake of at least magnitude 7.0 will occur on the southern or northern segments of the Hayward Fault within the next 30 years. The last earthquakes on those segments occurred in 1836 and 1868, respectively. The Hayward Fault is "probably the most built-on fault in the world" (BAREPP, n.d.). About 1.2 million people live within the epicentral region[5] of a potential 7.0 tremor on this fault, 10 times the population within a comparable distance of Loma Prieta (US Geological Survey Working Group, 1990, p. 4). *Ground shaking in the East Bay would be at least 12 times larger than during the Loma Prieta quake* (USGS, n.d.). It is thus likely that the East Bay will be a vortex of disaster during the next major Bay Area earthquake. Loma Prieta, which stretched emergency response to the limits, may in retrospect be viewed as merely a dress rehearsal.

The East Bay Hills firestorm of 1991

Two years after the Loma Prieta quake, the Bay Area mega-city was again struck by a major natural disaster. This time the area affected was localized and the victim population relatively wealthy. The East Bay firestorm of 20 October 1991 reminded disaster planners worldwide that natural hazards may be vastly aggravated by unwise building patterns and neglect of commonsense mitigation.

The cities of Oakland and Berkeley both contain prized residential sites on steep west-facing slopes of the East Bay Hills. The combination of bay views, accessibility to urban pursuits, and semi-wild surroundings has long attracted the wealthy. During the 1920s and 1930s, this district became interlaced with narrow winding roads lined with thousands of medium-sized homes on parcels whose average area was only 0.04 ha (5,000 sq. ft). Even by California standards it is an exceptionally hazardous district. The Hayward Fault marks the break of slope at the base of the hills and many public facilities lie directly over it, including the football stadium of the University of California at Berkeley. Landslides, flash-floods, and debris flows are additional geological hazards.

As in many other parts of California, the risk of fire along the boundary between wildlands and urban areas is a function of several related factors:

- **Vegetation**: The native oak and grassland ecosystem has been transformed by the introduction of decorative but highly flammable landscape trees, especially eucalyptus and Monterey pine. Unless removed, dead biomass from eucalyptus trees accumulates as an excellent source of fuel.
- **Fire suppression**: Fuel builds up because natural fires are controlled to protect local residences.
- **Drought**: Drought desiccates vegetation (in 1991 California was just emerging from a five-year drought).
- **Building materials**: Most homes are constructed of wood and are roofed with "shakes" (cedar shingles) in the characteristic style of California woodland architecture.
- **Narrow roads**: Narrow, steep, winding service roads impede the mobility of incoming police, fire, or medical vehicles and obstruct evacuation by car. During fires, abandoned cars easily block such roads.
- **Wind**: Santa Ana winds occur from time to time during the late summer and autumn in California. These hot dry winds are caused by high-pressure systems that form over the semi-arid interior Great Basin. They blow from the east and north-east towards the ocean at speeds that are amplified in narrow mountain passes and steep terrain.
- **Water supply**: It is difficult to maintain adequate water pressure in hilly areas. Failures of high-elevation reservoirs are common.
- **Arson**: It is widely believed that arson plays a role in igniting California wildland fires near urban areas.
- **Floods and debris flows**: Once a fire occurs, the burned areas may experience subsequent flash-floods, mudslides, and debris flows triggered by heavy precipitation on devegetated slopes.

The Oakland–Berkeley hills and California as a whole have a long history of urban/wildland fire disasters, and these have increased markedly in the 1990s (table 10.7). Since 1930, 14 fires have occurred in the Oakland Hills, including seven that erupted in the same place (Wildcat Canyon) where the 1991 firestorm began (Topping, 1992, p. 5). Today,

Table 10.7 Major California urban/wildland fires, 1923–1993

Year	County	Acres burned	Structures burned
1923	Alameda	130	584
1961	Mariposa	41,200	106
1961	Los Angeles	6,090	484
1964	Napa-Sonoma	71,601	174
1964	Santa Barbara	61,000	94
1967	Riverside	48,639	61
1970	San Bernardino	53,100	54
1977	Santa Barbara	804	234
1978	Sonoma	11,504	64
1980	San Bernardino	5,482	65
1980	San Bernardino	41,472	355
1981	Napa	22,000	69
1982	Ventura/L.A.	57,000	65
1986	San Diego	200	64
1988	Nevada	33,500	312
1988	Shasta	7,800	58
1990	Santa Barbara	4,900	641
1991	Alameda (Oakland)	1,600	3,349
1993	Southern Calif.	272,000	1,200

Source: Adapted from California Governor's Office of Emergency Services (1992), Annex A.

about 7 million people are estimated to live in regions of California that are susceptible to urban/wildland fires (FEMA, 1992).

Concern about these kinds of fires is both high and widespread in California. After a 1923 fire in Berkeley (Alameda County), a special commission offered more than 100 recommendations for mitigating future losses; most of these were ignored (Blakely, 1992). In 1980, following a series of southern California fires, a Governor's Task Force on Chaparral Fire and Flood Risk Management made recommendations affecting three areas of responsibility: (1) risk reduction, land use, and conservation standards; (2) vegetation management and fire-hazard reduction; and (3) fire command system improvements (Blue Ribbon Fire Prevention Committee, 1982, pp. 64–68).

Planning for natural disasters in mega-cities must be conducted at geographic scales that are commensurate with the scope of different hazards. Wildfires threaten many adjacent communities in the East Bay Hills but not the entire Bay Area. In 1982, a Blue Ribbon Fire Prevention Committee for the East Bay Hills was formed under the East Bay Regional Park District, which manages several large tracts of public open space in the Hills. The District served as coordinator for a comprehensive review of fire hazard in the entire Hills region. They issued a report

(Blue Ribbon Fire Prevention Committee, 1982) that focused primarily on vegetation management and the need to extend existing "fuel breaks" along the entire 41.1 km length of the urban/wildland boundary. It was also proposed that a new special assessment district be created to manage vegetation throughout the area at risk. Unfortunately, actions in support of these recommendations were not accomplished before the 1991 firestorm.[6]

The firestorm

The firestorm was the third most costly urban fire disaster in US history after the San Francisco earthquake of 1906 and the Chicago fire of 1871. Although the area burned (about 647 ha) was modest compared with other California fires, it was by far the most destructive urban/wildland fire in terms of damage to buildings. Twenty-five people died and over 150 were injured. In the space of nine hours, the fire destroyed or damaged 2,621 homes and 758 apartment and condominium units. About 5,100 people, many of them elderly, were left homeless. Damage amounted to at least US$1.5 billion (National Fire Protection Association, n.d.).

The fire erupted under textbook conditions. Daytime temperatures hovered around 30 °C, relative humidity was 17 per cent, and hillside vegetation was bone dry. Dead plant material littered the ground and trees overhung many homes, despite the long drought and warnings of fire danger. Hot, dry Santa Ana winds blew from the east on the morning of the conflagration. The State Department of Forestry had issued a "Red Flag" warning of potential fire hazard, but few residents took notice.

The blaze began on Saturday, 19 October as a small brush fire "of suspicious origin" high up Wildcat Canyon near the ridge line (fig. 10.4). It was apparently extinguished that day, but embers were re-ignited next day by Santa Ana winds blowing at medium strength. Flames rapidly spread downhill, leaping from house to house. Fire personnel who returned to the scene were overwhelmed. About 790 houses burned in the first hour. Turbulent winds generated by the fire formed a fire-storm that spewed burning material in all directions. Flames crossed an eight-lane highway (Route 24) and continued to consume homes and vegetation further downslope. Public orders to evacuate were difficult to communicate in the absence of sirens.

The presence of smoke was the primary warning to Hills residents. Many found it impossible to drive down obstructed roads. Cars by the hundreds were abandoned and destroyed as individuals literally ran for their lives. Few of the victims could do more than save themselves and

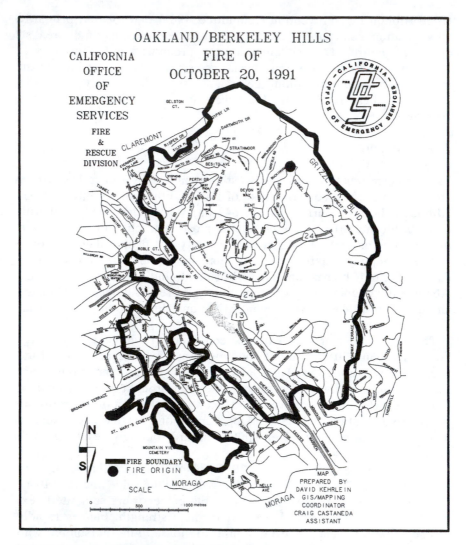

Fig. 10.4 Area affected by the Oakland firestorm, 20 October 1991 (Source: FEMA, 1992)

whatever they could carry. Sixty years of building on the hills had created a hazard that no one could individually undo. It was a classic "tragedy of the commons"; all were swept up in the common peril and personal consequences depended upon the fluke of the winds, not individual actions. Emergency response was massive but ineffectual. The city fire departments of Oakland and Berkeley were depleted by funding cuts, and were poorly trained and equipped for fighting a fast-moving urban/wildland fire. Both cities issued urgent requests for assistance from other cities and from federal and state agencies, pursuant to longstanding mutual-aid agreements.[7] Despite acts of great individual courage, fire-fighters were overwhelmed. Communication was hampered by overloaded radio and telephone circuits. Smoke and traffic congestion obstructed the mobility of the hundreds of emergency response vehicles and personnel. Even the Oakland hydrants were incompatible with the hose couplings of other municipal fire departments. That problem was however overshadowed by a loss of water pressure. Failure of the water distribution system stemmed from three causes: (1) all but 1 of 11 East Bay Municipal Utility District (EBMUD) reservoirs serving the area ran dry because of outflow through ruptured mains, melted pipes, and open faucets; (2) electrical power failed as transmission facilities burned thus disabling water pumping stations; and (3) EBMUD lacked emergency generators to operate the disabled pumps.[8] As in the 1906 San Francisco fire, failure of water pressure left hundreds of fire-fighters as helpless observers.

The conflagration was intensely hot and fast moving. Temperatures at the centre of the fire exceeded $1,000\,°C$, hot enough to boil asphalt (National Fire Protection Association, n.d., p. 12). Firestorm conditions began to develop within 15 minutes of the first building ignition. A thermal inversion formed at approximately 1,000 m; this trapped heat from the fire and intensified the pre-ignition heating of vegetation and structures. Flames consumed hundreds of structures simultaneously. Some houses appear to have exploded from within, perhaps due to the intense heating of interior spaces by picture windows that faced the flames (a form of greenhouse effect).[9] Propane and other flammable substances also contributed to the loss of many homes. Once a home was ignited, there was little possibility of saving it. All that remained next day were concrete foundations, chimneys, and charred tree trunks (photo 10.1). Fire victims lost personal mementos, pets, art and music collections, and the contents of computer disks – as well as their homes. Many such losses were not covered by fire insurance.

The flames reached within a few hundred metres of the opulent and historic Claremont Hotel. If the winds had continued, the fire could have consumed much of Berkeley, including the University of California campus. However, at about 7:00 p.m. on Sunday evening, the wind shifted and declined, nudging flames back over areas already burned. This allowed fire-fighters to contain and ultimately control the fire.

The recovery process: Private sector

Much support was provided by relatives, friends, and neighbours who welcomed victims as temporary house guests and helped them deal with

Photo 10.1 Concrete foundations were all that remained of hundreds of homes burned in the Oakland firestorm

the emotional trauma by listening to their tales of horror and their fears about the future (Adler, 1992[10]). Residents of neighbouring unburned hill areas in Oakland and Berkeley emerged with feelings of relief mixed with anxiety about future fires and perhaps guilt that they were spared.

It is usual for US public officials to vow to rebuild after disaster. The Mayor of Oakland followed this tradition. The day after the disaster he stated that the Hills community would be rebuilt. This was not simply a statement of solicitude for the victims but a clear-sighted recognition that the burned area provided some of Oakland's highest real estate tax revenues and that rebuilding would provide incomes to local merchants and contractors (Edward Blakely, personal communication, 12 August 1993). The alternative of leaving the burned area unbuilt and turning it over to an existing system of hillside regional parks was not seriously considered. A combination of high public acquisition costs and lost tax revenues provided powerful counter-arguments. Nevertheless, the failure to assess the economic and environmental impacts of not rebuilding represents a major shortcoming of the post-disaster recovery process.[11]

Despite official reticence to explore this alternative, private owners and public officials alike shared a pervasive determination to understand the causes of the disaster and, to the extent feasible, to rebuild more safely in light of risks from fires and earthquakes. Both professional and journalistic post-disaster assessments were produced in abundance. Seldom has a natural disaster been so thoroughly and competently examined from a multitude of disciplinary and personal perspectives. One of the earliest of these was a special series by the *Oakland Tribune* newspaper entitled "Firestorm: Bitter Lessons." Another journalistic outgrowth of the disaster was the appearance of a bi-monthly regional newspaper, the *East Bay Phoenix Journal*, published in Oakland. This began as a voice on behalf of fire victims who had to cope with issues of governmental policy, finance, and insurance as well as mounting a search for personal support services.

Unlike its role after the Loma Prieta earthquake, private insurance was a dominant force in financing rebuilding of the East Bay Hills. One year after the disaster, 3,954 claims amounting to US$1.4 billion (*c.* US$350,000 per household) had been filed with 49 insurance companies. Many of these were disputed, leading to frustration and often to lawsuits initiated by claimants. However, as settlements were reached on most claims, the fire victims received infusions of money. About two-thirds of the owners had purchased guaranteed replacement-cost insurance coverage. They were also reimbursed for costs of temporary housing and they enjoyed several kinds of relief from federal and state taxes. The average net worth of families who lost homes was expected to rise from 10 to 25 per cent compared with pre-firestorm levels (*San Francisco Examiner*, 31 October 1993, pp. E-1 and E-5).

Reconstruction of the Hills produced jobs in the building industry at a time when California was mired in recession. During 1993 some 11,000 new construction jobs were created. By 12 October 1993, building permits had been issued for about half the homesites burned in Oakland (1,368 out of 2,777) and for over half those burned in Berkeley (39 out of 69). Homes completed in these communities numbered 434 and 20, respectively (*East Bay Phoenix Journal*, 18 October 1993, p. 16). This represented over US$1 billion already invested in rebuilding, with several years of construction still ahead.

The recovery process: Public sector

Governmental institutions – city, county, special district, state, and federal – assisted the recovery process in many ways. These included emergency response and rescue, infrastructure repair, grants and tax subsidies to the victims, administration of rebuilding activities, and research on causes of the disaster. Research was especially important as a basis for reducing future vulnerability.

Because the disaster occurred in the heart of a highly developed megacity, there was no shortage of diverse expertise to address post-disaster

Table 10.8 Selected post-disaster reports following East Bay Hills firestorm of October 1991

- Federal Emergency Management Agency, *Hazard Mitigation Report for the East Bay Fire in the Oakland–Berkeley Hills* (Federal Disaster Declaration 919-DR-CA, 1992).
- Governor's Office of Emergency Services, *The East Bay Hills Fire: A Multi-Agency Review of the October 1991 Fire in the Oakland/Berkeley Hills.* Report prepared by the East Bay Hills Fire Operations Review Group (27 February 1992).
- Task Force on Emergency Preparedness and Community Restoration, "Final Report" (3 February 1992). The Task Force was established jointly by the mayors of Berkeley and Oakland and included committees on (1) Emergency Preparedness; (2) Communications; (3) Forestry and Vegetation; (4) Infrastructure and Development; and (5) Planning, Zoning, and Design.
- Ad Hoc Council on Replanting Needs (ACORN) (a project of the University–Oakland Metropolitan Forum), "Guide to Landscaping for Fire Safety" – collection of materials issued March 1992; second edition issued October 1992.
- Kenneth Topping, AICP, "Report on Oakland Hills Fire Prevention and Suppression Benefit Assessment District." (15 October 1992). Prepared as consultant to city of Oakland.
- Hills Emergency Forum, *1993 Report.* Ongoing intergovernmental forum consisting of two major components: (1) East Bay Fire Chiefs Consortium and (2) Vegetation Management Consortium.
- East Bay Municipal Utility District (EBMUD), *Firescape: Landscaping to Reduce Fire Hazard.* Public information document (1992).
- East Bay Municipal Utility District, *Preliminary Report on EBMUD's Response to the East Bay Hills Firestorm* (24 March 1993).
  ——— *Phase I Final Report* (15 July 1992)
  ——— *Phase II Final Report* (21 January 1993)
- National Fire Protection Association, *The Oakland–Berkeley Hills Fire, October 20, 1991* (1992).

issues. This included researchers and technical experts from the University of California at Berkeley, whose campus was situated directly in the path of the fire. Disaster response networks had already been primed by the Loma Prieta earthquake two years earlier and the City of Oakland had become particularly skilled in the art of disaster recovery. Post-disaster reports spanned many disciplines, including urban forestry, geology, urban planning and landscape design, fire-fighting, mental health, and water supply. Table 10.8 lists the most comprehensive of these.

The federal government officially entered the firestorm recovery process under a "major disaster declaration" for Alameda County issued by President Bush on 22 October 1991. For the second time in two years, FEMA established a disaster centre in Oakland and activated the many forms of aid available under the Federal Disaster Relief and Emergency Assistance Act. But the issues that confronted FEMA this time were dif-

ferent. Proportionally less public infrastructure and more private property were involved in the East Bay Hills than in communities affected by
the Loma Prieta earthquake. Most victims were affluent and well insured.
The fiscal burden thus fell more heavily on private insurers and less
on US taxpayers. The number of structures destroyed or badly damaged
was roughly comparable in both disasters: 3,957 "red-tagged" after the
earthquake and about 3,300 burned in the firestorm. But, by late 1994,
the total cost of federal assistance after the firestorm was US$513 million;
this is about 10 per cent of the funds so far committed to recovery from
the earthquake (FEMA, unpublished data). About four-fifths of federal
funding after the fire was spent on reimbursing state and local governments for their response and recovery costs

Part of the federal aid was used to fund operation of a Community
Restoration Development Center (CRDC), which was set up as a field
office of the Oakland city government. Situated in a former supermarket
close to the fire scene, the CRDC provided a "one-stop" facility where
fire victims could meet with representatives of city, state, and federal
agencies, utilities, insurance companies, and medical and social services
(photo 10.2). Planning staff at the CRDC reviewed proposed rebuilding
plans. Permit approval was expedited by a computerized geographical

Photo 10.2 Community Restoration Development Center, funded by FEMA,
provided convenient "one-stop" access to city officials, utility representatives, and
other services to victims

database that provided information about ownership, site, vegetation, utilities, and other characteristics of every parcel in the burned area (FEMA, 1992, p. 9).

Following established practice, FEMA convened an inter-agency hazard-mitigation team to identify ways that the risks of similar disasters might be reduced in the future, both in the East Bay Hills and in other areas subject to urban/wildland fires (e.g. southern California). The team made recommendations about vegetation management, roofing, road width, fire codes, flood hazards, and other matters (FEMA, 1992).[12] The California Governor's Office of Emergency Services also conducted a review of emergency response, particularly communications and fire-fighting. This agency urged, among other matters, that communities adopt compatible hydrant connectors so that fire equipment from any community could function in any other community, and it proposed installing a permanent auxiliary pumping system that would be capable of supplying water to all hill areas (California Governor's Office of Emergency Services, 1992). In California after urban/wildland fires there is an immediate upsurge of public anxiety about the risk of landslides, debris flows, and erosion on steep slopes during the next winter rainy season (McPhee, 1989). The City of Oakland spent US$5 million of federal money on aerial reseeding of the burned area and placed hundreds of straw-bale check dams in gullies and channels. But researchers found that such measures may have been unnecessary and were possibly even counter-productive (Booker et al., 1993).

The most comprehensive policy review was conducted by a regional Task Force on Emergency Preparedness and Community Restoration, created by the mayors of Oakland and Berkeley a week after the fire. This group was given eight weeks to prepare a detailed report that would guide recovery of the burned area and would protect other parts of the East Bay Hills. The Task Force included city and county officials, university faculty, utilities, local businesses, and representatives of the fire victims. It was organized in five work groups that were charged with different responsibilities:

1. *Emergency Preparedness* – to evaluate the emergency response systems of the cities, including fire stations, communications, early warning systems, and the use of volunteers.
2. *Communications* – to improve media and public information during a disaster.
3. *Forestry and Vegetation* – to develop landscape standards and other guidelines that reduce the potential for fire to spread.
4. *Infrastructure and Development* – to increase the quality of street systems and utilities (water, electricity, and telephone) for the fire areas.

Table 10.9 Task Force on Emergency Preparedness and Community Restoration: Key proposals and status, 1994

Proposal	Status
• Improve local emergency services through a major bond issue for new radio communications and fire protection equipment.	Adopted
• Increase the training and use of community volunteers to identify fire hazards and fight small fires.	Adopted
• Require all homes in designated fire hazard zones to remove wood roofs, use prescribed landscaping methods and materials, and incorporate fire control methods.	In progress
• Establish a fire protection district for the fire danger zone from funds allocated for fire prevention activities.	Adopted
• Move public utilities underground in the fire area, and improve water supply to the area.	In progress / In progress
• Revise building codes to require the use of sprinklers, and/or more fire-resistant building materials.	Not adopted / Adopted
• Improve street access for firefighting equipment, and relocate some fire equipment to fire-prone areas.	Not adopted / Adopted
• Limit the density of homes in fire-prone areas.	Not adopted

Source: Task Force on Emergency Preparedness (1992) as summarized in Blakely (1992). Status based on estimates by the author.

5. *Planning, Zoning, and Design* – to develop new zoning and design standards for the area to be rebuilt as well as for all potential fire areas.

Key policy proposals and outcomes are displayed in table 10.9.

Collaboration between city and university was especially fruitful with respect to forestry and landscaping issues.[13] A grant from the San Francisco Foundation funded the Ad Hoc Council on Replanting Needs (ACORN) – a city–university project that was designed to educate property owners about fire-resistant landscaping methods. The East Bay Municipal Utility District reinforced this effort with a public information document entitled *Firescape: Landscaping to Reduce Fire Hazard*, which was distributed as a public service to all of EBMUD's water customers in the Hills.

Oakland voters translated public education into political action during 1993 by approving an Oakland Hills Fire Prevention and Suppression Benefit Assessment District. The district encompasses all hill areas of Oakland, whether or not they were burned in the firestorm. All of the approximately 20,000 parcels of real estate in the district will be assessed at US$75 per year initially, which could rise to a maximum of US$300 per year. This revenue will be used primarily for reducing the amount of hazardous fuels on public and private lands; among other means, they will continue to employ herds of goats to graze public lands within the district. The programme will fund inspections of private lands to ascertain violations of the Uniform Fire Code and will generally strengthen the capability of Oakland to combat urban/wildland fires (Topping, 1992).[14]

The Task Force Committee on Planning, Zoning, and Design addressed the controversial issues of how the Hills should be rebuilt. It was firmly committed to rapid rebuilding and gave little public consideration to the alternative of buying burned areas for public open space (see endnote 11) or of limiting density by means of non-compensatory zoning regulations. Despite the very small size of house lots in the burned area, the Committee urged approval for reconstruction of homes within the same "footprint" (i.e. area covered by the previous building). It called for utilities to be placed underground and for the adoption of "innovative parking solutions" to avoid blockage of narrow roadways. (It did not, however, urge the widening of roads, presumably since that would further reduce the size of building lots and would be expensive.)

Fourteen months after the fire, the city of Oakland finally adopted a new zoning designation (S-14) for the burned area. It permitted enlargement of replacement structures by 10 per cent over previous building sizes and it exempted any plans already submitted, regardless of the size of structures proposed. Where it applies, the S-14 zone establishes a variety of rules concerning minimum lot setbacks and sideyards, heights of buildings, parking, and landscaping. These rules are specified in relation to specific slope and street frontage characteristics of different lots.

At the end of 1994 the physical results of this ordinance became visible on the hillsides of Oakland: hundreds of very large, free-standing homes of eclectic design, often sited within 3–5 m of each other on tiny lots (photo 10.3). Fire hazard has been reduced by using non-combustible roof materials, placing utilities underground, and limiting flammable vegetation. Off-street parking will be more abundant, but the roads are generally as narrow as before and larger homes may well produce more vehicles per household. Congestion of streets still remains as a public safety hazard. The proximity of homes, the lack of internal sprinklers capable of extinguishing common house fires, and the likelihood of street congestion all create a sense of unease about the potential consequences of an earthquake on the Hayward Fault during coming decades. Private home-owners may (incorrectly) assume that, if it were not safe to rebuild, the government would prevent them from doing so.

Conclusions

The Bay Area mega-city has been sorely tested by earthquake, wind, and fire since 1989, especially in the East Bay region. The Loma Prieta

Photo 10.3 The rebuilding process has produced eclectic variety in house styles, built in extreme proximity to one another

earthquake and the East Bay Hills firestorm each inflicted grievous – but not catastrophic – harm. Both disasters demonstrated the vulnerability of large mega-cities to lifeline failures – especially water, transportation, electrical, and medical facilities and their supporting systems. On the other hand, each displayed the advantages of possessing a well-organized "larger community," extending from the President of the United States to the professional staff of regional agencies, to whom private citizens and local city officials look for help in an emergency.

There is little cause for comfort in the fact that the Loma Prieta quake and the Oakland firestorm did not inflict heavier losses. As reported by FEMA, the emergency response capabilities of the entire federal, state, regional, and local governments were stretched to breaking point after the Loma Prieta quake. Differences in a few key variables could have produced a much worse outcome – if the epicentre had been closer to San Francisco, if the magnitude had been larger, if the wind had been blowing, if these events had occurred when workers were at their jobs and children were in school. Similarly, if the Santa Ana winds had continued to blow, the Oakland firestorm could have overwhelmed the Claremont Hotel, engulfed the University of California's Berkeley campus, and consumed many blocks of homes and businesses in Berkeley and Oakland.

The juxtaposition of two unrelated natural disasters in the same urban region within two years may be viewed as an unparalleled training exercise in disaster response and recovery. As suggested below, the implications for mega-cities elsewhere are salutory.

The importance of regional cooperation

Some American metropolitan areas, such as Denver, have established region-wide hazard-management agencies (Platt, 1986, 1987). Often, however, mega-cities are too large and individual cities are too small to serve as optimal geographic units for the purposes of disaster planning, disaster response, and disaster recovery. Mega-cities need to be divided into smaller sub-areas that are consistent with the functional requirements of particular services. Although the city of Oakland was a primary urban victim of both the 1989 earthquake and the 1991 firestorm, its leaders and citizens *could* not and *did* not face these trials in isolation. The "East Bay" provided a psychic and functional region that was drawn on to help Oakland recover.

Within mega-cities and their subregions, there is a high degree of interaction among institutions and individuals that have responsibilities for addressing common needs (e.g. government, corporations, non-profit organizations, and private citizens). Long before the Loma Prieta earthquake, California had evolved a strong tradition of mutual aid and intergovernmental cooperation. This is reflected in myriad intergovernmental agreements on public services, fire protection, and shared facilities. It also appears in strong and flexible institutions of regional governance. In the case of the East Bay, two key regional agencies are the East Bay Municipal Utility District and the East Bay Regional Park District. Other institutions crucial to the disaster recovery process have included the University of California at Berkeley, the Association of Bay Area Governments, Pacific Gas and Electric, and the Pacific Bell telephone company. Professional staff from these and other entities interact continuously on task forces, committees, and special projects such as the ACORN project of the University–Oakland Metropolitan Forum. Governmental, corporate, and philanthropic efforts are blended in an ongoing dialogue concerned with the welfare of the East Bay region.

Post-disaster hazard assessment

Both Bay Area disasters were followed by a deluge of post-disaster assessments drawn up under federal, state, regional, local, and private auspices. Some of these reports were prepared in accordance with Section 409 of the Federal Disaster Relief and Emergency Assistance Act. Section 409 evaluations are necessarily intergovernmental and interdisciplinary. Others focus on particular agency or corporate missions (e.g. urban forestry, soil stabilization, fire-fighting, medical care).

For post-disaster assessments to be useful, they must be: timely; scientifically accurate; prescriptive; and practical. But they must not be overly constrained by political considerations. Members of official post-disaster assessment teams should be charged to "tell it like it is" and make whatever recommendations seem desirable without fear of criticism for suggesting politically unpopular approaches.

In the case of the East Bay firestorm, the post-disaster assessments were professionally competent but politically cautious. The most prominent of the non-federal assessments, the Report of the Task Force on Emergency Preparedness and Community Restoration, articulated dozens of sound hazard-mitigation measures. But the ultimate mitigation measure was not mentioned; that is, the alternative of not rebuilding the Hills, or at least of reducing density and widening streets. Perhaps there was only a small chance that there would be significant public support for those alternatives but they should have received serious consideration (see endnote 11). Other communities have chosen to support public buy-out of disaster-affected communities (e.g. Rapid City, South Dakota; Times Beach, Missouri). The Mississippi Valley floods of 1993 prompted the federal government's purchase of several thousand damaged homes (Interagency Floodplain Management Review Committee, 1994; Myers and White, 1993).

Modest efforts to change land-use and building patterns in the Hills after the fire were ineffectual. Streets remain narrow and homes have become larger, albeit roofed with tile rather than shingle. The post-disaster assessments have led to improvements in public sector services, but little or nothing has been done by private owners. The possibility of an earthquake on the Hayward Fault was not a significant factor in deliberations by the Task Force on Emergency Preparedness and Community Restoration.

Lifelines

This chapter has emphasized lifeline vulnerability because it has been a central problem of past disasters in the Bay Area. It cannot be restated too often that there is no more significant issue for natural-hazard planning in mega-cities than lifeline reliability. In California, the seismic survivability of individual structures has been markedly improved as a result of changes made to building codes. Public education programmes have also provided individuals and households with information that is essential to ensuring better protection during a disaster. But, as clearly demonstrated by the Northridge earthquake of 17 January 1994, improvements in individual safety are in effect nullified by the increased vulnerability of public and private lifelines.

Twice found lacking, the East Bay Municipal Utility District is now vigorously seeking to improve the resilience of its water supply system against earthquake and fires. More difficult to upgrade, the Bay Area's ageing transportation arteries remain vulnerable to the kinds of failures that in 1989 led to the collapse of an interstate highway in Oakland and closed the San Francisco–Oakland Bay bridge. Where failure is likely, it must be anticipated and alternative means of coping must be devised. Before a disaster occurs, the public must be informed about appropriate actions to take if major highways and bridges are disabled.

The unexpected

Perhaps the most daunting challenge for disaster managers in mega-cities is to plan for the unexpected. The interconnectedness of modern urban regions triggers a cascading series of failures, not necessarily catastrophic individually, but cumulatively producing chaos and misery as well as economic hardship. Dealing with the unexpected requires trained professionals equipped with reliable technology for damage assessment, decision-making, and communication. Disaster response is both a professional skill and an art. Perhaps a disaster's only silver lining is that it can enhance the ability to prepare for the next one. Short of first-hand experience, the need to learn from disaster experience elsewhere is obvious. In that respect, the San Francisco Bay Area has much to teach its sister mega-cities.

Notes

1. Since 1950, the federal Bureau of the Census has designated "Metropolitan Statistical Areas" (MSAs) that encompass central cities and suburban areas. Periodically, MSA boundaries are revised (normally by adding more county units) or new MSAs are designated to reflect the outcome of continued urbanization. Where two or more MSAs are directly contiguous or functionally overlapping, the Census Bureau also designates "Consolidated Metropolitan Statistical Areas" (CMSAs) and refers to the individual MSAs within the CMSA as "PMSAs" (Primary MSAs).
2. The East Bay Municipal Utility District faces a similar threat. Aqueducts from Pardee Reservoir just north of Hetch Hetchy also cross the Hayward Fault.
3. The Chicago fire of 1871 cost US$1.96 billion in insured losses, and the Oakland firestorm of 1991 required US$1.75 billion in insurance payments (*San Francisco Chronicle*, 31 October 1993, p. B1).
4. Before 1990, earthquake insurance in California was available from private insurance companies for premiums on the order of US$2–4 per thousand dollars of coverage,

subject to a large deductible of 10 per cent of insured value. Thus home-owners had to pay a premium of several hundred dollars a year and were still not covered against moderate damage that did not exceed the 10 per cent deductible. Fewer than half the state's residents carried earthquake insurance. In 1990, after Loma Prieta, the state legislature established a mandatory programme of minimum earthquake coverage (Palm and Hodgson, 1992, chap. 2).

5. The epicentral region is the area of most intense ground motion on bedrock sites and was considered by the Working Group to extend for a radius of 10 km from the location of an earthquake fault rupture.

6. In view of the intensity of the 1991 firestorm, it has been questioned whether fuel breaks would have impeded the spread of the conflagration. The 1982 Blue Ribbon Committee perhaps overemphasized vegetation management to the exclusion of other mitigation approaches, such as coordination of fire-fighting amongst Bay Area cities (conversation with Tony Acosta, Oakland Park Department, 26 October 1993).

7. In response, an armada of fire-fighting forces converged on the fire: 440 local engine companies, 6 air tankers, 16 helicopter units, 8 communications units, 2 management teams, and 88 fire chiefs (California Governor's Office of Emergency Services, 1992, p. 31).

8. The FEMA (1990) Loma Prieta report had urged installation of on-site emergency generators by water supply managers, but the district had not done so (*Oakland Tribune*, 1991, p. 56).

9. Personal communication, Tony Acosta, Oakland Park District, 26 October 1993. News reports on the firestorms in southern California in October 1993 mentioned a similar phenomenon.

10. This collection of individual accounts of the disaster through photographs, essays, and poetry was published privately. Proceeds from the book are being donated to the Alta Bates Burn Center.

11. Acquisition might have been an economically feasible alternative to reconstruction. For example, the average price of lots sold in the burned area during October 1993 was approximately US$130,000 (*East Bay Phoenix Journal*, 18 October 1993, p. 17). If the approximately 3,000 vacant building lots could have been acquired at this price, the cost would have been around US$390 million less administrative costs and lost tax revenues. As compared with the public costs of reconstruction and restoring public streets and utilities to the area (in the order of US$100 million), this might not have been excessive. By this means the emotional costs of those who rebuilt under difficult conditions would have been eased. And the risk of future conflagrations would have been greatly reduced. However, Oakland relies on the Hills as its highest tax-generating district and the city would have required payments in lieu of taxes from an external source before it would be persuaded to adopt this approach.

12. The report reiterated the vegetation-management recommendations of the 1982 Blue Ribbon Fire Prevention Committee, whose executive summary was appended to the FEMA report.

13. The work group on this topic, chaired by Edward J. Blakely, Professor of City and Regional Planning at the University of California, Berkeley, was assisted by the University–Oakland Metropolitan Forum, a non-profit organization also headed by Blakely. Professor Blakely further served as Policy Advisor to the Mayor of Oakland and played a key role in coordinating local responses to both the earthquake and the firestorm.

14. The task of organizing regional responses to issues of vegetation management and fire response beyond the city of Oakland has been assumed by the Hills Emergency Forum,

a coalition of Oakland, Berkeley, East Bay Municipal Utility District, the East Bay Regional Park District, and the Lawrence Berkeley Laboratory of the University of California. All of these groups own and manage open lands in the Hills.

REFERENCES

ABAG (Association of Bay Area Governments). 1992. *Estimates of uninhabitable dwelling units in future earthquakes affecting the San Francisco Bay region.* Oakland, CA: ABAG.

Adler, P., ed. 1992. *Fire in the Hills: A collective remembrance.* Berkeley: Patricia Adler.

Bancroft, H. H. 1888. *History of California.* San Francisco: The History Company.

BAREPP (Bay Area Regional Earthquake Preparedness Project). n.d. *Living on the Fault.* Public information brochure. Oakland, CA: BAREPP.

Blakely, E. J. 1992. "This city is not for burning." *Natural Hazards Observer* 16(6): 1–3.

Blue Ribbon Fire Prevention Committee (for the East Bay Hills Urban-Wildland Interface Zone). 1982. "Report." Oakland, CA: East Bay Regional Park District, mimeo.

Bolin, R. 1993. "The Loma Prieta earthquake: An overview." In R. Bolin (ed.), *The Loma Prieta earthquake: Studies of short-term impact.* Program on Environment and Behavior Monograph No. 50. Boulder: University of Colorado Institute of Behavioral Science.

Bolin, R. and L. M. Stanford. 1993. "Emergency sheltering and housing of earthquake victims: The case of Santa Cruz County." In P. A. Bolton (ed.), *The Loma Prieta, California, earthquake of October 17, 1989 – Public response.* US Geological Survey Professional Paper 1553-B. Washington, D.C.: US Government Printing Office.

Booker, F. A., W. E. Dietrich, and L. M. Collins. 1993. "Runoff and erosion after the Oakland firestorm." *California Geology*, November–December: 159–173.

Bronson, W. 1986. *The earth shook, the sky burned.* San Francisco: Chronicle Books.

California Governor's Office of Emergency Services, East Bay Hills Fire Operations Review Group. 1992. *The East Bay Hills fire: A multi-agency review of the October 1991 fire in the Oakland-Berkeley Hills.* Sacramento: OES.

Dana, R. H. 1964. *Two years before the mast,* 2 vols. Boston: Little, Brown.

FEMA (Federal Emergency Management Agency). 1990. *Hazard mitigation opportunities for California.* State–Federal Hazard Mitigation Survey Team Report for Disaster Declaration FEMA 845-DR-CA. San Francisco: FEMA.

———— 1991. *The Loma Prieta earthquake: Emergency response and stabilization study.* Contract report by the National Fire Protection Association. Washington, D.C.: US Government Printing Office.

———— 1992. "Hazard mitigation report for the East Bay fire in the Oakland–Berkeley Hills." Prepared pursuant to Disaster Declaration FEMA 919-DR-CA. San Francisco: FEMA, mimeo.

French, S. 1990. "A preliminary assessment of damage to urban infrastructure." In R. Bolin (ed.), *The Loma Prieta earthquake: Studies of short-term impact.* Program on Environment and Behavior Monograph No. 50. Boulder: University of Colorado Institute of Behavioral Science.

Interagency Floodplain Management Review Committee. 1994. *Sharing the challenge: Floodplain management in the 21st century.* Washington, D.C.: US Government Printing Office.

Keeler, C. A. 1903. *San Francisco and thereabout.* San Francisco: The California Promotion Committee.

McPhee, J. 1989. "Los Angeles against the mountains." In *The Control of Nature.* New York: Farrar, Straus, Giroux.

Myers, M. F. and G. F. White. 1993. "The challenge of the Mississippi River flood." *Environment* 35(10): 6–9, 25–35.

Nash, R. 1982. *Wilderness and the American mind.* New Haven, CT: Yale University Press.

National Fire Protection Association. n.d. *The Oakland/Berkeley Hills fire – October 20, 1991.* Quincy, MA: NFPA.

Oakland Tribune. 1991. "Blocks could have been saved with more water." *Bitter Lessons* (special series by the *Oakland Tribune* published the week after the firestorm and reprinted as a special section).

O'Brien, P. W. and D. S. Mileti. 1993. "Citizen participation in emergency response." In P. A. Bolton (ed.), *The Loma Prieta, California, earthquake of October 17, 1989 – Public response.* US Geological Survey Professional Paper 1553-B. Washington, D.C.: US Government Printing Office.

Palm, R. and M. E. Hodgson. 1992. *After a California earthquake: Attitude and behavior change.* Geography Research Paper 233. Chicago: University of Chicago Press.

Plafker, G. and J. P. Galloway, eds. 1989. *Lessons learned from the Loma Prieta, California, earthquake of October 17, 1989.* US Geological Survey Circular 1045. Washington, D.C.: US Government Printing Office.

Platt, R. H. 1986. "Metropolitan flood loss reduction through regional special districts." *Journal of the American Planning Association* 52(3): 467–479.

——— ed. 1987. *Regional management of metropolitan floodplains.* Program on Environment and Behavior Monograph No. 45. Boulder: University of Colorado Institute of Behavioral Science.

——— 1991. *Land use control: Geography, law, and public policy.* Englewood Cliffs, NJ: Prentice-Hall.

Task Force on Emergency Preparedness and Community Restoration. 1992. "Final report," mimeo, 3 February.

Thomas, G. and M. M. Witts. 1971. *The San Francisco earthquake.* New York: Stein & Day.

Topping, K. C. 1992. "Oakland Hills Fire Prevention and Suppression Benefit Assessment District Report." Pasadena, CA: mimeo, 15 October.

USGS (US Geological Survey). n.d. *The next big earthquake in the Bay Area may come sooner than you think.* Menlo Park, CA: USGS.

US Geological Survey Working Group on California Earthquake Probabilities. 1988. *Probability of large earthquakes occurring on the San Andreas Fault.* Open File Rpt. 88–398. Washington, D.C.: US Government Printing Office.

———— 1990. *Probabilities of large earthquakes in the San Francisco Bay region, California.* USGS Circular 1053. Washington, D.C.: US Government Printing Office.

Vance, J. E., Jr. 1964. *Geography and urban evolution in the San Francisco Bay Area.* Berkeley: Institute of Governmental Studies, University of California.

Yanev, P. I. 1991. *Peace of mind in earthquake country.* San Francisco: Chronicle Books.

11

There are worse things than earthquakes: Hazard vulnerability and mitigation capacity in Greater Los Angeles

Ben Wisner

Editor's introduction

A former mayor of Newark, New Jersey, once claimed that, wherever American cities were going, his was going to get there before the rest. In the annals of urban hazard he might well have been speaking of Los Angeles. For here one can observe the interplay of many hazard variables that exist separately in cities elsewhere. Greater Los Angeles experiences a range of severe physical risks that is scarcely equalled by any other mega-city. It has deployed some of the world's most sophisticated large-scale science and engineering responses to environmental risk, as well as legions of professional and volunteer emergency personnel, and it continually throws up nascent self-help grass-roots hazard-management initiatives. The city region also exhibits sharp contrasts between first world affluence and third world squalor, exacerbated by deeply differentiating demographic, cultural, and political forces that are themselves a reaction to South California's version of the global post-industrial crisis. Moreover, the contradictory role of southern California's entertainment industry as dream factory in a setting of gritty reality is not to be discounted. It is a community conceit that almost anything might happen in Los Angeles, though experience suggests that this has not yet included many of the hazard adjustments that are so urgently needed.

375

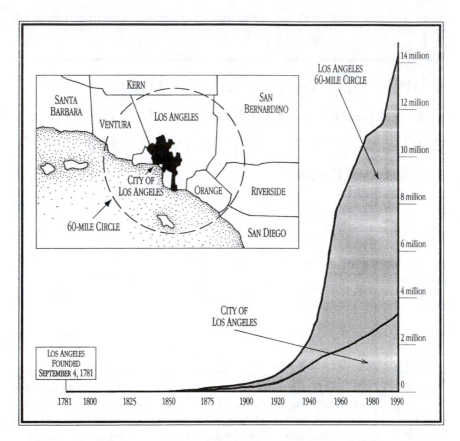

Fig. 11.1 Growth of population of Greater Los Angeles, 1781–1990 (Source: adapted from *The Economist*, 13 October 1990, p. 15)

Introduction

Los Angeles has grown enormously during the past 60–70 years, including a hefty addition of 2 million people in the 1980s (fig. 11.1). It currently includes over 15 million inhabitants, who occupy an area of 34,000 sq. miles (88,000 km²) spread over five counties in southern California. At its centre lies the city of Los Angeles (3.5 million in 1990), surrounded by the county of Los Angeles (combined population of 8.3 million in 1986), and both contained by the Los Angeles–Riverside–Orange County CSMA (Consolidated Metropolitan Statistical Area) (15,047,772 in 1992), herein referred to as Greater Los Angeles.

Greater Los Angeles is a mega-city composed of many communities and striking human contrasts. It includes 160 municipalities and accounts for roughly one-half of California's entire population. With a gross output of US$336 billion (1989), the economy of Greater Los Angeles is bigger than that of many nations. A large proportion of the residents of the city of Los Angeles (LA) are poor, young, unemployed, and non-White, whereas many suburban municipalities are islands of relative affluence and privilege. Twenty-six "edge cities" are located in the five-county region within 60 miles (96 km) of central Los Angeles (Garreau, 1991, p. 283). There, on the periphery of the mega-city, some 1.5 million jobs were created in the 1980s.[1] This is where the White working class migrated to during the 1970s and early 1980s, leaving the less mobile African-American working class and newly arrived Hispanic and Asian populations behind in LA city.

The population of this mega-city is highly stratified by race and income. Although Whites are numerically dominant, the African-American population has grown steadily since the Second World War. More recent immigrants have come from many parts of Asia and the Pacific, while existing Spanish speakers have been greatly augmented by arrivals from throughout the hemisphere, especially Mexico. The city of Los Angeles ranks second only to Mexico City in numbers of Mexican-born inhabitants. Population shifts in LA have been particularly dramatic. In 1990, the city's racial and ethnic composition was: White, 36 per cent; Hispanic, 40 per cent; African-American, 14 per cent; Asia and the Pacific, 10 per cent (Bureau of the Census, 1992, p. 36). A decade earlier there were considerably fewer members of minority groups (Bureau of the Census, 1988, p. 610): White, 61 per cent; Hispanic, 28 per cent; African-American, 17 per cent; Asia and the Pacific, 7 per cent. The process of uneven development, together with extremely rapid regional growth, has combined to increase risks of social, technological, and natural hazards in Greater Los Angeles. Vulnerability to the range of hazards is not evenly distributed across the population, but is differentiated by ethnicity and class. The ability of governments and local communities to mitigate hazards is severely curtailed by social, political, economic, and spatial factors.

"Top–down" hazard-reduction initiatives are limited by extreme fragmentation, decentralization, and complexity of governmental structure (federal, state, counties, municipalities, semi-autonomous bodies) and by the overwhelming influence of real estate and financial interests that have promoted weakly regulated growth for nearly a century. Economic recession and restructuring have eroded the tax base of government agencies thereby affecting their ability to plan, train, monitor, regulate, inspect, and deliver services. In addition, the city of Los Angeles has never undergone the democratic reforms that affected large cities of the eastern United States in the past few decades. Nearly 5 million people are represented by only 15 councillors, who control enormous precincts. By contrast, New York City has enlarged its council and other cities have created intermediate levels of government to encourage citizen participation (Davis, 1993b, p. 52). "Bottom–up" mitigation – action by communities on a local scale – is limited by social anomie and possessive individualism. Collective action is hindered by narrow self-interest (e.g. the NIMBY "not in my backyard" phenomenon), by the complex mosaic of ethnic and economic groups, and by inter-ethnic competition among the poor. There is also a long history of élite fear, which has fuelled the disruption of grass-roots organizations. This began with early "open shop" anti-union activities (Clark, 1983, pp. 281–282) and has since included: clandestine police surveillance of citizen groups (Davis, 1990, p. 298); deportation of Chicano labour organizers (Clark, 1983, p. 302); and police attacks on peaceful demonstrations of African-Americans (Martin, 1993), Hispanics (Acosta, 1973), and workers in the "Justice for Janitors" movement of the 1990s (Olney, 1993).

Recent economic and political changes have hindered public initiatives in a variety of ways. Economic restructuring has cost Greater Los Angeles at least 200,000 traditional, high-wage industrial jobs since 1990 (Davis, 1993b, p. 46). This economic shock has reduced the availability of tax revenues that might have financed "top–down" hazard-mitigation measures such as enforcement of air pollution regulations and may have decreased the capacity of the poor to cooperate with each other in community-based activities such as hazard assessment and mitigation. In the eyes of one observer, there has been a "virtual meltdown of local government" involving the abandonment of various social and health services in the wake of a recent 25 per cent "doomsday" reduction of LA County's budget (Davis, 1993b, p. 44).

Economic duress contributed strongly to a widely publicized rebellion of the poor in 1992 – the so-called "Los Angeles riots" (Williams, 1993). This is the latest in a long history of rebellions by LA minorities (African-American, Hispanic, and others) against perceived police brutality and injustice (e.g. the 1943 Zoot Suit riots; the 1965 Watts rebellion). Mexican-Americans and many other minorities are angry that they are excluded from all but minimum-wage employment, that they live in deteriorating conditions, and that they receive little respect from other Angelenos.[2] Poverty in the inner cities of the region is increasing at a rapid rate compared with the suburban ring (see table 11.1), and it is concentrated among members of racial and ethnic minority groups (Massey and Denton, 1989; Massey and Eggers, 1990).

Growth and uneven development in a geomorphically unstable setting (Cooke, 1984) have also created one of the most hazardous urban environments on the planet. Los Angeles expanded rapidly outward from its original site on the coastal lowland, filling in the flat-floored basins of San

Table 11.1 Economic welfare: Central cities compared with suburban ring

	Central cities' average as % of suburban cities' average	
	1980	1990
Households in poverty	360	650
% of suburban income	90	59

Source: Adapted from Davis (1993a), p. 14.

Fernando, San Gabriel, and San Bernadino (Wittow, 1979, p. 347). The extensive urbanized valleys and low alluvial plains are connected by a dense network of freeways and minor roads that carry more than 10 million motor vehicles.[3] A mountain chain (San Gabriel mountains, San Bernadino Mountain, and Mt. San Jacinto) acts as a topographic and climatic barrier that seals off this urban zone from the interior (Wittow, 1979, p. 347), and traps its infamous photochemical smog. Nearer the ocean, the Santa Monica mountains are somewhat cooler and less smoggy, but prone to wildfires, landslides, and beach erosion. This is a semi-arid area and extensive urbanization[4] has been possible only because massive amounts of water (and electricity) have been imported from distant sources. During the 1990s, the mega-city's insatiable appetite for water collided with regional-scale drought combined with competing demands on these sources, and a major crisis ensued. Ironically, flooding is also a significant LA hazard that has its origins in winter storms, steep topography, and large areas of impermeable surface. Storm discharges that exceed water treatment capacities, illegal disposal of hazardous chemicals, and a growing population have combined to create a serious coastal pollution problem.

The hazards notwithstanding, Los Angeles has grown rapidly for 70 years, and only now may be reaching the limits of what is socially and environmentally supportable. This chapter will discuss these twin limits to LA's growth. Challenges to social stability are deeply rooted and have been sharpened by economic restructuring and globalization during the 1980s. Different ethnic groups that exhibit extremes of wealth and poverty have been starkly juxtaposed here since early in the history of southern California. Polarization has increased in the 1980s and 1990s. Continuation of the rapid pace of urban growth, especially when combined with predicted global environmental changes including sealevel rise and atmospheric warming, can only increase the burden of many environmental hazards in Greater Los Angeles: seismicity, drought, wildfire, landslides, beach erosion, collapse of sewage-treatment facilities, coastal pollution, epidemic waterborne disease, air pollution, and increased cardio-respiratory diseases of various kinds. Given the complex pattern of government and the distribution of winners and losers in recent economic restructuring, it is unlikely that a general commitment to managed growth will emerge in the urban region as a whole, despite the fact that its physical infrastructure and social stability are so obviously interconnected. White suburbanites who fled the San Fernando valley into "edge cities" of Ventura County and who currently continue to settle in Orange, San Bernadino, and Riverside counties, hold tightly to the myth of separateness from LA and its social and physical problems. Ironically, it is the rapid growth of these edge cities, combined with an anti-urban political shift in state and federal legislative bodies, that has condemned the region's older core (the city of Los Angeles) to chronic conflict. Indeed, metaphors of warfare are often used to characterize the mega-city: struggles between urban governments and gangs. These deserve the label "low-intensity warfare," whereas a kind of "trench warfare" sets citizens against nature.[5]

The mega-city

The rise of Los Angeles to prominence can be partly explained by classic urban location theories of Christaller, Losch, Berry (1977), and others. These assume that systems of settlement are arranged in population hierarchies and spatial patterns according to principles of efficient travel by producers and consumers in search of markets and other services. In Los Angeles, comparative advantages over neighbouring communities were secured first when the railroad from Chicago reached it in 1885, later when a deepwater port was constructed in the vicinity and connected to the city by a narrow corridor of annexed territory (1890–1909), and again when the Panama Canal was completed in 1914 (Goetz, 1985, p. 312). But other factors, which do not normally feature in urban location theories, were also highly important to the mega-city's development. These include advertising, government investment, the availability of a non-unionized labour force, speculative building of transportation systems, LA's monopoly of the southern California water-supply system, and inter-racial tensions.

Sustained and flamboyant promotion of the local environment had a great deal to do with the city's early growth. Between the late 1880s and 1930 a quarter of all new arrivals came for health reasons or accompanied a relative who was drawn by the sunny, dry climate (Clark, 1983, p. 271, citing Baur). The hyperbolic promotion of LA's attractions (e.g. sea, mountains, health, and modernity) during this period foreshadowed the city's current postmodern identity crisis: confusion of image with reality, dream become nightmare. Twentieth-century growth spurts took place after petroleum was discovered in 1920 and triggered the creation of LA's refining and chemical industries. The Second World War brought a Kaiser steel mill to Fontana and sparked the building of aerospace plants, which eventually grew into a mighty empire of defence contractors during the Cold War. Between 1945 and 1965 the US government made US$100 billion worth of defence purchases in southern California (Clark, 1983, p. 285). Once established as a growth centre, Greater Los Angeles attracted a diversified range of additional industries, including those that migrated there in search of a non-union labour force (Bernard and Rice, 1983).

The transformation of Greater Los Angeles' natural environment and urban form owed much to electric trolley lines and automobile freeways. These were pushed out onto agricultural land ahead of urban builders, at first ensuring that the city of Los Angeles would grow, and then facilitating the creation of competing edge-cities that eventually superseded it. Much has been written about the exodus of LA's working-class and middle-class urban dwellers to suburbs and satellite towns (Berry, 1976; Steiner, 1981). The lure of suburban jobs and White fear of inner cities both contributed to this process. The edge cities contained new industries and 4,000 shopping malls. The working poor and the "underclass" were left behind, together with dying central business districts and old industrial facilities. To some extent, outward migration has been balanced by gentrification in the city of Los Angeles. This has produced a complex mosaic of affluent residential and commercial enclaves adjacent to poorer areas. These complement the historical refuges of privileged groups in the Santa Monica mountains, in the Hollywood Hills, and beside the beaches.

Very little investment, demolition, building, or rehabilitation takes place in any city without a complex interplay between government agencies and major economic interests (Dear and Scott, 1981; Harloe and Lebas, 1981; Harvey, 1973; Tabb and Sawers, 1978). Los Angeles would probably not exist as anything but a small agricultural service centre if federal and state governments had not provided financing for massive infrastructural investments: an elaborate system that brings in water from hundreds of miles away (see fig. 11.2), sewage-treatment plants, the deep-water port at San Pedro, the Boulder dam, the San Onofre nuclear power station, and freeways.

Moreover, decisions about the internal structural development of the city have not so much followed the dictates of economic rationality as stemmed from self-serving alliances between politicians and investors. For example, the powerful Community Redevelopment Authority (CRA) is composed of mayoral appointees not subject to public recall, most of

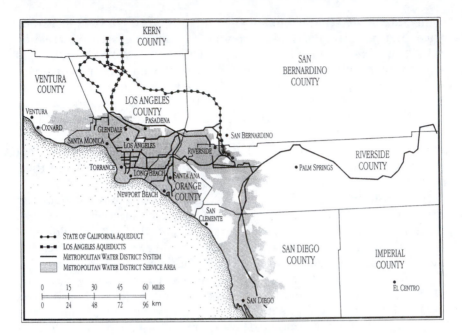

Fig. 11.2 Primary distribution system for water (Source: adapted from Nelson and Clark, 1976, p. 241)

whom come from the business community. They have power to designate redevelopment areas, to commandeer land by means of compulsory purchase, and to raise money. Tax dollars generated by CRA developments return to the CRA, not to the city of Los Angeles. The state of California insists that 20 per cent of this revenue be spent for low-income housing. On balance, however, poor and minority communities have fared badly because of massive displacement owing to a series of 18 redevelopment projects (Sudjic, 1993, pp. 100–102). These and other minority neighbourhoods – especially in Hispanic areas – have also been fragmented by newly constructed freeways.

The major influxes of African-American and Hispanic populations took place as an indirect effect of federal government policy. African-Americans from the US South came during the Second World War to find work in the rapidly expanding defence, construction, and transportation industries (Goetz, 1985, p. 309).[6] Many Mexicans eventually settled in Los Angeles as the indirect effect of the *bracero* programme launched by the US government, which recruited Mexican farm labourers for California's important agri-business sector to replace poor White farm labour that had shifted to higher-paying jobs in war-related industry. Civil wars in Central America and uneven development have continued

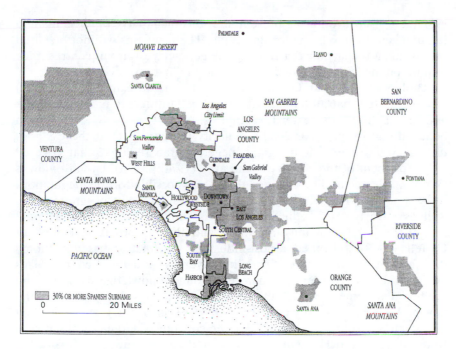

Fig. 11.3 Hispanic population in Greater Los Angeles (Source: adapted from Davis, 1990, pp. ix and x)

to drive Spanish-speaking people over the border, legally or illegally, so that now both the city and the county of Los Angeles are rapidly approaching a Hispanic majority (see fig. 11.3).

Los Angeles as a challenge to urban theorists

It has already been noted that explanations of the growth of Los Angeles do not readily follow from conventional urban location theory. This mega-city has also stimulated a variety of speculations about other aspects of urbanization. For example, Los Angeles casts new light on the long-running debate about links between the size and shape of cities and the quality of life of their inhabitants. Two opposing camps – the "garden city" decentralists, beginning with Ebenezer Howard (1898) and Camillo Sitte (1889), and the "modern" centralists, who hark back to Le Corbusier (1918) (Sudjic, 1993, pp. 10–32) – have taken up different positions in the debate. In LA, however, excessive urban size and ill-adapted urban form are not the main factors that undermine living conditions. Greater Los Angeles *is* clearly too big, as evidenced by the crippling pressure it places on distant water sources and the damage it has done to the regional environment. But, even at its present size, this mega-city *could* be made a fulfilling and livable home for its people. The problems will not be cured by curtailing the size or rearranging the geometry of urban centres; they require political and economic reforms in the first instance.

Scholars such as Jean Gottmann (1961, 1983) and Emrys Jones (1990) have suggested that all human settlement will ultimately take on the size and general form of Greater Los Angeles. As early as 1968, Kevin Lynch believed that by the year 2000 a majority (60 per cent) of the US population would live in four "giant megalopolitan regions" that covered only 7 per cent of the country's land area (1971, p. 523). Lynch's view is an optimistic one and he believed that it was possible for the megalopolis to be a new kind of city that would lift its residents out of ignorance as well as poverty. His "educative city" is similar to the "transactional metropolis" of Gottmann and Jones and the "informational city" of Castells (1989). All reflect a post-industrial, information-based economy that facilitates and requires flexibility, diversity, and rapid change. This new world urban order is organized around knowledge-workers who are based at home or in satellite offices in edge cities. Jones believed the new patterns were already apparent in places such as the great *obijo-toshi* that stretches from Tokyo to Osaka, the corridor of large cities that runs from Glasgow and Hamburg to Lyons and Rome, and the "Boswash," "Chipitts," and "Sansan" urban regions of North America (1990, pp. 133–139). Doxiades extrapolated the process to produce a global "ecumenopolis" – a single world city (Jones, 1990, p. 135).

Mega-city theorists have not delved deeply into the economic and policy factors that encourage cities to "evolve" in this direction. But they are aware of the gap that separates their idealized formulations from reality, especially the existence of serious urban environmental problems and equally serious problems of race and class (Jones, 1990, pp. 163–212). Nevertheless, I believe that the situation in Greater Los Angeles is more lethal and urgent than most theorists have realized. With high school drop-out rates approaching 80 per cent in South Central LA, the "educative" city takes on an ironic meaning. During the 1980s, most of the institutions in what Davis (1990, p. 75) calls the "museum archipelago"[7] expanded and redecorated, at the same time that the physical infrastructure of the city, including its major sewage treatment plant, deteriorated. The Malibu firestorm of 1993 neatly encapsulates the paradox. Probably set by arsonists, it nearly devoured the Getty Museum. So much for the educative city!

This does not mean that the megalopolitan theorists are entirely wrong. Globalization is real (Angotti, 1993, pp. 22–25; Harvey, 1989; Palloix, 1977; Taylor and Thrift, 1986). However, the new world order is, in fact, also a new world *disorder* (Thrift, 1986). What the theorists have left out of their analyses is the historical fact of uneven development (Smith, 1984). Global integration has rapidly and brutally shifted industrial jobs from region to region, provoked indebtedness, and encouraged corruption and militarization. Within mega-cities such as Greater Los Angeles that are the fruit of globalization, uneven development has produced zones of dereliction and despair, as well as thousands of mini-malls.

Contemporary urban issues and trends

The factors that are most likely to bring major change to the Los Angeles region over the next two decades are economic, political, and environmental. They operate at scales from the local to the global. Only some of them are widely recognized and debated.

Successive legislatures and governors of California have tended to view the state's large cities as tax burdens. One reflection of this is Proposition 13, a measure that was passed by voters in 1978. It rolled back and capped property taxes, thereby precipitating a major fiscal crisis for many local governments, including, especially, those in San Francisco and Los Angeles. A further expression of unwillingness to aid cities occurred after the second Los Angeles uprising in 1992, when the state legislature not only refused to fund disaster relief for affected businesses and individuals but mandated further cuts in state funds for cities and counties. (This contrasted sharply with the compassion it showed to victims of the 1989 Loma Prieta earthquake.) The results were predictable; during the fiscal year 1984/85, LA City and County owed creditors a total of US$10.5 billion. The city was already spending 22 per cent of its budget on police, 12 per cent on sewerage and sanitation, and less than 1 per cent on health and hospitals (Bureau of the Census, 1988, pp. 54 and 615). This continuing crisis raises painful questions for Greater LA. If sufficient money cannot be found to fund human services, infrastructural maintenance, and routine governance, is it likely that there will be *new* expenditures for hazard mitigation and preparedness? Where will the political support for bond issues and other means of raising revenue come from when the political climate is anti-urban and specifically unsympathetic to the city of Los Angeles?

The financial plight of city government also receives little attention from the federal government. Nationally, there has been a marked tendency over the past three US presidencies to disinvest in the cities, especially Los Angeles. In his provocative and thoughtful essay, "Who Killed LA: The Verdict Is Given," Mike Davis (1993b) presents federal funding figures (as a percentage of cities' budgets) that speak for themselves. The federal contribution to LA's budget has shrunk to almost nothing at a time when the states have also been labouring under federally mandated requirements to meet national standards for a wide range of environmental pollutants (e.g. the Clean Air Act). It might be asked how the federal government could tolerate the social and physical decline of the country's second-largest population centre? Although only a minority of scholars and activists are presently putting the issue in these terms, over the next two decades it will clearly become the focus of a new debate about urban policy.

At the regional level, Greater Los Angeles is embroiled in continuing disputes with the state of Arizona, with much of northern California, with residents of the Owens Valley, and with the Hopi Indians of New Mexico about preferential access to various natural resources that sustain the mega-city. These include: water supplies from the Colorado River, the Sacramento River, and the Owens Valley;[8] and coal that provides Greater LA with cheap electricity. A new wave of environmental consciousness has brought with it a willingness to charge resource-users the "external" costs of land destruction, air and water pollution, and other impacts that are felt at the point of production. It is likely that such actions will significantly add to the economic burden of Los Angeles over the next 20 years.

Internationally, the most important political and economic factors that will affect Greater Los Angeles involve the mobility of capital, labour, technology, and commodities. Long before the North American Free Trade Agreement (NAFTA) was ratified by the US Congress, Greater LA was losing high-wage, unionized, industrial jobs to plants relocated in the Caribbean and Mexico, elsewhere in Central America, and in the Pacific Rim. In the three years from 1990 to 1993, 27 per cent of all US job losses took place in Greater LA. One-fifth of the industrial workforce (nearly 200,000 workers) have lost their jobs (Davis, 1993b, p. 46). Many of these (85,000) were unionized workers; meanwhile, non-union, minimum-wage jobs continue to account for an increasing proportion of the employment that remains. The percentage of male workers earning US$20,000 or less has tripled since 1973, and the ratio of low-wage service jobs to high-wage industrial jobs grew from 2 : 1 in 1970 to 5 : 1 in 1990 (ibid., p. 47). The burden of these job losses has fallen disproportionately on African-American and Hispanic workers (see map of location of plant closings superimposed on ethnicity, fig. 11.4).

Not only have jobs been relocated outside the country; they have also been shifted to the edge cities, where they are beyond the reach of inner-city residents. Moreover, imported cars, electronics, textiles, and other consumer goods have undermined local production. Foreign capital accounts for 90 per cent of the investment in the new high-rise developments of LA's historic downtown and elsewhere (Century City, San Gabriel Valley) (Davis, 1992, 1993a, 1993b). Although these investments

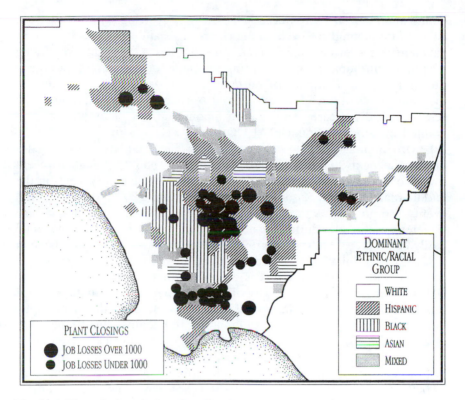

Fig. 11.4 Plant closings in Los Angeles County, 1978–1982 (Source: adapted from Oliver et al., 1993, p. 124)

have created some construction jobs, they have also displaced large numbers of poor and lower-middle-class residents, continuing the process of "urban removal" initiated by the freeway system.

Opinion is divided over whether NAFTA will decrease or increase Mexican immigration to Greater LA; the population of LA City is already over 40 per cent Spanish speaking. At the same time that foreign investments in edge-city business complexes are placing growing demands on infrastructure (roads, water, power, solid waste, etc.), the social service needs of the predominantly poor Hispanic population also have to be met. The crises in health care and education are so great that California's Governor has proposed denying citizenship to any child born of illegal immigrant parents and denying even emergency medical care to anyone who is an undocumented alien. In November 1994, the California State electorate passed a ballot initiative, known as Proposition 187, that would do precisely that. Its implementation is, however, now caught up in litigation. Even more recently (1996), federal legislation has been adopted that would cut back on the rights of legal immigrants to welfare payments and certain other public programmes. Given the continuing reshuffling of global economic arrangements, another major set of questions for Greater Los Angeles during the next two decades will concern the joint effects of NAFTA, the General Agreement on Tariffs and Trade (GATT), and increased cooperation among nations of the Asian Economic Development Council (AEDC).

Environmental changes will dramatically affect Greater Los Angeles during the next 20 years. Pollution is already bad (although a recent survey of 75 US cities misleadingly places the city of Los Angeles in the middle ranks – World Resources Institute, 1993, p. 220). Extreme natural events also pose acute threats. The probability of a major earthquake in Greater Los Angeles during the next 20 years is very high. Damage could easily be greater than from the 1994 Northridge earthquake, estimated at US$15 billion dollars (Place and Rodrigue, 1994, p. 1). Far more people could be killed and injured. It is estimated that there are as many as 50,000 older structures in the city of Los Angeles that pre-date the 1935 seismic building codes. Many of these are densely inhabited multiple-family dwellings. Large numbers of injuries and deaths must be expected in areas such as the Mexican *barrio* of East Los Angeles. Room occupancy is very high in other low-rise areas such as the unincorporated enclave of Lennox, near Los Angeles International Airport, where some 27,000 Mexican immigrant tourism and hotel workers "are sardine-packed into tiny bungalows and stucco hovels under an aircraft noise barrage so extreme that none of the community's schools can have windows" (Davis, 1993b, p. 47). Such areas will be very difficult to evacuate in the case of a large earthquake and the secondary risk of fire may be higher because of the dense, squalid living conditions.

One of Greater Los Angeles' more vibrant economic sectors also contributes to future hazard potential. Chemical production and transhipment plants have not suffered significant economic stress. The LA Standard Metropolitan Statistical Area (SMSA) has the second-highest number and concentration of chemical facilities in the United States. Many of these are located near earthquake faults. The probability of multiple chemical releases in the case of an earthquake is considered quite high (Showalter and Myers, 1992, p. 14, citing Tierney and Anderson, 1990).

Longer-term environmental changes are also likely to affect the region within the next 20 years. Global warming may exacerbate the region's water problems (Hanson, 1988, pp. 260–262). Associated sealevel rise could combine with sinking coasts to inundate considerable portions of the littoral plain. Worldwide, there is a trend toward more rapid mutation and diffusion of new viruses and other known diseases. Continued large-scale immigration (legal and illegal) could combine with an impoverished health-care system, over-burdened water and sewage-treatment facilities, and poor living standards to create conditions ripe for epidemics of disastrous proportions.

Many of the preceding concerns have made their way onto the agenda of planners and managers. The catastrophic collapse of the main sewage-treatment plant in 1987 eventually led to historic City Council limitations on real estate development in 1988 (Davis, 1990, pp. 200–203). The implications of urban expansion for traffic and air quality have been a matter of concern for many decades, and continue to be a central focus. More recently there has been a debate over the deterioration of the quality of LA County schools, the secession of San Fernando schools from the county system, the problems of homeless youth and of health care, and increasing unemployment.

Unfortunately, all of the above have been overshadowed by vociferous public debate about drugs, crime, and street gangs. Law enforcement already eats up more than 20 per cent of the LA City budget and 7 per cent of the LA County budget (Bureau of the Census, 1988, pp. 54 and 615). Greater Los Angeles has some of the largest and best-financed security forces in the United States. These are complemented by a vast army of private security guards in the employ of wealthier residents who live behind walls in the edge cities and in "fortress LA" (Davis, 1990, pp. 223–263). Yet the general consensus in southern California seems to be that more police and more money for the police are needed. Cities and counties squeezed to finance crumbling infrastructure and to combat third world-scale hunger and disease among their poorest residents are likely to be forced by White voters to spend the little they have on additional deadly force. The November 1994 state elections in California saw passage of a referendum requiring mandatory life imprisonment after conviction for a third serious crime. The cost of the additional prisons and life-long maintenance of prisoners will eat further into state budgets.

The 1992 uprising and media-inflamed images of deadly and irrational "gangbanging" youth have driven majority White voters of the edge cities furious with anxiety and resentment. Their feelings were expressed through representatives in the California legislature when they voted against post-disaster aid for South Central LA. It was feared that the victorious candidate for Mayor of LA might dismantle the multiracial, liberal coalition that had dominated LA government since 1973 because he believed that police power was the answer to curbing drug use, membership in gangs, and civil unrest – not community economic development, jobs, youth training, or similar schemes. However, this does not appear to have occurred and he was returned to power in a subsequent election (1997).

Responses to the multiple public crises of Greater Los Angeles

What approaches have been taken to address such enormous urban problems? What have managers, politicians, and citizens tried to do about the rapid rate of growth, about the widening disparity between rich and poor, about city and county finances, about vital natural and physical systems – air, water, sewage treatment, roads – stretched to the breaking point?

There has been a complex and often contradictory mosaic of approaches. Though many problems that are intimately interconnected have sometimes been treated as though they were separate, energy distribution, water, air quality, and traffic have received relatively well coordinated and thorough attention for many years. The South Coast Air Quality Management Zone, the Metropolitan Water District of Southern California, Southern California Edison, and CALTRANS (the state transportation agency) are large, professional bureaucracies. As good as these agencies may be, they can deal only with the effects of deeper-seated problems. Political decision-making and direction are required to manage the twin challenges that threaten the entire region: unregulated growth and a lack of jobs that pay a living wage. Here again the pattern is varied. Some California cities, such as Santa Barbara, have tried strictly to regulate new urban development. LA City itself attempted new growth regulations in the 1980s when water supply, sewage treatment, and traffic flow all threatened to collapse. On the whole, however, the history of the region shows the political dominance of banking and real estate interests, for whom "growth" is synonymous with profit.

In an earlier, simpler era, the problems faced by planners, boosters, and managers of Los Angeles were large but straightforward. The budding mega-city lacked a water supply, a deepwater port, and an energy supply for industry (Clark, 1983, p. 273). LA's early leaders set out to overcome these deficiencies in a direct manner, mixing bold modernist engineering with bare-knuckle politics. Boundaries were redrawn to give the city access to the harbour. A political battle was fought against the Southern Pacific Railway, which wanted federal support for a deepwater port at Santa Monica, where it monopolized the land (ibid., pp. 273–274). Equally impressive was the engineering required to move water some 223 miles (357 km) from the Owens Valley[9] (while simultaneously generating electricity as it descended over 2,000 feet [667 m]) – all in the face of often violent protests by the valley's ranchers (Kahrl, 1982).

By comparison, the problems facing the current generation of *Angelenos* are much more complex. Engineering solutions no longer suffice and the mega-city's politics have become multifaceted and subtle. In 1906, President Theodore Roosevelt could dismiss opponents of the Owens

Table 11.2 Hazards affecting the Los Angeles region

Geophysical/climatic	Biological	Technological	Social
Earthquake	HIV infection	Air pollution	Civil uprising
Tsunami	Drug overdose	Hazardous waste	Violent crime
Sinking coast	Childhood cancer	Chemical fire	Child poverty
Sealevel rise	Marine pollution	Dam failure	Homelessness
Hot winds	Heat exhaustion	Nuclear accident	
Drought	Waterborne disease		
Wildfire			
Gale-force ocean wind			
Flood			
Landslide			

Valley aqueduct as a "few settlers ... [who] ... must unfortunately be disregarded in view of the infinitely greater interest to be served by putting the water in Los Angeles" (quoted by Clark, 1983, p. 275). No present-day President, Governor, or Mayor would publicly dismiss the suffering of those who are now affected by Los Angeles' transition from Sun Belt industrial city to its new role in the greater Pacific Rim.

Hazards

People who live in the LA region are at risk from at least 25 diverse hazards that interact in complex ways (table 11.2). These hazards can be classified as geological, climatic, biological, technological, and social. This list of LA hazards is longer and more varied than those that have customarily been compiled by other writers (Cooke, 1984; Nelson and Clark, 1976; Wittow, 1979). Previous accounts have largely ignored a vast number of hazards that originate in the industrial production process and in the organization of the region's infrastructure. The importance of environmental hazards in Greater LA has grown in recent years. Two periods, over half a century apart, illustrate these changes: 1928–1938; 1983–1993. The first decade brackets most of the Great Depression; the second encompasses another major societal transformation – the end of the Cold War.

Two historical windows

1928–1938

Among the noteworthy disasters that marked the 1928–1938 period were one dam collapse, several river floods, a major earthquake, and the onset of the Great Depression. In 1928, between 400 and 450 people were killed and the town of Oxnard was evacuated when a 172 foot (52 m) dam failed, sending a wall of water, mud, and debris down the Santa Clara River to the sea between Oxnard and Ventura (Caughey and Caughey, 1976, p. 293; Outland, 1963). A year later, failure of the New York stock exchange ushered in the Great Depression and a period of privation, homelessness, and immigration to Los Angeles by those seeking employment or refuge. The city of Los Angeles sent police to turn back migrants at the state border. Large numbers of Mexican nationals were deported from Los Angeles in this period of social unrest (Caughey and Caughey, 1976, p. 294).

In 1933, the city of Long Beach and adjacent areas were severely damaged by a Richter scale 6.3 earthquake whose epicentre lay offshore on the Newport–Inglewood Fault. It struck in the afternoon after schools had emptied for the day. This helped to keep down casualties, but 120 died none the less and damage of more than US$50 million (1933 dollars) was sustained. Lessons learned from the physical damage to schools and other structures were translated into stronger building codes by 1935 (Caughey, 1976). Finally, extensive floods during 1934 killed 40 and inflicted US$15 million in damages, and again in 1938, when the tolls were 59 dead and US$62 million (Wittow, 1979, pp. 364–365).

1983–1993

The decade from 1983 to 1993 was marked by an upsurge of hazard events. Winter 1983 brought torrential rain and gale-force storms that severely eroded the Greater Los Angeles coast. Destructive earthquakes were a frequent problem (1983, 1987, 1992). In addition to the associated deaths and damage, they provided a drum beat that measured out the region's slow march toward ecological collapse and social upheaval as a result of unregulated development. Indeed, the 1987 Whittier quake actually cleared a path for developers anxious to buy up real estate in the old downtown for redevelopment (Davis, 1992). Los Angeles also felt the grip of a drought that lasted five years (1988–1993).[10] According to Hanson (1988, pp. 263–264), global warming is likely to add to existing water problems via negative feedback from increased water consumption in southern California and elsewhere, via contamination of Greater Los Angeles' own limited groundwater sources, and via reduced ability to recharge groundwater from surface waste water (as is the present practice, see Nelson and Clark, 1976, p. 241). Finally, in 1993 there occurred the most destructive and widespread brush fires ever throughout much of the region.

Failures of technological systems contributed other important hazards.

A massive breakdown of the Hyperion sewage-treatment plant in 1987 sent millions of gallons of waste into Santa Monica Bay. The same rapid urban growth that had produced the extra burden of sewage had also produced more runoff from impermeable surfaces that threatened to flood the flat urbanized basins and coastal plains (Davis, 1990, pp. 198–199). In 1987, studies revealed that as much as 40 per cent of LA's groundwater was contaminated by hazardous chemicals. This was extremely bad news for a city that is already dependent on limited heavily used water sources.

It was during this period that the deadline for compliance with federally mandated clean air standards passed. Modest but significant air-quality improvements that had been achieved in Greater Los Angeles through widespread use of catalytic converters and other measures were being cancelled out by the ever-growing number of cars on the road. Monitoring of industrial compliance with air emission standards was also declining.[11] Moreover, voters had rejected several bond issues that would have financed rapid transit systems during the 1980s. Both traffic and air quality were becoming progressively worse.[12]

The ongoing AIDS pandemic, which affected Greater Los Angeles as much as any Western mega-city, raised the spectre of other potentially catastrophic disease outbreaks. These included: a new cholera pandemic, which swept through much of South and Central America but stopped – for now at least – short of Greater Los Angeles; drug-resistant tuberculosis; and the return of preventable childhood diseases in large numbers of Los Angeles' children too poor to obtain routine immunizations.[13] In the years between 1990 and 1993, economic recession hit the region as hard as any destabilizing event since the Great Depression and seriously reduced the tax base for county and city revenues. Combined with reductions in state support of county and city budgets, this meant that there was even less money than usual available to deal with a concatenation of environmental crises. As in the previous period of severe economic disruption (the 1930s), Mexican immigrants were again the focus of intense nativist hostility, with the Governor of California recommending that "illegals" be denied emergency medical care.

For many, perhaps the most important hazard of the second period was a social one – the uprising of April 1992. In addition to the deaths, injuries, suffering, and property losses that occurred, at least 71 of the 671 buildings burned contained asbestos or other hazardous material that contaminated rubble, complicating its removal and disposal (*Environmental Reporter*, 15 May 1992, p. 411). Finally, although it did not take place during the decade under inspection, the Northridge earthquake (1994) provides a dramatic coda to the unfolding theme of recent increases in hazard.

Overview of hazards

An inspection of hazards during these two periods quickly reveals the contrasts. Matters have become much worse recently (see table 11.3). The tight coupling between overloaded natural systems and mismanaged technological ones is particularly evident. With the possible exception of flood control, little progress has been made toward preventing or miti-gating even the more straightforward natural hazards. Not only is the ratio of technological hazard events to natural hazard events increasing, the dividing line between these two subtypes is often blurred because of complex interactions between society and Nature in this urban region. For example, road-building operations at the head of an unstable slope inaugurated the massive landslide at Portuguese Bend; extraction of oil triggered subsidence beneath the Baldwin dam and its eventual failure during a storm; the breakdown of a sewage-treatment plant is inter-connected with the changing ecology of Santa Monica Bay; traffic con-gestion and airborne industrial waste are connected with the epidemiol-ogy of respiratory disease; and the release of asbestos from buildings damaged by earthquakes is connected with a suite of secondary health hazards (Gruenwald, 1988). In other words, *amalgams* of natural and technological hazard are increasing. For example, in 1987 the Whittier Narrows earthquake (Richter 5.9) displaced a 1 ton chlorine tank that was being filled, resulting in the release of a half-ton toxic cloud. Power failure caused by the earthquake disabled the company's siren, and mal-functioning telephones made it impossible to warn authorities (Showalter and Myers, 1992, p. 14).

There is probably no hazard or no mix of hazards that would surprise planners who work in southern California. But whether there is an insti-tutional capacity to plan and budget for the increasingly complex Nature–technology interactions that characterize newer hazard amalgams is an open question. Undoubtedly the answers have more to do with lead-times and financing arrangements than with issues of psychology and intellectual style. The history of Los Angeles is one of rapidly alternating economic booms and busts (Rieff, 1991, p. 56). Planners and politicians have often reacted slowly to changes that challenged existing planning procedures.

The expansion of LA County is a case in point. Although growth rates have been modest by national and international standards, the addition of new homes and businesses in Greater Los Angeles places a dis-proportionately heavy burden on the environment. Unlike the faster-growing squatter populations of Rio de Janeiro, Mexico City, or Lagos, *Angelenos* make disproportionately heavy demands on water, sewer, and electricity connections, on the absorptive capacities of roads and local airsheds, and on other natural processes.

Table 11.3 Selected hazards in Greater Los Angeles

Damaging earthquakes	Reservoir failures	Flooding	Large landslides	Brush fires	Ocean gales & coastal erosion
1872 Owens Valley (R8.3), 60 k	1928 San Francesquisto Dam, 400 k	1914 & 1916 $14 million	1956–7 Portuguese Bend, $10 million, 150 p	1932	1978 Malibu
1933 Long Beach (R6.7), 120 k, $50 million	1963 Baldwin Hills Dam	1934 40 k, $15 million	1969	1953	1983 El Niño storms
1940 Imperial Valley (R7.1), 9 k, $12 million	1971 Van Norman Dam (severely damaged)	1938 59 k, $62 million	1978 10 k	1961	Winter 1992
1952 Kern County (R7.7), 12 k, $63 million	1978 Big Tujunga Dam	1969 92 k, $62 million		1970 295 p	
1971 San Fernando Valley (R6.6), 65 k, $439 million		1978		1979	
1983 Coalinga (R6.5), $33 million		1983		1991	
1987 Whittier Narrows (R5.9)				1993 1,000 p	
1992 Yucca Valley (R7.4)					
1994 Northridge (R6.7), 57 k, $15 billion					

k = number killed
$ million = cost of losses
p = properties lost

What has possibly been most surprising to the planning professionals is the shifting attitudes and political behaviour of citizens who perceive threats to their way of life. The "taxpayers' revolt" of the late 1970s that marked the beginning of the current fiscal crisis did, indeed, inspire "wonder and amazement" at its scale and rapid development. (At one point 1.5 million signatures were gathered on behalf of Proposition 13.) In retrospect, the sudden rejection of central planning and finance was compounded by a desire on the part of the White middle class to isolate itself from the social and physical problems of the inner cities. Although this was a common trend in the United States at the time, it clearly made the tasks of coordination, conflict-resolution, and strategic planning all the more difficult, if not impossible. The pulling of neighbourhoods and communities back into themselves, behind protective walls, may give the illusion of security, but the vital task of hazard mitigation is made more difficult, and, in the long run, security for all declines.

There have been two major uprisings in Los Angeles. Soldiers of the National Guard (state militia) were on the streets in 1965. In 1992, there were federal troops freshly returned from the Gulf War. The post-1992 period has at times approximated a state of war. In South Central LA clashes among gangs and between the community and the police have resembled other urban wars in places such as Kabul, Sarajevo, Belfast, Gaza, Johannesburg, and Mogadishu. The death rate among young people in this part of the city, especially young African-American men, is very high. Inter-gang truces have held, but the background conditions of violence are sobering. Drive-by shootings, revenge killings, and the marketing of addictive and debilitating "crack" cocaine continue. In inner-city neighbourhoods of LA, police tactics have terrorized many blameless people: low-flying helicopters shine blindingly bright searchlights in the night, youth are subject to random interrogations, and there are shows of paramilitary force directed against alleged "crack houses." Some observers speak of the "'Ulsterization' of riot control in the inner-city" and "creeping 'Israelization' of residential security in affluent hillside and valley districts" (Davis, 1993b, p. 32).

Of course, the region is also subject to more conventional natural hazards. Apartment buildings and freeways were among the more common casualties of the 1994 Northridge earthquake. Variables such as design, age, level of maintenance, and specific earthquake characteristics make identification of particularly dangerous sections of freeway difficult, but not impossible.[14] Landslides often block portions of the Pacific Coast Highway as well as narrow canyon roads that serve the Santa Monica mountains, while brush fires and their smoke can temporarily block secondary roads in hill areas where alternative routes are more difficult to plot than on the dense, geometrical "grid" of the urban plains. Power outages and disruption of telephone service have accompanied earthquakes. Water, sewage, and solid waste removal services have seldom been disrupted for long, but future quakes could affect them more severely. If this happened, there would be a cascade of further public health hazards.

Police, fire-fighters, emergency medical technicians and emergency room staff, and California Emergency Services staff are all trained for immediate response to short-onset natural and technological disasters such as earthquakes, explosions, brush fires, and large-scale transportation accidents. But existing budgetary crises and an intense preoccupation with "policing" as an all-purpose answer to urban unrest may undermine disaster-readiness among these professionals. In any event, what they provide is crisis management and relief, not mitigation of hazard causes. Short-term relief is coordinated by the state in combination with the Federal Emergency Management Agency (FEMA). Relief for smaller, local events remains in the hands of city and county social services and voluntary agencies such as the Red Cross. On the whole, this means that there is more local flexibility, and less "red tape" is required to secure grants and loans (Popkin, 1990).

Longer-term recovery is another matter altogether. Here the private insurance sector is the most important actor. Many banks will not provide a mortgage if home-owners in floodplains fail to purchase available flood insurance. The purchase of earthquake insurance is less often required, and studies have found that many Californians decline to buy it (Palm, 1990; Palm and Hodgson, 1992). Apart from the fact that the poor are less likely to afford insurance of any kind, some low-income people, especially those belonging to ethnic minorities, may have more difficulty negotiating the recovery-assistance bureaucracy. There is evidence that they recover more slowly, if at all, from major disasters. Since the 1992 LA uprising, there had not been any major programme for rebuilding South Central and reviving its locally owned business despite presidential campaign promises, abortive attempts to legislate aid in both national and state legislatures, and an ambitious-sounding business-sponsored initiative called "Rebuild LA" (Davis, 1993a, pp. 4–6). Similarly, for a year or more after the Northridge earthquake, demolitions of damaged buildings were occurring much more slowly in Crenshaw, an African-American area that was among the more severely damaged, than in Northridge itself.[15]

This section has provided a general overview of hazards in Greater Los Angeles. Now let us turn to a more detailed accounting of trends in four different dimensions of hazard: risk (probability and physical characteristics of the hazard event); exposure (measure of the population at risk); vulnerability (probability of sustaining loss); and response (preparedness, mitigation, emergency action, short-term relief, and recovery).

Systematic analysis of hazards

Risk

With the possible exception of flooding, the risks of all hazards in Greater Los Angeles have increased over the past 60 years (Wittow, 1979, pp. 365–366). Earthquake risk has risen because of the length of time that has passed since a major release of tension in the "locked" portion of the San Andreas Fault near the mega-city. Other risks (e.g. drought, coastal erosion, landslide, brush fire) also have increased because of a combination of urban growth and probable long-term climatic change. During the same time-period, entirely new hazards have been created (e.g. potential nuclear accidents, HIV infection, violent crime, and drug overdose deaths of epidemic proportion).

Exposure

Exposure has universally increased owing to very rapid expansion of the urban area and increased population density. Development of exposed sites has become common (hillsides, beach-front locations, canyons, air traffic corridors, gang war zones, locations near chemical plants and other hazardous activities).

Vulnerability

Vulnerability has both decreased and increased in a complex pattern that is explained by technological innovations, governmental regulation and enforcement, changes in socio-economic well-being, and shifts of political power. For example, anti-seismic building codes that date from 1935 have reduced vulnerability to earthquakes in the built environment, but many low-income people, particularly Hispanics, live and work in pre-1935 structures. Buffer areas and other measures that are designed to reduce the risk of brush fires around homes have been offset by other trends such as the sheer magnitude of residential development in exposed locations, the use of fire-susceptible landscaping close to homes, and the size of resulting firestorms. A circum-Pacific tsunami warning system has probably reduced the vulnerability of persons to loss of life and injury near the beaches, but the amount of real estate exposed has increased.

Response

Response has also showed a complex pattern of change since the 1930s. Warnings have improved considerably for tsunami, brush fire, floods, and some types of landslides. Conversely, a reliable system of earthquake prediction and warning is still not available, despite hundreds of millions of dollars invested in research. Hazard-mapping has also become very detailed, including large-scale maps showing even small earthquake faults, landslide probabilities, and flood recurrence intervals. Communities have become active generators and users of hazard data, particularly those relating to pollution of air, soil, and water. As noted earlier, some aspects of emergency response (emergency medical care, search and rescue, fire-fighting) have become highly professionalized and effective, but the pattern of short-term relief and long-term recovery is mixed. Considerable controversy surrounds the distribution of relief and recovery funds and services by some government agencies and the private sector – especially insurance companies.

Changes in natural hazards

Earthquakes

Earthquake remains a major hazard in Greater Los Angeles. Those listed in table 11.3 cost a total of 323 lives. The most recent large earthquake was a Richter scale 6.7 event in January 1994 that affected many northern and central parts of the megacity: 57 people were killed, 16 of them in the spectacular collapse of an apartment building in Northridge; thousands were injured; several major highways were blocked, including the Santa Monica freeway (I-10), "I-5," which serves commuters in Antelope Valley, and Route 188 into Simi Valley and Grenada Hills. The total cost of this earthquake has been estimated at US$15 billion (Place and Rodrigue, 1994).

Earthquake *risk* is a controversial subject in Greater LA. The San Andreas Fault is believed to be "locked" for about 100 miles north and south of Palmdale. There has not been a major quake here since 1857. A large event has been expected for many years, but ground surface uplifts have not yet been followed by a major seismic shock (fig. 11.5). However, a series of smaller earthquakes may have reduced stress and along with it the risk of a "big one" (Jaume and Sykes, 1992).

Unfortunately, the theoretical models of earthquake causality upon which predictions are based are disputed (Robinson, 1993, pp. 57–65; Savage, 1991). Deformed strata at Pallet Creek (34 miles north-east of Los Angeles) indicate that earthquakes of Richter magnitudes 8 and above have occurred, on average, once every 160 years (Alexander, 1993, pp. 77–78). Past quake activity suggests that earthquakes greater than Richter 6 are likely to occur on various parts of the San Andreas Fault system by the year 2018. Probabilities range from 10 per cent for the Carrizo section to 30 per cent for the Anza section, 40 per cent for the

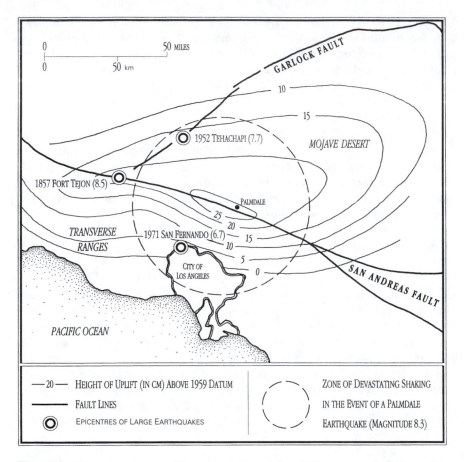

Fig. 11.5 Ground surface uplift around Palmdale, California, and the possible effects of a major earthquake on Los Angeles (Source: adapted from Wittow, 1979, p. 352)

Coachella Valley section, and greater than 90 per cent for the section around Parkfield (Alexander, 1993, p. 78, citing USGS, 1988; Ayer, 1992, p. 36, citing the same source). Unfortunately the recurrence interval is so variable for specific sites (e.g. 55–275 years for Pallet Creek), that it is impossible to "predict" actual events in any meaningful way (Robinson, 1993, p. 72). For example, it was predicted that there was a 95 per cent probability of a Richter 6+ magnitude earthquake at Parkfield by the end of 1992 (Bakun and Lindh, 1985; Robinson, 1993, p. 72). As this chapter goes to press (late 1997) scientists are still waiting to witness this event!

Nevertheless, available evidence suggests that the risk of a major earthquake on the San Andreas Fault is high. So too is the risk of earthquakes on other fault systems that criss-cross the Greater Los Angeles area. Some of these faults have generated very destructive quakes in the recent past (Long Beach in 1933, San Fernando Valley in 1971) (Caughey, 1976; Griggs and Gilchrist, 1983, pp. 43–47), including the most recent, Northridge (1994).

Earthquake *exposure* is generally high and growing, but it varies in response to factors of location, depth, and magnitude of the energy release. Since the area circumscribed by a circle of 20 mile radius would contain around 15 million people, no matter where it is placed in the mega-city, any major earthquake will affect a very large population. Locations near faults are more exposed than those some distance away, but topography, coastal location, or other locational factors could be associated with increased exposure to secondary landslides, liquefaction (Greene et al., 1991), fire, or release of hazardous material.

Vulnerability is often a function of income level; in Greater Los Angeles income is in turn highly correlated with ethnicity, race, and age. The homes and workplaces of poor, non-White populations are particularly susceptible to loss. Somewhere between 20,000 and 50,000 pre-1935 structures house low-income residents and are places of work for others, including undocumented foreign workers (Alexander, 1993, p. 337; Lee and Shepherd, 1984).[16] Also, low- and moderate-income groups inhabit areas downstream from major dams (e.g. Van Norman) that could collapse in a major earthquake (Ayer, 1992, p. 27; Griggs and Gilchrist, 1983, p. 45; Wittow, 1979, pp. 350–351). Finally, many low-income families also live along the Newport–Inglewood Fault, in the vicinity of the country's second-largest concentration of chemical industries (Showalter and Myers, 1992, p. 14), and near major oil refineries.

Response to earthquakes in southern California is limited by the lack of a precise short-term warning system. Automated systems for shutting down natural gas delivery systems, oil pipelines, etc. have been put in place, but the length of time that would be required to evacuate the area's very large populations – and the attendant health and logistical problems – make it unlikely that a reliable and practical warning system for humans can be developed in the foreseeable future. Fortunately, the emergency preparedness and response system is highly organized. The planning context for earthquake preparedness (and mitigation) is provided by the Joint Committee on Seismic Safety of the California Legislature and the California Master Plan. The latter incorporates estimates of the economic significance of different geological hazards, which combine data on property values, infrastructure, and lives at risk in each urban area (Alexander, 1993, p. 349). Every department of city and county government has an earthquake plan, and a coordinating body brings together all of these plans for different geographical units (e.g. the

City of Los Angeles Emergency Operations Organization). Schools have their own earthquake plans, as do many businesses and industries. Business and industry associations encourage earthquake preparedness (e.g. the Business and Industry Council on Emergency Planning and Preparedness and the Downtown Emergency Preparedness Action Council) (Mattingly and Melloff, 1992, p. 286). Considerable efforts have also been made to understand how laypeople respond to messages about preparedness and warnings (Dooley et al., 1992; Mulilis and Lippa, 1990; Palm and Hodgson, 1992; Turner et al., 1986). Some of these research projects have highlighted important cultural differences in disaster response (Johnson and Covello, 1987; Turner and Kiecolt, 1984; Vaughan and Nordenstam, 1991; Vaughan, 1993).

The earthquake-vulnerability of schools and other public buildings is comparatively small because most of these have slowly come into compliance with the strengthened building codes that were adopted in 1935 after the Long Beach earthquake. Building codes for housing are strict and were updated following the 1989 Loma Prieta earthquake (American Society of Civil Engineers, 1990). New construction is required to observe easements or "setbacks" away from faults and fault traces. In addition, ambitious programmes of hazard abatement have been carried out by bracing existing structures. Fortunately, until recently there has been relatively little high-rise construction in Greater Los Angeles and the light wood and stucco design of many residences has stood up well to past shaking. Newly constructed skyscrapers are designed to withstand earthquakes, but they have not yet been tested by a major event.

Other natural hazards

Over the past 60 years there have been: solid successes in responding to some natural events, such as the creation of a tsunami early warning system; some questionable mixtures of success and possibly long-term failure, such as the almost complete containment of the Los Angeles River within concrete flood-control works (figs. 11.6 and 11.7); and some dismal failures. The greatest failure is the management of brush fire (Pyne, 1982, pp. 405–423).

The climax vegetation of Greater LA's coastal hills is oily chaparral, which produces abundant litter that is very flammable. Vast areas are susceptible to destruction, especially when hot, dry Santa Ana winds blow seawards from the Mojave Desert. Fire risks have traditionally been reduced by periodic controlled burning of leaf litter. As more people invade the hills, smoke from controlled burns has also become a nuisance and residents have demanded an end to the practice. This has set the stage for ever-larger fires. Table 11.3 above lists the dates of some of the

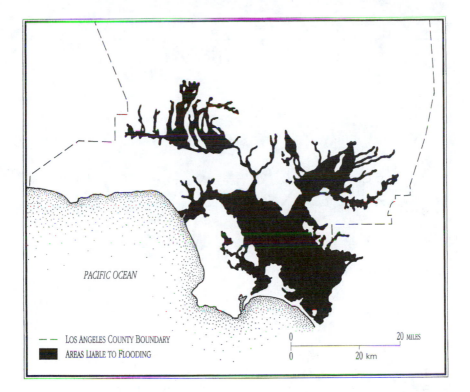

PACIFIC OCEAN

— — LOS ANGELES COUNTY BOUNDARY

▣ AREAS LIABLE TO FLOODING

0   20 MILES

0   20 km

Fig. 11.6 Flood hazards in Los Angeles County (Source: adapted from Wittow, 1979, p. 365)

largest burns. Figure 11.8 shows the extent of historical fires from 1919 to 1973.

Between 1950 and 1965, 242 fires larger than 100 acres (40 ha) consumed around 332,000 acres (135,000 ha) of vegetation and homes. The Bel Air fire of 1961 cost US$24 million, and the fires stretching from Newhall to Malibu in 1970 burned 180,000 acres (72,900 ha) and destroyed 295 homes (Nelson and Clark, 1976, p. 244). Since then the problems have accelerated. The fires of 1993 burned more than 200,000 acres (81,000 ha), destroyed over 1,000 homes, and caused losses of more than US$500 million, although only a few people were killed (*New York Times*, 31 October 1993). During the last days of October and early November 1993, 13 separate fires burned in six counties over a distance of 200 miles (322 km) from Ventura County to the Mexican border. Further urban encroachment into the hills combined with possible climate changes is likely to increase the hazard of future fires. There is, in short, no sign yet of success in mitigating the brush-fire hazard.

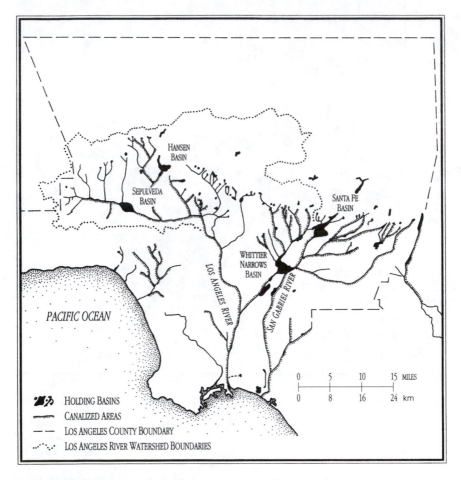

Fig. 11.7 Los Angeles County flood-control structures (Source: adapted from Nelson and Clark, 1976, p. 243)

Changes in technological hazards

The past 60 years have seen a wide array of new technological hazards. New risks, ever-greater exposure, more differentiated vulnerability, and more professionalized (and usually narrowly technocratic) responses have been the general pattern. Success in reducing technological risk through mitigation has been less than that achieved with natural hazards. Although air quality is better than it was in the 1950s, the air is still lethal.

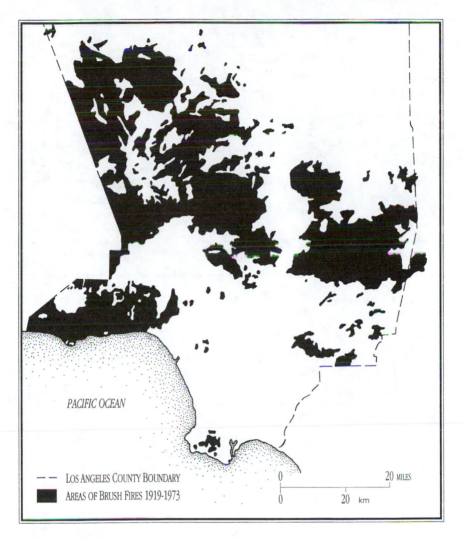

Fig. 11.8 Areas of brush fires in Los Angeles County, 1919–1973 (Source: adapted from Wittow, 1979, p. 362)

Air pollution

In Greater Los Angeles there is reason to question Goethe's time-honoured assertion that "city air makes people free"; here it is more likely to make them sick. This mega-city is the largest producer of several classes of air pollutants in the United States.[17] A combination of climatic and topographic factors exacerbates the underlying problem of vehicular

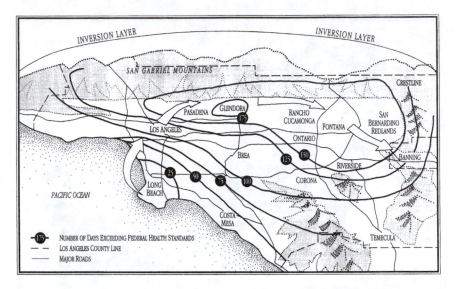

Fig. 11.9 Ozone smog in southern California (Source: adapted from Labor/ Community Watchdog, 1991, p. 7)

and industrial emissions. Pollutants from these and other sources remain trapped in the air over the mega-city (Labor/Community Watchdog, 1991, p. 6). Upward dispersal is stopped by a thermal inversion; lateral dispersal is blocked by mountain walls to the north and east. Photochemical smog forms as the stagnant stew is cooked by solar radiation (fig. 11.9). LA's smog is destructive to lung tissue. Even the young exhibit serious damage. Of youths between the ages of 15 and 25 who died from traumas in 1990, 80 per cent had "notable lung abnormalities" and 27 per cent had "severe lesions on their lungs" (Labor/Community Watchdog, 1991, p. 5, citing Sherwin, 1990).

The *risk* of damaging air pollution can be defined at two levels: (1) the probability of a day during which federal air-quality standards are not met (i.e. an air pollution "event"), or (2) the probability of a day when residents are advised to refrain from strenuous outdoor exercise, while infants and the elderly stay indoors (i.e. an air pollution "alert"). Risks become larger inland from the coast, with the worst air trapped in the San Gabriel valley (fig. 11.9). There are also diurnal variations: higher pollutant concentrations occur in the late afternoon and evening, lower ones at night. It has been estimated that air pollution in Greater LA has caused 1,600 premature deaths each year (Lents and Kelley, 1993, p. 38).[18] Although many of these are among older people and others who already suffered from respiratory and heart diseases, the gross human toll is greater than any inflicted by earthquakes, floods, or other natural disasters. Moreover, almost anyone at any location in Greater LA is at risk, although the trend has been downwards in recent years as a result of improved emission controls. Whether this will continue is problematic. The number of cars on the road in Greater LA doubled from 4 million in 1978 to 8 million in 1985, and is now approaching 11 million (Labor/ Community Watchdog, 1991, p. 7; Lents and Kelley, 1993, p. 32; Wittow, 1979, p. 370).

Exposure can be calculated by estimating the number of people living within various zones of smog risk. For example, in the 1980s there were about 4–5 million within areas that exceeded federal air-quality standards on more than 150 days a year (fig. 11.9). As with other hazards in Greater Los Angeles, population growth and the spread of building into inland valleys has considerably increased exposure.

Vulnerability to air pollution is a function of age, sex, health status, and activity. Children are more vulnerable because they frequently play out of doors during the afternoon and early evening, when smog levels peak; small body weight and small lung surface area also allow children to receive higher pollutant doses per breath than adults (Kilburn et al., 1992; Kleinman et al., 1989; Labor/Community Watchdog, 1991, p. 22, citing Roan, 1990; National Academy of Sciences, 1988).[19]

Pregnant women are especially vulnerable. Carbon monoxide pollution has been associated with low birth weight and sudden infant deaths (Labor/Community Watchdog, 1991, p. 23, citing Kleinman et al., 1989). People with respiratory diseases such as chronic bronchitis, emphysema, or asthma are also more vulnerable than others. This group contains many frail elderly and AIDS sufferers (Labor/Community Watchdog, 1991, pp. 23–24). Indeed, AIDS victims may be at a special disadvantage; preliminary animal studies suggest that nitrogen oxides – one of the more common pollutants – can accelerate the pathological effects of AIDS viruses (ibid., p. 24). Finally, those who are very active outdoors – athletes and people engaged in construction, gardening, or other heavy labour outdoors – are also particularly vulnerable to air pollution (ibid., p. 26).

Response has taken many forms. The problem of air quality was recognized in the 1940s, and the Los Angeles County Air Pollution Control District was established in 1947. Early attempts at mitigation included banning burning in open dumps of Los Angeles County (1948). At the time there were 54 such open-burning rubbish dumps, mostly municipally operated (Gordon, 1963, p. 72).[20] The use of backyard solid-waste incinerators was outlawed in 1957. Regulation of factory emissions by permit began in the 1960s. Other early measures included banning of certain industrial solvents (1966), shifts to natural gas or low-sulphur oil for electricity generation, and a series of automobile exhaust restrictions from 1966 onwards (Gordon, 1963, p. 73; Lents and Kelley, 1993, pp. 33–37; Nelson and Clark, 1976, p. 247). Anti-pollution agencies began to appear after 1947. The Los Angeles Health Department monitors the air and issues warnings when ozone levels reach critical levels. "Level II" emergencies are declared infrequently and industries are asked to reduce activity by 10–20 per cent while commuters are requested to car pool. The department has never exercised its statutory authority to declare a "level III" emergency, which would shut down industry and halt traffic.

The South Coast Air Quality Management District (SCAQMD) may require operators of fleets with 15 or more vehicles to use clean fuels such as natural gas, propane, ethanol, or methanol. The district currently requires companies with more than 100 personnel to offer cash incentives for using car pools on the journey to work (*The Economist*, 7 April 1990, p. 39). This measure eliminated 90,000 trips a day and increased car occupancy from 1.13 to 1.24 between 1987 and 1992 (Lents and Kelley, 1993, p. 38). Since 1976, the SCAQMD has required all new or expanding industries to use the cleanest available technology. However, loop-holes allowed many firms to avoid these provisions by setting up smaller branch units (ibid., p. 37). Citizens' groups such as Mothers of East Los Angeles and Concerned Citizens of South Central LA play significant roles in research, awareness-raising, and lobbying. For example, from 1985 to 1987 the Concerned Citizens of South Central LA campaigned against city plans to build a trash incinerator in their neighbourhood. As a result, the city withdrew this plan and started a recycling programme instead (Labor/Community Watchdog, 1991, p. 32).

Taken together, all of the preceding efforts at mitigation reduced peak ozone concentration from 680 parts per billion (ppb) in 1955 to 300 ppb in 1992 (Lents and Kelley, 1993, p. 32). But the results may not be enough to bring the air in Greater LA within federal health standards. The SCAQMD estimates that, to achieve this goal, hydrocarbon emissions would have to be cut by a further 80 per cent, nitrogen oxides by 70 per cent, sulphur oxides by 62 per cent, and particulates by 20 per cent (ibid., p. 38). However, planners continue to try. A complex three-stage plan is now in effect that identifies 135 measures that can be accomplished with existing technology. If implemented by 1996, the plan anticipates an 85 per cent reduction in automotive emissions by the following decade. Stage one assumes expanded use of electric cars. Stage two involves large-scale use of alternative fuels (natural gas, ethanol, methanol, etc.). Stage three requires commercial development of new technologies. The SCAQMD has so far contributed US$40 million in seed money toward the development of these technologies (ibid., p. 39).

The current strategy is resolutely committed to so-called tech-fixes. Growth controls – such as have been enacted by Santa Barbara and some northern California communities – are rarely talked about by public leaders. By contrast, local community groups such as the Labor/Community Watchdog insist that "air quality and social justice must go hand in hand." This means that market forces alone should not be allowed to shape the growth of Greater Los Angeles or to determine such decisions as those that bear on industrial location, choice of technology, and levels of emissions. Instead, the wishes of citizens, workers, and industries should be equally weighted (Labor/Community Watchdog, 1991, p. 65). This kind of tripartite model is not a rhetorical political demand but an increasingly accepted risk-management model that has been endorsed by the United Nations (Fiorino, 1990; UNEP, 1988, 1992).

Other technological hazards

Hundreds of thousands of people now live under the flight paths near 11 airports. Similar numbers live within a few miles of oil refineries and chemical plants capable of catastrophic explosions, fires, and toxic releases. For example, Los Angeles possesses 24 of the highly toxic "Superfund" sites designated by the federal government.[21] Even without emerging new natural and technological (na-tech) releases, the "normal" environmental load caused by spills and releases of hazardous materials is cause for concern. Nuclear power stations also pose significant risks in Greater Los Angeles. The San Onofre nuclear power station is a particularly instructive example. After operating on temporary licences since 1967, one reactor was finally closed in 1992 when a suit was filed by community groups. Two other reactors were commissioned in 1982 and have been in full operation since 1983. The Nuclear Regulatory Commission (NRC, 1979) estimated that 3.6 million people lived within 50 miles (80 km) of the plant in 1970. Since then this area has witnessed some of the fastest growth in the mega-city. Projections indicate that there may now be 5.6 million or more people there. At present there are only two nuclear power stations still functioning in California: San Onofre and the even more controversial Diablo Canyon station in northern California. Apart from concerns raised by na-tech releases during a possible earthquake, there exists a problem of small, intermittent releases (including "controlled" airborne releases of radioactive gases) and disposal of radioactive waste in a region fast filling up with every possible sort of waste. Finally, there remains the problem of possible terrorist attacks – a not inconceivable risk, especially amid the troubled politics of Greater Los Angeles (Whittaker, 1978).

Changes in biological hazards

The statistical *risk* of biological hazards such as cancer, AIDS, drug overdose, heat exhaustion, and waterborne diseases is more difficult to quantify than the risk of air pollution. Some risks are involuntary and the agents (e.g. carcinogens) are widely pervasive in air, water, and food chains. For example, Los Angeles County has the highest rate of excess deaths from childhood cancers in the United States (Goldman, 1991, p. 133).[22] Other risks are voluntary, in that they are associated with specific behaviours. Thus, HIV infection is primarily connected with intravenous drug use (needle sharing) and unprotected sex. In 1989, an estimated 112,000 people were HIV positive in LA County (Labor/Community Watchdog, 1991, p. 26). This is equivalent to a population-wide rate of 1.4 per cent, but most of the risk is actually concentrated in specific vulnerable groups (see below).

Exposure to biological hazards is growing almost everywhere in Greater LA because population is increasing along with the prevalence of poverty, homelessness, drug abuse, and other conditions that are associated with high-risk behaviours and high-risk environments.

Vulnerability is highly dependent on age, gender, income, race, and ethnicity. In the 1980s, 40 per cent of the children in LA County lived at or below the poverty line, a high proportion of them in single-parent households headed by their mother. Females headed 14 per cent of all households in LA City in 1980 and 12 per cent of LA County households (Bureau of the Census, 1988, pp. 43 and 611). A disproportionate number of these family units were African-American or Hispanic (Davis, 1990, p. 306). The percentage of LA County residents living below the poverty line increased from 11 per cent in 1969 to 13 per cent in 1979 and to 16 per cent in 1987 (Labor/Community Watchdog, 1991, p. 7). Of those living below the poverty line, 28 per cent receive no welfare benefits (Davis, 1993a, p. 25, citing Kennedy, 1992), and the value of both the minimum wage and the median welfare benefit for families with dependent children (AFDC) has declined by 40 per cent since 1970.[23] These are the young women who cannot afford prenatal care, who seek primary health care from hospital emergency rooms, and who cannot have their children immunized against the same childhood diseases that UNICEF identifies as characteristic of the third world (i.e. measles, tuberculosis, polio, diphtheria, pertussis, tetanus). Young people from such families are the ones most at risk from drug overdose, death by violence, and HIV infection. These families are also the ones at greatest risk from epidemics of waterborne disease if and when the precarious water and sewage-treatment infrastructure breaks down again. Many of the homeless street children and youth are gays and lesbians who have been turned out of family homes elsewhere and have migrated to LA in search of refuge and new lives. Likewise, the frail elderly living on small, fixed incomes and those who have fallen into homelessness are much more vulnerable to the stress of LA's heat waves and bad air.

Response has taken the form of an elaborate social welfare system dating from the 1960s' and 1970s' era of the "Great Society" and "Model Cities" expenditures of state and federal money on programmes for the poor. This approach is rapidly disappearing, the social safety net unravelling, as limitations on property taxes and the decline in federal contributions have cut over US$1 billion from the heart of such programmes (Davis, 1993b, p. 44). The other approach has been technocratic. The Metropolitan Water District (MWD) is another large, highly professional bureaucracy like the South Coast Air Quality Management District discussed earlier. Although designed to accepted standards of water quality and sewage treatment, rapid growth of population within the system has exceeded its capacity. The 1988–1993 drought, complex legal and financial negotiations to try to increase water supply, and the breakdown in 1987 of the Hyperion treatment plant are signs of stress on a system that is vital to the public health of many millions of people. As with many of the hazard responses reviewed in this chapter, there appears to be an over-reliance on engineering solutions and not enough attention is paid to the larger issue of growth control, or even to the many management issues that might be called the "software" counterpart to infrastructure's "hardware" (Herman and Ausubel, 1988, p. 1). For example, if the MWD were to charge the marginal price of additional service to its new customers (and not its old customers), there would be both more funding available for maintenance, hazard mitigation, and "surprises" as well as a significant disincentive to growth (balanced by incentives for conservation) (Hanson, 1988, pp. 265–266).

Solid-waste disposal is also a problem of possible significance for the future of public health, and another to which the response has been a technical solution. Greater Los Angeles creates 80,000 tons of garbage a day. The 1991 City of Los Angeles Solid Waste Plan predicted a shortfall of 22,000 tons per day in nearby landfill capacity. As noted earlier, there is community resistance to incineration. In the face of this large problem, the technological/commercial solution is a partnership between Waste Management of North America (the largest waste-management corporation the world) and the Atchison and Topeka and Santa Fe Railroad. They are investing US$100 million in an integrated system of recycling recovery centres along existing rail lines throughout the Los Angeles area, where non-hazardous recyclables will be separated from the waste stream and sold locally to industry, and a long-distance disposal site some 225 miles from Los Angeles in the high desert of San Bernardino County for the rest. Each of the recovery sites is to have a capacity of 3,000 tons per day. The high desert site has a capacity of 200 million tons in a very arid zone planners claim is seismically stable. Waste Management promises to run the facility in an ecologically sound manner, covering each small cell, and eventually returning the site to its natural condition (Jacobson, 1993, pp. 34–35).

Whatever the pros or cons – and there have been some objections to the giant scheme – one has to ask why the plan revolves around only one of three basic waste-management "R's": recycling should be the last resort once attempts to *reuse* material (internally within industries, for instance) and to *reduce* it have been tried. Here again, as in the cases of air quality, drought, and brush fire, the fundamental issue such technical fixes avoid is the eventual need to control and regulate growth.

Changes in social hazards

The *risks* of many social hazards, like some of the biological ones, are small in the aggregate but are increasing. A practical distinction must be made between the "social problems" of individuals and families and the "social hazards" that affect a large proportion of a subgroup such as the African-American or Hispanic population. Civil unrest of the kind seen in the first[24] and second Los Angeles uprisings is both the cause and effect of social hazards. To the extent that economic recession, youth unemployment, and disinvestment in social services are major causes of rioting, the future is likely to hold more uprisings (Williams, 1993). The vicious circle of poverty, degraded urban environment, anger, despair, drugs, and refuge in alternative paramilitary solidarity (the "hood") and alternative economic systems ("gangbanging") leads toward more and more confrontation with rival gangs and with the police. Escalation of violence is the necessary result.

Exposure is also growing. Gangs are now present throughout the Greater Los Angeles region, and alienated youth from poor White, Hispanic, Chinese, Vietnamese, Cambodian, and other ethnic groups have become features of the mega-city landscape alongside the original African-American gangs (Davis, 1990, pp. 267–322).

Vulnerability to social hazards is age and race dependent. Los Angeles City had a "violent crime" rate of 9,239 per 100,000 population in 1985.[25] These violent acts are disproportionately suffered by young African-American and Hispanic people (Sorenson et al., 1993) at the hands of other young people of colour. This is the second-highest rate in California, after the city of Oakland in the Bay Area, and one of the highest in the United States. The rate for the county of Los Angeles is only somewhat lower at 7,189 per 100,000 (Bureau of the Census, 1988, p. 45). A little less vulnerable are those youths who stay in school. However, the highest drop-out rates in both junior and senior high school are in the South Central district (Oliver et al., 1993, pp. 128 and 129).

The predominant *response* to this complex of social hazards has been the use of deadly force and imprisonment. This includes the well-known "war on drugs," which really is experienced as a war by citizens caught in the cross-fire. In April 1989, the Los Angeles Police Department began "Operation Hammer" by sweeping 10 square blocks of South Central Los Angeles, arresting more African-American youth than at any time since the Watts uprising in 1965. One Mayor of Los Angeles has called the gangs "the Viet Cong abroad in our society" (Davis, 1990, p. 268). Although many public health and social service workers believe it is possible to prevent youths joining gangs by dealing with socially embedded causes (Lasley, 1992; Ostos, 1991), the Los Angeles Police Department appears committed to the notion that some youth are irreparably evil and must simply be "weeded" out of the community (Gates and Jackson, 1990). Construction of prison space is a second prong of the predominant response to social hazards. It is equally technocratic, narrow in its social vision, and self-defeating because the scarce city and county resources devoted to prison construction and policing have been extracted from the programmes for job-training and school retention, drug rehabilitation, family counselling, and economic welfare that might do something about the root causes of the problems (Davis, 1993a, pp. 23–25; 1993b, pp. 41–45). More constructive and humane is the approach of the Los Angeles Department of Children's Services. During the 1992 uprising it activated an earthquake contingency plan in children's shelters, provided additional counselling, and engaged in immediate relief activities (Patterson and Boehm, 1992).

Summary of hazard vulnerability

Of the four principal hazard dimensions discussed here (risk, exposure, vulnerability, response), patterns of vulnerability provide most insight into the complex interaction of social, technological, and natural processes in Greater Los Angeles. These are summarized in table 11.4.

The socio-economic and ethnic group that is most vulnerable to the first five hazards listed in table 11.4 is poor and non-White. These people are most likely to inhabit one of the 20,000–50,000 older buildings that pre-date earthquake building codes (Alexander, 1993, p. 337). A large number of immigrants from Mexico and locally born Mexican-Americans occupy such structures in East Los Angeles. Poor, non-White families are also more likely to live in the floodplain of the Los Angeles River, and they are more likely to live in areas where air pollution is not abated by

Table 11.4 The distribution of vulnerability in Greater Los Angeles

	Earthquake injury	Flood	Bad air exposure	Chemical fire	Home fire	Violence	Drug overdose	HIV	Bad air injury	Heat exhaustion	Landslide	Nuclear accident	Brush fire	Coastal storm & beach erosion
Poor, non-White	×	×	×	×	×									
Poor, non-White youth	×	×	×	×	×	×	×	×						
Poor, old									×	×				
Homeless, old						×			×	×				
Homeless, youth						×	×	×	×					
Rich, White											×	×	×	×

coastal breezes. Likewise, low-income people – including many people of colour – live in heavily industrial areas, especially those near oil refineries in such places as Wilmington, El Segundo, Carson, and Torrance (Labor/ Community Watchdog, 1991). These are the areas that are most at risk from chemical fires and these are the people most likely to die in such blazes (Bureau of the Census, 1980, p. 208). More recent statistics, local-ized specifically for Greater LA, may well reveal that the poor, non-White population is more vulnerable to home fires.[26]

Youthful members of families in the foregoing group are vulnerable to all of the hazards just discussed, but are also particularly vulnerable to the next four hazards in the table. Children are more vulnerable to the health effects of air pollution. Their older siblings are at risk of violence (assault, murder), drug overdose, and HIV infection, because of shared needles and sex with drug-using partners.

The elderly population, whatever its ethnicity, is physiologically vul-nerable to heat exhaustion and to the ill-effects of air pollution on pre-existing cardiopulmonary conditions if they do not have enough money to live in the cooler and less polluted hills or at the beach or to afford air-conditioning. Those of their number who become homeless are also vul-nerable to street violence, in addition to the stresses of malnutrition and exposure to heat and to some of the worst air, which collects around the old downtown area where many homeless congregate (Macey and Schneider, 1993).

It is estimated that there are as many as 50,000 homeless youths on the streets of Los Angeles County.[27] They are subject to diverse hazards of the street, including violence (e.g. rape and sexual exploitation), HIV, drug overdose, malnutrition, and chronic diseases such as tuberculosis, compounded by the effects of air pollution.

Beach and hillside locations render upper-middle-class and rich *Ange-lenos* more exposed to landslides, coastal storms, beach erosion, and brush fires. These are natural hazards not generally visited upon the poor of Greater LA. Although spared the worst of the bad air and proximity to hazardous chemical industries, a large number of the rich live well within reach of the San Onofre nuclear power complex. A Chernobyl-scale event could spread a radioactive cloud well into the Mexican *barrio* of East Los Angeles (just 57 miles from San Onofre), but smaller (and more likely) accidents could well mar the "perfect" Orange County lifestyle in, for example, Irvine Spectrum or John Wayne, CA.

Opportunities for intervention

This chapter has emphasized the contribution of pathological social pro-cesses to hazard-vulnerability in Greater Los Angeles. It is necessary to focus on these pathologies not simply because social (and biological) hazards are important components of the mega-city's hazard portfolio, but also because a clear understanding of social pathology is crucial to the formulation of effective "bottom–up" strategies of hazard reduction. Given the shortcomings of existing "top–down" strategies, bottom–up strategies offer alternative paths toward a less hazardous future. These reasons are amplified below.

First, "biological" and "social" risks threaten large numbers of people. In terms of contributions to overall rates of suffering and death, their magnitude is greater than all the "natural" hazards combined, with the possible exception of a very large earthquake.

Secondly, social and biological hazards bear most heavily on specific groups. For the African-Americans and Hispanics who are most affected, they are more than random events; they are perceived as further products of racial and ethnic discrimination. Therefore they threaten to add to social disintegration as well as to induce direct harm.

Thirdly, the potential is strong for interaction between social and biological hazards on the one hand and natural and technological ones on the other. Continued racial tension, economic polarization, and social disintegration may produce more arson, sabotage, and environmental terrorism.

Fourthly, denial and flight from burgeoning social and biological hazards have set in motion political processes that stripped inner-city LA and LA County of the funds necessary for mitigating all kinds of hazards.

Fifthly, in Greater Los Angeles comprehensive community-based hazard mitigation depends on the willingness of people from all ethnic, social, and economic backgrounds to work collaboratively for the common good. Continued polarization perpetuates short-term parochial interests and frustrates this goal.

Bottom–up hazard mitigation

Ideally, communities should be able to assess the hazards that affect them and to plan and execute effective mitigation programmes (Anderson and Woodrow, 1989; Blaikie et al., 1994; Hardoy and Satterthwaite, 1989; Maskrey, 1989). Despite anomie, economic polarization, and spatial and social fragmentation, this opportunity still exists in Greater Los Angeles. Groups such as Mothers of East Los Angeles, Concerned Citizens of South Central, and Labor/Community Watchdog have shown that it is possible. Unfortunately, such initiatives are exceptions. Many people remain locked behind their security fences and trust that large bureaucracies will "solve" hazards by means of technology: police helicopters and computers; fire-fighting air tankers that drop retardant chemicals in

250 gallon (946 litre) batches; a Rail Cycle that hauls garbage into the desert; more and fancier treatment plants; perfect "clean" fuels for cars.

What is needed is more active community participation in decisions that affect people's lives, and, specifically in hazard identification and mitigation. Models for this kind of planning have existed for a long time (Clavel, 1986; Gans, 1968; Wisner et al., 1991). The obstacles – lack of funding, economic collapse of many communities, racial separation, and fear – are very great, but the effort has to be made. As shown by Puente in chapter 9 in this volume, Mexico City has achieved considerable progress in community-based disaster reconstruction, and the same principles could be applied more widely to other hazard-management initiatives. Not only did the Mexican programmes strengthen the city's capacity for mitigating, coping with, and recovering from any future disaster; locally based programmes created 115,000 new construction jobs that were filled almost exclusively by locally recruited workers (Pantelic, 1991, pp. 92 and 93). Do the social networks that support poor people in Mexico City not exist in the Los Angeles *barrios* (Lomnitz, 1977; Romo, 1984)? Is it not possible to organize recovery, from riots and fires as well as from natural disasters, in ways that strengthen the social and economic capacity of the local community in the long run? For a start, it is necessary to *listen* to what people in such communities are saying (Madhubuti, 1993; Martinez, 1993).

Conclusion: What future for Greater Los Angeles?

Casting about for futures, I am drawn to the past. Catullus and Plutarch provide descriptions of flood, fire, and redevelopment in Rome *c.* 50–40 B.C.:

"The low parts of the city were subject to periodic floods, and the collapse and conflagration of buildings were common occurrences. In the Principate it is said that not a day passed without a serious fire, yet then there were 7,000 vigiles to put them out, in the Republic only a small force of publicly owned slaves. Crassus had a gang of five hundred builders, and bought up houses that were afire or adjacent to a blaze at knock-down prices with a view to rebuilding on the sites." (Brunt, 1966, p. 12)

This pessimistic view of the urban future is one that has been shared by many who have written about Los Angeles. In one novel, the mega-city of 2052 is a place where water and gas are rationed, disease flourishes, and everyone is regarded as a stranger (Kadohata, 1992). In the movie *Bladerunner* (Harvey, 1989, pp. 308–323), the picture of 2010 is no less inviting: the county's population of 10 million contains 5 million living in poverty; there are half a million homeless; and commuters spend two hours in freeway traffic travelling at 15 mph (24 kph), surrounded by air so poor that it is a major cause of death for 100,000 people each year.

These portraits are hardly just speculative science fiction. They can be arrived at by modest projections of current trends (Labor/Community Watchdog, 1991, p. 9). However, by the time conditions deteriorate to the levels envisaged above, the last historic chance to put "planning as debate" or "participative conflict planning" into place, both from the top–down and from the bottom–up, will be lost. Despair, anger, and fear will seal the haves and have-nots off from each other. Such a future might bring sabotage of water and sewage-treatment plants, nuclear terrorism, or brush fires set by arsonists, combined with police sweeps of suspected communities. Or it might involve an intensification of "natural" disasters and "normal" accidents (Perrow, 1984), or some combination of socially instrumental retaliatory hazard and "normality."

Is such a future avoidable? It must be. Deyan Sudjic probably had something of the sort in mind when he wrote:

Unfettered capitalism, when it meets the absolutes of photochemical smog and water shortages, finds itself having from necessity to think about its future just like any social-democratic, Northwest European utopia – particularly when the city's dispossessed begin to take the law into their own hands and start burning down shopping centres. (1993, p. 102)

Since the election of Mayor Riordon in 1994, there are signs that the sheer magnitude of the multiple hazards facing Los Angeles have deflected the ideology of growth and the marketplace toward more pragmatic directions. For example, Mayor Riordan did not support his fellow-Republican, Governor Pete Wilson, when he campaigned for Proposition 187 – the ballot initiative that would deny public services such as health and education to illegal immigrants and their children. The new Mayor was also impressive in a technical way as he pushed agencies and contractors to repair the Santa Monica freeway after the Northridge earthquake. He succeeded two months ahead of schedule.

On a gloomier note, in December 1994 it was announced that the city of Los Angeles would not be awarded federal "enterprise zone" grants to rebuild after the riot of 1992 and the earthquake of 1994. This was another financial blow to the mega-city, following on the heels of an announcement that Orange County had lost US$2 billion in risky investments. If the dystopia is to be avoided, what is needed, above all, is financial stabilization combined with closer integration and coordination of county and municipal governments. As this chapter has emphasized, the economic force of globalization and the political force of possessive individualism both tend to work against integration and coordination. Yet many competent and caring professionals and energetic community groups in Greater Los Angeles are working for a safer and more humane future. In these people is the hope of the mega-city.

Notes

1. Davis (1993a, p. 16, fn. 31) defines the "edge-city rim" slightly differently and comes up with the figure of 2 million new jobs between 1972 and 1989 in a region that is just 2 per cent Black (versus LA County with 11 per cent Black population and LA City with 14 per cent Black).
2. Davis (1990, p. 316) identifies 230 Black and Latino gangs, and another 81 Asian gangs, including those composed of Filipino, Vietnamese, and Cambodian youth, as well as young Chinese from Pasadena "unwilling to spend lifetimes as busboys and cooks."
3. The total urban street and freeway system in the city of Los Angeles alone is 24,906 miles (40,000 km). This is the second-largest urban transportation system in the United States (behind the New York–Newark urban region) (Carpenter and Provorse et al., 1993, p. 354).
4. The city of Los Angeles has a population density of only 7,426 persons per sq. mile (2,867.2/km^2), covering 469.3 sq. miles (181.2 km^2), compared with New York City's density of 23,701 persons per sq. mile (9,151/km^2) and the Manhattan density of 52,415/ sq. mile (20,237.5/km^2) (Bureau of the Census, 1992, p. 36).
5. War metaphors about LA are frequently used by the Mayor, police officials, urban planners, and gang members. Analogies are drawn with Bosnia, Viet Nam, Beirut, Lebanon, and Israel.
6. A similar migration in the San Francisco Bay Area during the war led to the development of one of the most stable and affluent Black working-class communities on the West Coast, West Oakland, which grew up around the high-wage jobs in the rail and port complex there.
7. A narrow swathe of territory running from the J. Paul Getty Museum near the Pacific Coast Highway in the west to the Huntington Library and Art Gallery in the east in Pasadena, with the UCLA campus, Los Angeles County Museum of Art, and Southwest Museum in between.
8. In 1963 and again in 1985, Arizona won cases in the Supreme Court giving it a larger share of the river's water, but, with its own Sun Belt cities such a Phoenix growing rapidly, it will want more (Clark, 1983, p. 276; Goetz, 1985, p. 312).
9. To bring water from the Owens River, 142 tunnels totalling 52 miles were required. The city was dependent on water from the Los Angeles River until the Owens Valley aqueduct was completed in 1913. In 1930 the aqueduct system was extended another 105 miles (169 km) further north into the Mono basin on the eastern side of the Sierra Nevada mountains. Construction on an aqueduct from the Colorado River began in 1931, adding water from this source in 1941 (Goetz, 1985, p. 312; Clark, 1983, pp. 275–276; Rolle, 1992, p. 756).
10. There is a long history of drought in southern California going back at least to the catastrophic drought of 1862–1865, which marked the end of the dominance of the large cattle ranches in the region (Goetz, 1985, p. 311; Cleland, 1951). There was also a severe urban drought in 1976–77 (Goetz, 1985, p. 312). In Los Angeles County, a full 35 per

cent of all water use was devoted to landscaping irrigation in either public spaces or private homes (Rieff, 1991, p. 252).

11. Inspections of air pollution sources declined by 48 per cent between 1989 and 1992 in the South Coast Air Quality Management District, according to an audit by the California Air Resources Board (*Environmental Reporter*, 26 August 1993, p. 605).

12. In 1992 there were 33 State I smog alerts and 96 days in which ozone levels exceeded federal health standards, up from 21 alerts and 70 unhealthful days in 1991 and 29 alerts and 89 ozone days in 1990 (*Environmental Reporter*, 20 August 1993, p. 1299).

13. California had the highest rate of measles in the United States in 1992: 42 per 100,000 of the population. This is a good indicator of very poor immunization coverage (Carpenter and Provorse et al., 1993, p. 33).

14. Several contributions to the National Academy of Engineering workshop on "Cities and their Vital Systems" bear on the question of freeway security; see Ausubel and Herman (1988); Ibbs and Echeverry (1988); Marland and Weinberg (1988).

15. Personal communication from the research team, Drs. Chrys Rodrigue and Susan Place, California State University Chico. These scholars note elsewhere (Place and Rodrigue, 1994) that even the name of the January 1994 earthquake reveals bias against the poor. Although the epicentre of this earthquake was in Reseda, it was named by the media and governmental officials the "Northridge" earthquake. These authors note that the per capita income of Reseda is only US$15,177 (below the US$16,409 average for California), whereas Northridge is more affluent, with an average of US$23,308 per capita. Place and Rodrigue suggest that this naming "may possibly reveal demographic bias on the part of the media, compounded by initially confusing pronouncements by the California Institute of Technology" (1994, p. 5; also see Rovai and Rodrigue, 1994).

16. Another study cited by Griggs and Gilchrist (1983, p. 55) found 14,000 buildings in Los Angeles "deficient in terms of earthquake design."

17. The US government's Environmental Protection Agency (EPA) has developed a Pollutant Standards Index (PSI) that takes into account ambient levels of sulphur dioxide, nitrogen oxides, particulate matter, carbon monoxide, and ozone as well as the differing standards and monitoring techniques used from city to city. The World Resources Institute assembled average PSIs for 75 cities in the USA for 1990. Los Angeles was by far the worst, with an average PSI of 73. Eight other cities had PSIs in the 50s. All the rest were 49 or below.

18. Lents and Kelley (1993, p. 38) review a series of studies of the health effects of photochemical smog in the general population of the South Coast Air Quality Management District. One of these estimates that meeting federal air-quality standards would save US$9.4 billion in health costs each year and prevent 1,600 premature deaths annually. Also see Ostro et al. (1993).

19. Another child health issue related to air pollution that the community group Concerned Citizens of South Central Los Angeles is trying to get the city to investigate is the likelihood that years of traffic using leaded fuel passing overhead on the multiple interchanges that traverse East Los Angeles have built up considerable lead in the soil. This lead may be causing brain damage to local children.

20. In an early example of the fragmentation and isolationism that continue to plague attempts to mitigate hazards in Greater Los Angeles, the city of Whittier nearly seceded from the county because it did not want to spend the additional money necessary to dispose of solid waste by other methods (Gordon, 1963, p. 72).

21. LA is fourth behind New York with 101 Superfund sites, Philadelphia with 77, and San Francisco–Oakland with 34 (World Resources Institute, 1993, p. 205).

22. "Excess deaths" are those observed in excess of what would be expected if the whole of the USA had the same "average" death rate from childhood cancers, with no variation

from county to county, once the effect of random variation has been accounted for (Goldman, 1991, pp. 17–19).

23. Davis (1993a, p. 24), who continues (pp. 24–25, citing the US Congress, House Ways and Means Committee, *Green Book*, Washington D.C., 1992) to the effect that the median welfare benefit for a family of three is now barely equal to one-third of the poverty threshold.

24. Six days of looting and burning in Watts, at the centre of South Central LA, began on 11 August 1965; 34 people died, 1,032 were wounded, and 3,952 were arrested (Goetz, 1985, p. 309).

25. "Violent crime" is defined as murder, non-negligent manslaughter, forcible rape, and aggravated assault (Bureau of the Census, 1988, p. 611).

26. My guess is that the high numbers for people of colour on a national basis is the result of an epidemic of arson perpetrated by the owners of dilapidated apartment buildings in many big cities, who had them burned for the insurance money (so-called "insurance fires") and in order to get rid of "sitting tenants" and clear the way (albeit illegally) for more lucrative property redevelopment.

27. Personal communication from Gabe Kruks, acting Executive Director, Gay and Lesbian Community Services, West Hollywood, 1992.

REFERENCES

Acosta, O. 1973. *The revolt of the cockroach people*. New York: Vintage.

Alexander, D. 1993. *Natural disasters*. New York: Chapman & Hall.

American Society of Civil Engineers. 1990. "California stiffens earthquake measures." *Civil Engineering* 60 (September): 14.

Anderson, M. and P. Woodrow. 1989. *Rising from the ashes*. Boulder, CO, and Paris: Westview and UNESCO.

Angotti, T. 1993. *Metropolis 2000: Planning, poverty and politics*. London: Routledge.

Ausubel, J. and R. Herman, eds. 1988. *Cities and their vital systems: Infrastructure past, present, and future*. Washington, D.C.: National Academy Press.

Ayer, E. 1992. *Earthquake country: Travelling California's fault lines*. Frederick, CO: Renaissance House.

Bakun, W. and A. Lindh. 1985. "The Parkfield, California, earthquake prediction experiment." *Science* 229 (August): 619–624.

Bernard, R. and B. Rice, eds. 1983. *Sunbelt cities: Growth and politics since World War II*. Austin: University of Texas Press.

Berry, B., ed. 1976. *Urbanization and counter-urbanization*. Beverly Hills, CA: Sage.

——— 1977. *Growth centers in the American urban system*. Cambridge, MA: Ballinger.

Blaikie, P., T. Cannon, I. Davis, and B. Wisner. 1994. *At risk: Vulnerability to disasters*. London: Routledge.

Brunt, P. 1966. "The Roman mob." *Past & Present* 35 (December): 3–27.

Bureau of the Census (US Department of Commerce). 1980. *Social indicators III*. Washington, D.C.: US Government Printer.

———— 1988. *City and county data book*. Washington, D.C.: US Government Printer.

———— 1992. *Statistical abstract of the United States*. Washington, D.C.: US Government Printer.

Carpenter, A., C. Provorse, et al. 1993. *The world almanac of the U.S.A.*. Mahwah, NJ: World Almanac (Funk & Wagnalls).

Castells, M. 1989. *The informational city*. London: Basil Blackwell.

Caughey, L. 1976. "The Long Beach earthquake." In J. Caughey and L. Caughey (eds.), *Los Angeles: Biography of a city*. Berkeley: University of California Press, pp. 301–304.

Caughey, J. and L. Caughey. 1976. "A time of catastrophe." In J. Caughey and L. Caughey (eds.), *Los Angeles: Biography of a city*. Berkeley: University of California Press, pp. 293–295.

Clark, D. 1983. "Improbable Los Angeles." In R. Bernard and B. Rice (eds.), *Sunbelt cities: Politics and growth since World War II*. Austin: University of Texas Press, pp. 268–308.

Clavel, P. 1986. *The progressive city: Planning and participation 1969–84*. New Brunswick, NJ: Rutgers University Press.

Cleland, R. 1951. *The cattle on a thousand hills: Southern California, 1850–1880*. San Marino, CA: Huntington Library.

Cooke, R. 1984. *Geomorphological hazards of Los Angeles*. London: Allen & Unwin.

Davis, Mike. 1990. *City of quartz*. New York: Vintage.

———— 1992. "Chinatown revisited? The internationalization of downtown Los Angeles." In D. Reid (ed.), *Sex, death and god in L.A.* New York: Pantheon, pp. 19–53.

———— 1993a. "Who killed Los Angeles? A political autopsy." *New Left Review* 197 (January/February): 3–28.

———— 1993b. "Who killed Los Angeles? The verdict is given." *New Left Review* 199 (May/June): 29–54.

Dear, M. and A. Scott, eds. 1981. *Urbanization and urban planning in capitalist society*. London: Methuen.

Dooley, D., R. Catalano, and S. Mishra. 1992. "Earthquake preparedness: Predictors in a community survey." *Journal of Applied Social Psychology* 22 (March): 451–470.

Fiorino, D. 1990. "Citizen participation and environmental risk: A survey of institutional mechanisms." *Science, Technology, and Human Values* 15(2): 226–242.

Gans, H. 1968. *People and plans*. New York: Basic Books.

Garreau, J. 1991. *Edge city: Life on the new frontier*. New York: Anchor.

Gates, D. and R. Jackson. 1990. "Gang violence in L.A." *The Police Chief* 57 (November): 20–22.

Goetz, P., ed. 1985. "Los Angeles." *The New Encyclopaedia Britannica*, 15th edn. Chicago: Encyclopaedia Britannica, vol. 23, pp. 307–312.

Goldman, B. 1991. *The truth about where you live: An atlas for action on toxins and mortality*. New York: Times Books.

Gordon, M. 1963. *Sick cities*. New York: Macmillan.

Gottmann, J. 1961. *Megalopolis: The urbanized north eastern seaboard of the United States*. Cambridge, MA: Harvard University Press.

———— 1983. *The coming of the transactional society*. Baltimore, MD: University of Maryland Press.

Greene, H., J. Gardner-Taggart, and M. Ledbetter. 1991. "Offshore and onshore liquefaction at Moss Landing spit, central California – result of the October 17, 1989, Loma Prieta earthquake." *Geology* 19 (September): 945–949.

Griggs, G. and J. Gilchrist. 1983. *Geological hazards, resources, and environmental planning*, 2nd edn. Belmont, CA: Wadsworth Publishing.

Gruenwald, G. 1988. "Asbestos levels in a Los Angeles building before and after an earthquake." *American Industrial Hygiene Association Journal* 49 (February): A87–88.

Hanson, R. 1988. "Water supply and distribution: The next 50 years." In J. Ausubel and R. Herman (eds.), *Cities and their vital systems*. Washington, D.C.: National Academy Press, pp. 258–277.

Hardoy, J. and D. Satterthwaite. 1989. *Squatter citizen*. London: Earthscan.

Harloe, M. and E. Lebas, eds. 1981. *City, class and capital*. London: Edward Arnold.

Harvey, D. 1973. *Social justice and the city*. Baltimore, MD: Johns Hopkins University Press.

———— 1989. *The condition of postmodernity*. Oxford: Basil Blackwell.

Herman, R. and J. Ausubel. 1988. "Cities and infrastructure: Synthesis and perspectives." In J. Ausubel and R. Herman (eds.), *Cities and their vital systems*. Washington, D.C.: National Academy Press, pp. 1–21.

Ibbs, C. and D. Echeverry. 1988. "New construction technologies for rebuilding the nation's infrastructure." In J. Ausubel and R. Herman (eds.), *Cities and their vital systems*. Washington, D.C.: National Academy Press, pp. 294–311.

Jacobson, T. 1993. *Waste management: An American corporate success story*. Washington, D.C.: Gateway Business Books.

Jaume, S. and L. Sykes. 1992. "Changes in state of stress on the southern San Andreas fault resulting from the California earthquake sequence of April to June 1992." *Science* 258 (November): 1325–1328.

Johnson, B. and V. Covello, eds. 1987. *The social and cultural construction of risk*. New York: D. Reidel.

Jones, Emrys. 1990. *Metropolis*. London: Oxford University Press.

Kadohata, Cynthia. 1992. *In the heart of the valley of love*. New York: Penguin Books.

Kahrl, W. 1982. *Water and power: The conflict over Los Angeles' water supply in the Owens Valley*. Berkeley: University of California Press.

Kilburn, K., R. Warshaw, and J. Thornton. 1992. "Expiratory flows decreased in Los Angeles children from 1984 to 1987: Is this evidence of effects of air pollution?" *Environmental Research* 59(1): 150–158.

Kleinman, M., S. Colome, and D. Foliart. 1989. *Effects on human health of pollutants in the South Coast air basin*. Final report to South Coast Air Quality Management District. Los Angeles: SCAQMD.

Labor/Community Watchdog. 1991. *L.A.'s lethal air: New strategies for policy, organizing, and action.* Los Angeles: Labor/Community Strategy Book Center.

Lasley, J. 1992. "Age, social context, and street gang membership." *Youth and Society* 23 (June): 434–451.

Lee, D. and R. Shepherd. 1984. "Hazardous buildings: Aspects of the Los Angeles earthquake problem." *Earthquake Engineering and Structural Dynamics* 12 (March/April): 149–167.

Lents, J. and W. Kelley. 1993. "Clearing the air in Los Angeles." *Scientific American*, October: 32–39.

Lomnitz, L. 1977. *Networks and marginality: Life in a Mexican shanty town.* New York: Academic Press.

Lynch, K. 1971. "The possible city." In L. Bourne (ed.), *Internal structure of the city: Readings on space and environment.* New York: Oxford University Press, pp. 523–528.

Macey, S. and D. Schneider. 1993. "Deaths from excessive heat and excessive cold among the elderly." *Gerontologist* 33(4): 497–500.

Madhubuti, H., ed. 1993. *Why L.A. happened: Implications of the '92 Los Angeles rebellion.* Chicago: Third World Press.

Marland, G. and A. Weinberg. 1988. "Longevity of infrastructure." In J. Ausubel and R. Herman (eds.), *Cities and their vital systems.* Washington, D.C.: National Academy Press, pp. 312–332.

Martin, T. 1993. "From slavery to Rodney King: Continuity and change." In H. Madhubuti (ed.), *Why L.A. happened: Implications of the '92 Los Angeles rebellion.* Chicago: Third World Press, pp. 27–40.

Martinez, R. 1993. *The other side: Notes for the new L.A., Mexico City and beyond.* New York: Vintage.

Maskrey, A. 1989. *Disaster mitigation: A community-based approach.* Oxford: Oxfam.

Massey, D. and N. Denton. 1989. "Hypersegregation in the U.S. metropolitan areas: Black and Hispanic segregation along five dimensions." *Demography* 26(3): 373–391.

Massey, D. and M. Eggers. 1990. "The ecology of inequality: Minorities and the concentration of poverty, 1970–1980." *American Journal of Sociology* 95(5): 1153–1188.

Mattingly, S. and V. Melloff. 1992. "International collaboration for local earthquake response and recovery planning." In A. Kreimer and M. Munasinghe (eds.), *Environmental management and urban vulnerability.* World Bank Discussion Paper 168. Washington, D.C.: World Bank, pp. 285–286.

Mulilis, J-P. and R. Lippa. 1990. "Behavioral change in earthquake preparedness due to negative threat appeals: A test of protection motivation theory." *Journal of Applied Social Psychology* 20 (May): 619–638.

National Academy of Sciences, ed. 1988. *Air pollution, the automobile and public health.* Washington, D.C.: National Academy Press.

Nelson, H. and W. Clark. 1976. "The Los Angeles metropolitan experience." In J. Adams (ed.), *Twentieth century cities.* Contemporary Metropolitan America, vol. 4. Cambridge, MA: Ballinger Publishing.

NRC (Nuclear Regulatory Commission). 1979. *Demographic statistics pertaining to nuclear power reactor sites.* NRC Report No. 0348, October. Washington, D.C.: NRC.

Oliver, M., J. Johnson, and W. Farrell. 1993. "Anatomy of a rebellion: A political-economic analysis." In R. Gooding-Williams (ed.), *Reading Rodney King/Reading urban uprising.* New York: Routledge, pp. 117–141.

Olney, P. 1993. Presentation by Peter Olney, representative of Service Employees' Union, Local 399, at the Third Annual Conference on Social Ecology: Burning Issues: Social Ecology and the Urban Crisis. Santa Monica College, Santa Monica, CA, 28–30 May.

Ostos, T. 1991. "Alternatives to gang membership." *Public Health Reports* 106 (May/June): 241.

Ostro, B., M. Lipsett, K. Mann, A. Krupnick, and M. Harrington. 1993. "Air pollution and respiratory morbidity among adults in southern California." *American Journal of Epidemiology* 137(7): 691–700.

Outland, G. 1963. *Man-made disaster: The story of St. Francis Dam.* Glendale, CA: Arthur H. Clark.

Palloix, C. 1977. "The self-expansion of capital on a world scale." *Review of Radical Political Economy* 9(2): 1–28.

Palm, R. 1990. *Natural hazards: An integrative framework for research and planning.* Baltimore, MD: Johns Hopkins University Press.

Palm, R. and M. Hodgson. 1992. *After a California earthquake: Attitude and behavior change.* Chicago: University of Chicago Press.

Pantelic, J. 1991. "The link between reconstruction and development." In A. Kreimer and M. Munasinghe (eds.), *Managing natural disasters and the environment.* Washington, D.C.: World Bank, pp. 90–94.

Patterson, K. and S. Boehm. 1992. "How can human service agencies be ready? First, prepare for an earthquake..." *Public Welfare*, Fall: 7–8.

Perrow, C. 1984. *Normal accidents: Living with high-risk technologies.* New Haven, CT: Yale University Press.

Place, S. and C. Rodrigue. 1994. "Media construction of the 'Northridge' earthquake in English and Spanish language print media in Los Angeles." Manuscript, Center for Hazards Research, Department of Geography and Planning, California State University Chico.

Popkin, R. 1990. "The history and politics of disaster management in the United States." In A. Kirby (ed.), *Nothing to fear.* Tuscon: University of Arizona Press, pp. 101–129.

Pyne, S. 1982. *Fire in America: A cultural history of wildland and rural fire.* Princeton, NJ: Princeton University Press.

Rieff, David. 1991. *Los Angeles: Capital of the third world.* New York: Touchstone Books.

Roan, S. 1990. "Air sickness: Evidence mounts of dramatic, permanent damage to lungs of children." *Los Angeles Times,* 4 March.

Robinson, A. 1993. *Earth shock: Climate, complexity and the forces of nature.* London: Thames & Hudson.

Rolle, A. 1992. "Los Angeles." *Encyclopedia Americana.* New York: Encyclopedia Americana, vol. 17, pp. 748–757.

Romo, R. 1984. *East Los Angeles: History of a barrio.* Austin: University of Texas Press.

Rovai, E. and C. Rodrigue. 1994. "The (mis?)construction of reconstruction: Mental maps and timelines of recovery from the Los Angeles earthquake of 17 January." Manuscript, Center for Hazards Research, Department of Geography and Planning, California State University Chico.

Savage, J. 1991. "Criticism of some forecasts of the National Earthquake Prediction Evaluation Council." *Bulletin of the Seismological Society of America* 81 (June): 862–881.

Sherwin, R. 1990. "Findings." Unpublished paper presented at the International Specialty Conference on Tropospheric Ozone and the Environment, Los Angeles, California, 21 March.

Showalter, P. and M. Myers. 1992. *Natural disasters as the cause of technological emergencies: A review of the decade 1980–1989.* Natural Hazards Research Working Paper 78. Boulder, CO: Natural Hazards Research and Applications Information Center, University of Colorado.

Smith, N. 1984. *Uneven development: Nature, capital and the production of space.* Oxford: Basil Blackwell.

Sorenson, S., B. Richardson, and J. Peterson. 1993. "Race/ethnicity patterns in the homicide of children in Los Angeles, 1980 through 1989." *American Journal of Public Health* 83 (May): 725–727.

Steiner, R. 1981. *Los Angeles: The centrifugal city.* New York: Kendall/Hunt.

Sudjic, D. 1993. *The 100 mile city.* London: Flamingo.

Tabb, W. and L. Sawers, eds. 1978. *Marxism and the metropolis.* New York: Oxford University Press.

Taylor, M. and N. Thrift, eds. 1986. *Multinationals and the restructuring of the world economy.* Beckenham: Croom Helm.

Thrift, N. 1986. "The geography of international economic disorder." In R. Johnston and P. Taylor (eds.), *A world in crisis? Geographical perspectives.* Oxford: Basil Blackwell, pp. 12–67.

Tierney, K. and R. Anderson. 1990. *Risk of hazardous materials release following an earthquake.* Preliminary Paper 152, Disaster Research Center. Newark, NJ: University of Delaware.

Turner, R. and J. Kiecolt. 1984. "Responses to uncertainty and risk: Mexican American, black, and Anglo beliefs about the manageability of the future." *Social Science Quarterly* 65 (June): 665–679.

Turner, R., J. Nigg, and D. Paz. 1986. *Waiting for disaster: Earthquake watch in southern California.* Berkeley: University of California Press.

UNEP (United Nations Environment Programme). 1988. *APELL: Awareness and Preparedness for Emergencies at Local Level: A process for responding to technological accidents.* Nairobi: UNEP.

——— 1992. *Hazard identification and evaluation in a local community.* Paris: UNEP Industry and Environment Programme Centre.

USGS (United States Geological Survey). 1988. "Probabilities of large earthquakes occurring in California on the San Andreas Fault." *US Geological Survey Open File Report 88–398.*

Vaughan, E. 1993. "Individual and cultural differences in adaptation to environmental risks." *American Psychologist* 48(6): 673–680.

Vaughan, E. and B. Nordenstam. 1991. "The perception of environmental risks
among ethnically diverse groups." *Journal of Cross-Cultural Psychology* 22(1):
29–60.

Whittaker, H., ed. 1978. *State comprehensive emergency management.* Final report
of the Emergency Preparedness Project, Center for Policy Research, National
Governors' Association. Washington, D.C.: US Government Printing Office.

Williams, R. 1993. "Accumulation as evisceration: Urban rebellion and the
new growth dynamics." In R. Gooding-Williams (ed.), *Reading Rodney King/
Reading urban uprising.* New York: Routledge, pp. 82–96.

Wisner, B., D. Stea, and S. Kruks. 1991. "Participatory and action research
methods." In E. Zube and G. Moore (eds.), *Advances in environmental
behavior and design.* New York: Plenum Press; Washington. D.C.: US Environmental Design Research Association, vol. 3, pp. 271–295.

Wittow, J. 1979. *Disasters: The anatomy of environmenal hazards.* Athens:
University of Georgia Press.

World Resources Institute. 1993. *The 1993 Information Please Environmental
Almanac.* New York: Houghton-Mifflin.

12

Environmental hazards and interest group coalitions: Metropolitan Miami after hurricane Andrew

William D. Solecki

Editor's introduction

On the wall of a well-known fast food restaurant in Miami Beach is a locally painted mural that features a great ocean wave rearing up and appearing to curl over high-rise hotels that line the shorefront. Whether the artist intended to communicate a more benign message about sea, sand, and relaxation is not known, but the image draws attention to the precarious situation of this low-lying mega-city on the edge of a frequently turbulent ocean. It is a view that is shared by many authors, from the novelist Joan Didion (1987) to the cartographer Mark Monmonier (1997). But the generally available literature on environmental hazards in Miami is surprisingly thin, despite the recent dire experience of hurricane Andrew. More attention is paid to the city's high crime rate and to its ethnic and social divisions than to its natural environmental risks. Yet all are part of the package of hazards that confronts residents and visitors alike and all periodically become mutually intertwined in public policy issues. As Bill Solecki shows, a joint approach to human and natural components of vulnerability can have significant implications and benefits for both scholars and managers. Also important in Miami's case is the potential for extending emerging notions of socially sensitive hazard management from the context of more developed mega-cities to less developed mega-cities, where they are perhaps most needed. Miami's role as de facto capital of the Caribbean Basin assures it high visibility as a model suitable for emulation elsewhere.

428

Introduction

Megacities are often defined as cities that contain populations in excess of 10 million. With more than 3 million residents, greater Miami does not qualify as a true mega-city but it shares many characteristics of its larger brethren, especially a remarkable human heterogeneity. Like most contemporary big cities, Miami contains a jumbled mix of contrasting populations whose daily lives and outlooks are quite different. Because of this social fragmentation it is often described as a city "on the edge" – behaviourally as well as locationally (Portes and Stepick, 1993). Some analysts believe that social cleavages and contending social visions weaken the urban fabric by promoting deprivation and marginalization (Davis, 1995; Ezcurra and Mazari-Hiriart, 1996; Linden, 1996). Others argue that diverse populations and outlooks provide the energy that fuels continued growth and makes institutional reform possible (Cohen et al., 1996; Sassen, 1993). For students of environmental hazards and disasters these interpretations are not just the opposite sides of an academic debate; the effectiveness of mega-city programmes for disaster recovery, reconstruction, and reduction hangs in the balance.

This chapter examines the capabilities of subpopulations and institutions for coping with hazards and disasters in south Florida's Dade and Broward counties (fig. 12.1). The underlying thesis is that vulnerability to hazards is socially constructed. Societal processes make certain places and certain people more vulnerable to hazards than others; such inequities are crucial contributors to the increasing vulnerability of mega-cities. But vulnerability is not fixed, especially in big cities. The social structures of mega-cities are always developing inequities, undergoing progressive reform, fracturing, and then achieving new cohesion. Here the dynamics of social heterogeneity are examined through the experience of interest group coalitions that emerged in the wake of hurricane Andrew (1992). Attention is directed to how these coalitions were differentially affected by the hurricane as well as to their roles in the response and recovery process and the consequent effect on Miami's overall vulnerability. The presentation is organized in five sections. First, I elaborate the concept of mega-city vulnerability. I then discuss interest group coalitions and their effects on urban change, and provide information about the development of coalitions in metropolitan Miami. After detailing the local history of environmental hazards, I finally examine coalition interactions in the aftermath of hurricane Andrew, as well as resulting changes in vulnerability.

Fig. 12.1 The state of Florida

Megacities: Big places, big problems

Broadly speaking, there are two interconnected sets of explanations for the growth of mega-cities. The first sees them as products of processes that are internal to specific countries – processes that drive the concentration of people, wealth, and power. The second attributes their growth to the operation of global capitalism, which has created a supervening world system that requires certain functions to be performed in very large cities (Knox and Taylor, 1995; Sassen, 1994; *Urban Geography*, January–February 1996).

Although big cities and smaller ones both suffer from similar types of social and environmental problems (e.g. crime, pollution, inequity, infra-structural decay), the scale of mega-city problems is distinctive. For a start, sheer size begets complexity and other so-called diseconomies of scale. Western liberal scholars and planners often describe these cities as too large to manage in a coherent fashion (Cohen et al., 1996; Haughton and Hunter, 1994; World Resources Institute, 1996). Huge mega-city populations and rapid rates of growth pose potentially overwhelming problems for managers who are responsible for regional resource bases and for the physical environment. Secondly, mega-cities experience a heightened degree of social, economic, and political fracturing. Particularly important are divisions of income (rich and poor), race and ethnicity (often exacerbated by recent immigration), and political affiliation (multiple jurisdictions) (Clark and McNicholas, 1995; Mitchell, 1995).

Mega-cities are also more vulnerable to environmental hazards than are smaller cities. It is inherently difficult to develop a hazard-management plan for a mega-city (Mitchell, 1995). Size, dynamism, societal dissonance, and lack of basic knowledge about local hazards are particular barriers. Moreover, the impact of a mega-city disaster may spread to affect adjacent rural areas and even an entire country. In poor nations, the potential for deaths, injuries, and other losses is heightened because so much of the country's population and wealth is usually located in its mega-cities. The psychological and economic impacts of devastation may also extend well beyond the local level, especially for the international banking, financial, and trade institutions that typically cluster in mega-cities. For example, if the central economic institutions of Miami were to be devastated, the implications for business and commerce would be global in scope.

Vunerability and its opposite – resilience – are central topics of mega-city hazards research (Blaikie et al., 1994; Dow, 1996; Hewitt, 1984; Liverman, 1990a, 1990b; Timmerman, 1981). Although formal definitions of these terms are elusive, vulnerability can be defined as the likelihood that an individual or group will be exposed to and adversely affected by a hazard (Cutter, 1993). Resilience can be defined as the ability of an individual or group to respond to – and recover from – a hazardous event. It is possible to make a distinction between vulnerable places and vulnerable people; different, yet potentially overlapping, sets of factors contribute to both types of vulnerability. These include biophysical characteristics (e.g. earthquake frequency, climatic variation) and societal characteristics (e.g. social inequity, access to education, poverty).

Several attempts have been made to identify different components of vulnerability. Liverman (1990a) presents one summary list. This includes: environment (e.g. climatic variability, biodiversity, deforestation); technology (e.g. infrastructure, energy use, indigenous knowledge); social relations (e.g. income, gender, race, ethnicity), demographics and health (e.g. population growth rate, age structure, nutrition); land use and ownership (land tenure, property regimes); and economic processes and institutions (e.g. access to markets, price structure, effectiveness of government policies). Kasperson (1994) identifies three broad categories of factors that are important in determining relative vulnerability. These are: ecosystem sensitivity (e.g. local and regional ecological shifts associated with environmental change); economic sensitivity (e.g. cascading of environmental change impacts through an economic system and its ability to respond); and social structure sensitivity (e.g. response capabilities of particular social groups to environmental change). Blaikie et al. (1994) articulate a three-step progression of societal conditions and processes that explain vulnerability: (1) root causes (e.g. access to power, dominant social ideology) affect (2) dynamic contextual pressures (e.g. lack of local investments, rapid population growth), which have outcomes in the form of (3) unsafe conditions (e.g. failing local infrastructure, low income levels).

It is widely believed that vulnerability changes gradually in step with modifications of societal structures. But other scholars have begun to focus on specific moments when vulnerability changes dramatically (Dow, 1996) and on broader enabling changes in Nature–society relations. Merchant (1989), for example, presents a history of ecological revolution in New England that involves three shifts of so-called "production paradigms" over a period of 400 years. This approach provides a broad context for understanding other changes in hazard perceptions and hazard-management practices that can be traced to single events (Cooke, 1984; Solecki and Michaels, 1994). The latter studies argue that aftermaths of disasters often facilitate the adoption of new hazards policies and plans by individuals and/or coalitions with vested interests. Such initiatives can, in turn, change the vulnerability of society to future events.

Interest group coalitions in mega-cites

Interest group coalitions are "temporary alliances [of interest groups] for limited purposes ... which are constructed to pool limited resources and coordinate strategies" (Knoke, 1990, p. 209). They attracted the interest of resource geographers and other social scientists during the 1970s and 1980s (Clarke, 1990; Cox and Mair, 1988; Jonas, 1992, 1993; Logan and Molotch, 1987; Mollenkopf, 1983; Molotch, 1976). A central research issue has been the role of coalitions in developing and maintaining social inequities. Much early work focused on the politics of urban growth coalitions, which were made up of members of local élites who were attempting to perpetuate their own privileged economic and political positions. Most of this research adopted a political economy perspective, focusing first on class conflicts and then on ethnic and racial conflicts. More recently, natural resource conflicts have become a focus of study (Blaikie and Brookfield, 1987; Roberts and Emel, 1993).

Several common themes have emerged from the literature. A primary finding is that most coalitions focus on control, power, and capital accumulation – whether driven by individual or collective ambitions. Secondly, coalitions are quite flexible and can form or dissolve quickly as requirements change, especially in advanced capitalist Western countries. Thirdly, a coalition intersects with other coalitions at varying spatial scales (i.e. local, regional, state, national, and international). In summary, both nested relationships and alliances among coalition partners are common, and fluid connections are the rule. Conversely, such fluidity also permits frequent conflicts about interests or ideology among former coalition partners. Coalitions are particularly important in cities of the postmodern era because such communities are fractured by processes of change and spatial mobility, especially along lines of class, income, and ethnicity (Mollenkopf and Castells, 1991; Soja, 1995). People who live in these cities depend on coalitions to make local politics work. Three types of coalition are common: ethnic- and/or racial-based coalitions; place-based coalitions; and production-based coalitions. Ethnic- or race-based coalitions are organized around different groups (e.g. African-Americans, Italian-Americans, or Korean-Americans; business, political, or social groups). Place-based coalitions are defined by bounded spaces (real or subjective) such as neighbourhoods, municipalities, regions, or types of landscape (e.g. downtown business associations, municipal chambers of commerce). Production-based coalitions form around particular methods or modes of production, such as primary, secondary, or tertiary industries (e.g. home-builder leagues, farm cooperatives), or around other defining factors, such as site-based versus situation-based industries (e.g. tourism vs. trade) or fixed capital industries versus mobile capital industries (e.g. pulp and paper industry versus telecommunications).

All these types of coalition are present in metropolitan Miami. Different ethnic and racial groups coalesce and spin apart as their needs change. Although African-American, Cuban, and White populations represent the best-known groups, the city also has many other well-defined ethnic and racial entities (e.g. Haitians, Jamaicans, Jews). In addition there are marked land-use and occupancy interest groups (e.g. urban, suburban, exurban, and agricultural), with differing and often competing interests or agendas. These may come together or split apart around different questions at different times. The city also contains some coalitions that reflect different political aspirations of industries that are closely tied to a location (e.g. tourism) and others that are more mobile (e.g. banking, finance, insurance). Together, the coalitions and their activities both take their form from, and contribute to, the complex and dynamic character of social relations in Miami.

Development of metropolitan Miami

Birth of a city

Until the end of the nineteenth century much of south-east Florida remained a settlement frontier.[1] With the opening of a coastal railway to Miami in 1896, the area became accessible to tourists and potential residents. In 1900, apart from distant Key West, which had a population of 18,000, only 5,000 people inhabited the seven counties south of Lake Okeechobee. The largest node was Miami (300), founded in 1896 (Chapman, 1991). Fewer than 10 named settlements were present in what would eventually become metropolitan Miami. All were located close to the coast, no more than a mile from estuaries and beaches. Little land was in agricultural production and yields were correspondingly low (Snyder and Davidson, 1994).

By 1930, the population of Dade and Broward counties had grown to approximately 163,000, as retirees, tourists, and other settlers flooded in. Growth rates in the region during this time were some of the highest recorded by the US Census, exceeding 100 per cent per decade from 1900 to 1930 (see table 12.1). The population influx to coastal areas encouraged the construction of major water- and flood-control systems that made possible additional growth. But the vast majority of new settlers remained tightly clustered within a few miles of the Atlantic coast where they valued the natural amenities. Only near Miami did significant settlement occur further inland. Encouraged by a state law that permitted groups of 25 registered voters or freeholders to establish municipalities, many cities were incorporated during this early period, thereby laying the basis for the fractured political landscape of today (fig. 12.2). In 1990, there were 56 incorporated municipalities in metropolitan Miami, with the largest (the city of Miami) containing approximately 10 per cent of population. Moreover, roughly a third of the population live in unincorporated areas, outside municipal governmental control (table 12.2). By 1930, agriculture was also a firmly established economic activity. In Dade and Broward counties, two large agricultural districts had developed on lands just beyond the urban fringe. Winter vegetables, such as string beans, tomatoes, potatoes, and celery, were dominant crops south of Miami and in a discontinuous zone stretching north between the cities from Miami to Palm Beach County.

Population growth slowed somewhat in the 1930s and 1940s but the resident population reached 580,000 by 1950. By 1950 the general regional pattern of land use had not changed, although urban areas were beginning to expand (fig. 12.3). Much of the increased residential growth occurred as filling-in of unoccupied areas or agricultural land around Miami and Ft. Lauderdale.

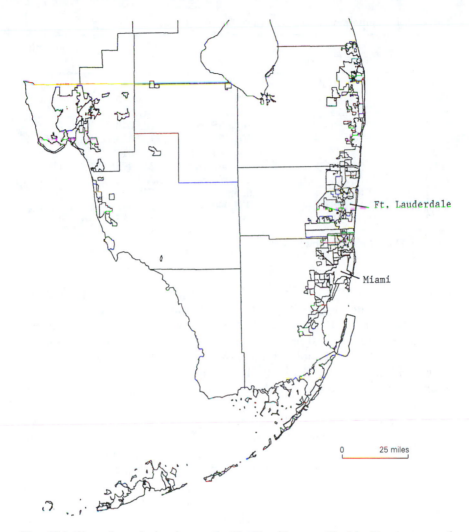

Fig. 12.2 Place boundaries in south Florida (Source: Florida Resources and Environmental Analysis Center and the Florida State University Department of Geography)

Post–Second World War expansion

Metropolitan Miami has grown enormously since 1950, propelled by several forces. Chief among these in the 1950s and 1960s were (1) increased national demand for winter vegetables and fruits, and (2) shifts in tourism and residential preferences, particularly among retirees. Strong non-local

Table 12.1 Population of metropolitan Miami, 1900–2000

Year	Broward County	Dade County	Total	% growth per decade
1900	–	500 est.	500 est.	n.a.
1910	–	11,933	11,933	n.a.
1920	5,135	42,753	47,888	301.3
1930	20,094	142,955	163,049	240.5
1940	39,794	267,739	307,533	88.6
1950	83,933	495,084	579,017	88.3
1960	333,946	935,047	1,268,993	119.2
1970	620,100	1,267,792	1,887,892	48.8
1980	1,018,257	1,625,509	2,643,766	40.0
1990	1,255,488	1,937,094	3,192,582	20.8
2000	1,474,228	2,083,029	3,557,257	11.4

Source: US Bureau of the Census.

markets for Florida's winter vegetables and fruits developed throughout the United States as the city's transportation and product distribution system became more integrated into Northern and Midwestern markets. The value of agricultural product sales in south Florida as a percentage of all farm sales in the United States more than doubled during the period from 1949 to 1968 (Winsberg, 1991). In the meantime, south Florida had become the destination of choice among a very large section of the nation's retirees.

Another factor that affected growth patterns was a series of massive, hurricane-related floods in the late 1940s. These gave increased impetus to the expansion of regional water projects. As water-management goals shifted from drainage to flood control, the United States (federal) government built many dykes and levees. A prominent example was the Eastern Protective Levee, which was designed to prevent flood waters in the Everglades from flowing east towards developed areas. The levee became a physical and psychological boundary that separated the remaining natural Everglades from developed areas. It in effect moved the location of the normal flood-free area 5 to 10 miles further west and established a boundary beyond which lay large tracts of the former natural hydrological system that were barred to farming and urban development.

By the late 1960s, Greater Miami had taken on the character of a large diverse metropolitan region. It possessed a fully developed infrastructure, including a system of major highways and canals that facilitated continued expansion and intensification. It also acquired some other trappings of big-city life in America, such as a major professional sports team (Miami Dolphins, in 1966), a national political convention (Republican

Table 12.2 Population of metropolitan Miami by municipality type, 1940–1989

Year	Miami	Miami Beach	Ft. Lauderdale	Other municipalities	Total	Unincorporated area	
						Broward County	Dade County
1940	172,172	28,012	17,996	47,064	45,289	3,738	41,551
1950	249,276	46,282	36,328	122,587	124,534	14,796	109,738
1960	291,688	63,145	83,648	134,879	490,433	138,216	352,217
1970	334,859	87,072	139,590	664,615	661,757	124,464	537,293
1980	346,681	96,298	153,279	1,080,744	966,764	167,711	799,053
1989	371,444	98,047	150,631	1,349,502	1,145,902	157,682	988,220

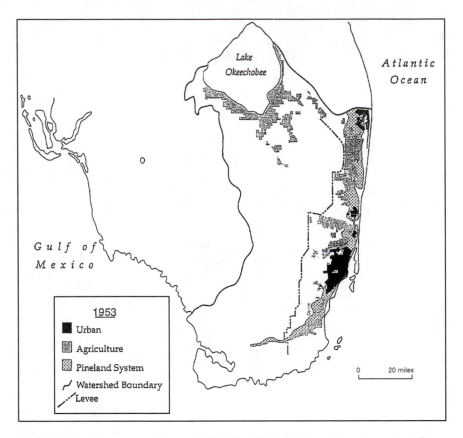

Fig. 12.3 Land use/cover in south Florida, 1953 (Source: Department of Geography, Florida State University)

Party, in 1968), and race riots. A new urban political structure began to emerge characterized by lack of an entrenched patrician class, a marginalized African-American community (Mohl, 1995), and rapid upward economic and political mobility for wealthy in-migrants (Portes and Stepick, 1993).

Agricultural expansion and increased demand for warm weather retirement living have remained important components of regional growth during the past two or three decades but they have now been joined and overtaken by other stimuli. Metropolitan Miami has increasingly become a destination for international migrants, and a centre for international banking, finance, and trade, especially within the Caribbean and Central America. The Cuban Revolution provided an initial catalyst that set off mass migration of mainly middle- and upper-income Cubans to Miami. When Cuba's new communist leader, Fidel Castro, came to power in 1959, Hispanics accounted for 5.3 per cent (50,000) of Dade County's population; in 1990, they comprised 47.5 per cent (916,000). The Cuban impact on metropolitan Miami, particularly Dade County, has been profound. Building on their former class status, many Cubans quickly became prominent members of the local economic and political system.

By the late 1970s, metropolitan Miami was composed of several different population subgroups: a large Cuban community, northern US transplants (including a substantial Jewish population), southern US African-Americans and Whites (Boswell, 1991; Mohl, 1982a, 1982b; Moore, 1994; Portes and Stepick, 1993). During the 1980s and 1990s, the ethnic and racial make-up changed again as new streams of migrants arrived. These came mostly from the Caribbean basin and Latin America. By the early 1990s, sizeable numbers of people from Colombia, the Dominican Republic, Ecuador, Haiti, Honduras, Jamaica, Nicaragua, Panama, Peru, and Puerto Rico had taken up residence. The degree of spatial clustering varied among and within these groups. Generally, higher-income families moved to class-defined neighbourhoods rather than ethnically defined ones. Whereas some groups – such as the Haitians and Nicaraguans – subsequently remained tightly clustered in well-established enclaves, other migrant groups – such as the Colombians – dispersed more widely throughout the region. Regardless of residential location, members of different nationalities today usually maintain strong intra-ethnic social and political ties, which have served as bases for the numerous ethnically organized coalitions that are present in the region.

Continued rapid population growth during the past several decades has put additional pressures on the local environment and on social systems. By 1990, the population of Dade and Broward counties had almost reached 3.2 million. Most people lived in land-intensive, low-rise, single-family dwellings. In-migrants, both retired and working, increasingly sought out lower-rent, inland locations. Tourists still clustered near the coasts, and non-residential developments with detached homes and landscaped lots near amenities such as golf courses began to increase dramatically. By 1973, an almost continuous strip of urban development stretched along the Atlantic coast. As Atlantic coast rural land was converted to urban use, other parcels in inland locations were converted to agricultural uses. Agriculture was increasingly confined to isolated pockets situated between the urban fringe and publicly owned conservation lands to the west (including the Everglades) (fig. 12.4).

Changes of land use and land cover are evidence of increased competition for space between agricultural and urban interest groups in metropolitan Miami (Walker et al., 1997). As the urban land cover area has increased, agricultural interests have moved operations to newly opened

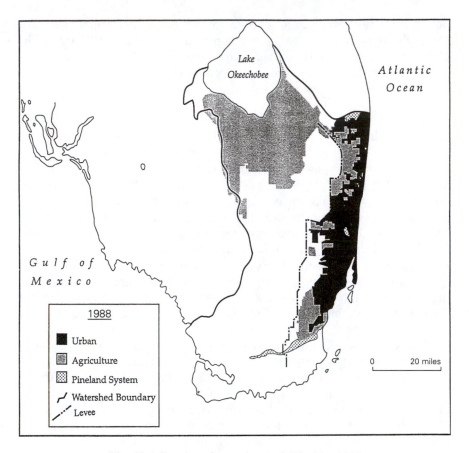

Fig. 12.4 Land use/cover in south Florida, 1988

areas, particularly interior locations. Other tensions between the agricultural and urban interests also have emerged. Points of contention include: water pollution from agricultural chemical runoff; pesticide spraying; and allocation of water resources. These problems, along with rising costs and increased foreign competition, have placed metropolitan Miami farmers under tremendous pressure. Continued land conversion and increased land speculation have raised concerns about the long-term viability of farming in the region (Dunlop, 1995; Winsberg, 1991). Only a relatively small amount of land in Dade County still remains in cultivation, mostly in the south (table 12.3). Broward County too is rapidly losing its remaining agricultural land and the value of agriculture (both land and products) is an increasingly small contributor to the regional economy (tables 12.3 and 12.4).

Table 12.3 Farming in 1992

County	No. of farms	Land in farms (acres)	Change in land, 1987–1992	Land in cropland (acres)	Irrigated land (acres)	Size per farm (acres)	Market value per farm (US$)	Total market value of agricultural product (US$)
Broward	393	23,735	−12,174	(D)	3,388	60	315,376	34,742,000
Dade	1,891	83,681	+620	68,795	52,363	44	389,694	356,967,000
Total	2,284	107,416	−11,554	n.a.	55,751	n.a.	n.a.	391,709,000

Source: *Census of Agriculture: State and County Data, Florida*, US Department of Commerce, Census Bureau.
D = not disclosed because of confidentiality restrictions

441

Table 12.4 Land use in 1993: Assessed value (US$ million)

County	Residential	Commercial	Industrial	Agricultural	Institutional	Miscellaneous
Broward	40,658.00	8,338.00	2,722.00	114.47	1,129.92	3,563.18
Dade	47,886.72	13,096.60	4,444.63	520.96	1,923.59	8,187.95
Total	88,544.72	21,434.60	7,166.63	635.43	3,053.51	11,751.13

Source: Department of Revenue, State of Florida, unpublished data; printed in *Florida Statistical Abstract 1995*, Bureau of Economic and Business Research, College of Business Administration, University of Florida, Tallahassee, FL: University Press of Florida.

Expansion of the economic base

By 1970, the economy of the city had three mains components: agriculture, tourism, and federal government transfer payments to retirees and others. This is one of the few areas of the United States where crops can be grown in winter. Retirees were attracted on the one hand by warm winters, beaches, and a slow, pleasant pace of life and on the other by low tax rates and a concentration of services for senior citizens, particularly health care.

Since 1970 the economy of the region has been transformed to capitalize on its situational characteristics. Most important is Miami's relative location *vis-à-vis* markets in Latin America and North America. Market advantages operated for both conventional and unconventional economic products. For example, a narcotics boom occurred in the mid-1970s and Miami became a drug capital for much of the world. It was estimated that more than 70 per cent of the US supply of heroin, cocaine, and other illegal substances flowed through the region. This traffic brought drug-related crimes and wealth to Miami. An influx of "hot" dollars quickly made Miami a major financial centre by the early 1980s, with banks and multinational corporations being lured there by the huge cash flows. Cash came directly from Latin America and from large retail sales of drugs in cities such as Chicago and New York. Throughout the 1980s, the Federal Reserve Bank of Miami reported surpluses of US$4–6 billion per year. Money laundering and cash surpluses were conservatively estimated to have added between US$1 billion and US$2 billion dollars to the Miami economy every year (Cartano, 1991).

Another major source of capital came from exporters, retailers, and realtors who catered to the more than 2 million Latin Americans who visited Miami each year from 1976 to 1983. Fearful of volatility in their home countries, élites used Miami as a safe haven for their money as well as an entertainment and shopping centre. After the collapse of many Latin American economies in the early 1980s, the rate of growth slowed significantly. None the less, by 1990 Miami had become a major international centre of trade. Air traffic (both passengers and cargo) was the most significant factor in this change. Whereas the Miami customs district handled only 2.1 per cent of US trade in 1990 (US$19.1 billion), it processed approximately 40 per cent of the trade to Central and South America and the Caribbean (Nijman, 1996). The aspiration to make Miami a world-class trading centre has continued to be expressed and acted upon throughout the 1990s. For example, the most recently elected (24 July 1996) Mayor of Miami, Joe Carollo, has called for construction of a world trade centre to capitalize on the city's desire for prominence in global trade as well as trade with Latin America. He has argued that

Miami might take over Hong Kong's role after that city returns to Chinese rule in 1997 (Anon., 1996a).

Fire, insurance, and real estate (FIRE) industries have lately assumed high prominence in Miami, but tourism still plays an important role in the local economy and the employment sector. For example, Broward and Dade counties together maintained 76,641 hotel and motel rooms (Florida Department of Business and Professional Regulation, 1994). More than 25,000 people were employed by hotels and motels and another 20,000 were employed in amusement and recreation services. Federal money transfers are also still important to the local economy. Metropolitan Miami garnered close to US$13 billion in federal government transfers and grants during 1992. The almost 400,000 retirees in Broward and Dade counties together receive more than US$262 million of social security payments every month.

Hazards of metropolitan Miami

Water-related hazards abound in Miami because water is ubiquitous here and the boundaries between land and water are blurred (Blake, 1980). For example, most of the metropolitan area receives more than 60 inches of rain per year and the highest spot is just a few feet above sea level. But water is dynamic; according to Craig (1991), before large-scale water use began in the early twentieth century, the hydrostatic head of fresh water coming down to the coast via the western Everglades system was so great that artesian springs existed at numerous offshore locations in the shallows of Key Biscayne. These no longer function but the entire region remains almost totally surrounded by water. To the east lies Biscayne Bay and the Atlantic Ocean, to the west are the seasonally inundated Everglades, and to the south is Florida Bay and the Florida Straits. The pre-settlement water table was close to the surface and underlying rock formations are honeycombed with karst features. The visually inconspicuous Atlantic coastal ridge is the only above-ground solid geological stratum that connects Miami with the rocks of the US mainland. As Craig's comments about the now-defunct artesian springs make clear, human impacts on Miami's hydrosphere have been extensive. In particular, construction of the dense network of canals and levees has dramatically changed local hydrology, draining some areas and making others subject to increased flooding (Blake, 1980).

The watery character of Miami is reflected in misleading characterizations of the city as a tranquil paradise. Instead it is highly vulnerable to violent meteorological events that are characteristic of its subtropical climate. Most people live on what is actually a long thin peninsula rising just a few feet above the water and jutting out into the path of severe storms, especially hurricanes. A few miles inland the eastern Everglades are distinguished by intense heat, heavy rain, and swarms of mosquitoes. Nineteenth-century explorers described it as a desolate region that should be avoided (Derr, 1989). Although hurricanes are the most significant local environmental hazard, the region is also subject to lightning, floods, tornadoes, waterspouts, drought, and freezes. The history of Miami has been punctuated by extreme events such as the great freeze of 1895, the 1926 hurricane, the drought of the early 1960s, and hurricane Andrew (1992). These and other extreme events are discussed next (table 12.5).

Hurricanes

Dozens of hurricanes have swept through south Florida in the past 125 years. Before 1992, the most significant was that of 1926, which directly struck Miami Beach and Miami. Then national headlines proclaimed, "Miami Beach is wiped out" (Chapman, 1991). That storm killed hundreds, injured thousands, caused more than US$100 million worth of damage, left 47,000 homeless, and established a benchmark of loss that was not surpassed for two-thirds of a century (Parks, 1986).

Between 1926 and the arrival of hurricane Andrew, the region's vulnerability to hurricanes changed. Risks of death or injury declined dramatically as warning systems and evacuation procedures improved, but the economic costs of damage continued to rise. At around US$30 billion, economic losses attributable to hurricane Andrew confirmed this trend, although the storm was not the worst that could be envisaged. It is believed that the so-called "big one," a large hurricane that directly strikes a heavily populated urban coast, could cause up to US$50 billion in losses. Hurricane Andrew gave some clues about the likely effects of such a superstorm. It struck in August 1992, cutting across a relatively narrow swathe of southern Dade County. As a category 4 storm (on the Saffir/Simpson scale) it might well have caused extensive water damage and significant flooding in low-lying coastal areas. In fact, wind caused most of the problems. Andrew destroyed or rendered uninhabitable 80,000 homes, made more than 250,000 people homeless, and destroyed or damaged 82,000 businesses. In Dade County alone, 15 deaths were directly attributed to Andrew, and another 25 perished from indirect causes. More than 700,000 people were evacuated as it approached. Local crops, particularly limes and avocados, suffered heavy damage (Pielke, 1995; Tait, 1993).

In many respects, Andrew was an unusual hurricane. It was compact and relatively fast moving. The area of extreme damage was quite small, extending outward about 10–15 miles from the point of landfall (fig.

Table 12.5 Selected hazards and disasters affecting metropolitan Miami, 1900 to present

Event	Year	Killed	Injured	Damage (US$ million)	Description[a]
Hurricane	1904	n.a.	n.a.	n.a.	Category 1
Hurricane	1906	n.a.	n.a.	n.a.	Category 2
Hurricane	1906	129	n.a.	n.a.	Strikes Florida Keys (category 2)
Hurricane	1909	n.a.	n.a.	n.a.	Category 3
Hurricane	1916	n.a.	n.a.	n.a.	Category 2
Hurricane	1924	n.a.	n.a.	n.a.	Category 1
Tornado	1925	5	35	n.a.	Miami's deadliest (F3 intensity) skirts north-east edge of city
Hurricane	1926	n.a.	n.a.	n.a.	Category 2
Hurricane	1926	243	n.a.	115.1	Severe damage in Miami area (category 4)
Hurricane	1928	1,836	n.a.	26.2	Focused on Lake Okeechobee region
Hurricane	1929	n.a.	n.a.	n.a.	Category 3
Drought	1930–31	n.a.	n.a.	n.a.	Extensive fires throughout Everglades resulting from drought and water table lowering
Muck fires	mid-1930s	n.a.	n.a.	n.a.	
Hurricane	1935	n.a.	n.a.	n.a.	Category 2
Hurricane	1935	408	n.a.	11.5–40.0	"Labor Day" hurricane – record low barometric reading. Keys hit hard (category 5)
Hurricane	1941	n.a.	n.a.	n.a.	Category 2
Hurricane	1945	4	n.a.	54.1	Damage in Dade County (category 3)
Hurricane	1947	17	n.a.	51.9	Damage in Broward County/Ft. Lauderdale (category 4)
Hurricane	1947	n.a.	n.a.	n.a.	Category 1
Hurricane	1948	n.a.	n.a.	n.a.	Category 2
Hurricane	1949	2	n.a.	45.0	Gold Coast hurricane. West Palm Beach (category 3)
Hurricane King	1950	3	n.a.	31.6	Damage in Miami (category 3)
Freeze	1957–58	n.a.	n.a.	n.a.	Freezing temperatures throughout south Florida
Hurricane Donna	1960	11	n.a.	305.0	Hits extreme southern Florida and the Keys (category 4)
Drought	1961–63	n.a.	n.a.	n.a.	

Event	Year				Description
Hurricane Cleo	1964	0	n.a.	n.a.	Along east coast (category 2)
Hurricane Betsy	1965	8	n.a.	139.3	Extensive flood damage in Miami and Keys (category 3)
Hurricane Inez	1966	n.a.	n.a.	n.a.	Category 1
Tornadoes	1968		n.a.	n.a.	5 funnels strike Miami
Race riots	1968	3	n.a.	n.a.	Liberty City (Miami)
Drought	1970–71	n.a.	n.a.	n.a.	South Florida experiences worst water deficiency in 200 years
Freeze	1977	n.a.	n.a.	n.a.	Snow in Miami; agricultural disaster
Hurricane David	1979	<5	n.a.	5.0 approx.	Along east coast – category 2 when hit Florida
Race riots	1980	19	hundreds	n.a.	Liberty City (Miami)
Drought	1981	n.a.	n.a.	n.a.	
Race riots	1982	n.a.	n.a.	n.a.	Overtown (Miami)
Flooding	1983	n.a.	n.a.	n.a.	High rain in south Florida
Freeze	1984	n.a.	n.a.	n.a.	Extensive throughout south Florida
Drought	1985	n.a.	n.a.	n.a.	
Brush fires	1985	n.a.	n.a.	n.a.	
Die-off – sea grass	1987	n.a.	n.a.	n.a.	Extensive die-off of sea grass first noticed in Florida Bay
Hurricane Floyd	1987	n.a.	n.a.	n.a.	Category 1
Drought	1988–91	n.a.	n.a.	n.a.	
Freeze	1989	n.a.	n.a.	n.a.	
Race riots	1989	n.a.	n.a.	n.a.	Miami
Biological	1989	n.a.	n.a.	n.a.	Biologists claim 200 exotic plant species have invaded region
Muck fires	1989	n.a.	n.a.	n.a.	Smoke produces unhealthy conditions in Miami
Hurricane Andrew	1992	17	n.a.	20,000–30,000	Southern Dade County, 250,000 homeless (category 4)

Sources: Doig (1996); Fernald and Patton (1984); Henry et al. (1994); Nash (1976); Winsberg (1990).
a. Saffir/Simpson Damage Potential Category.

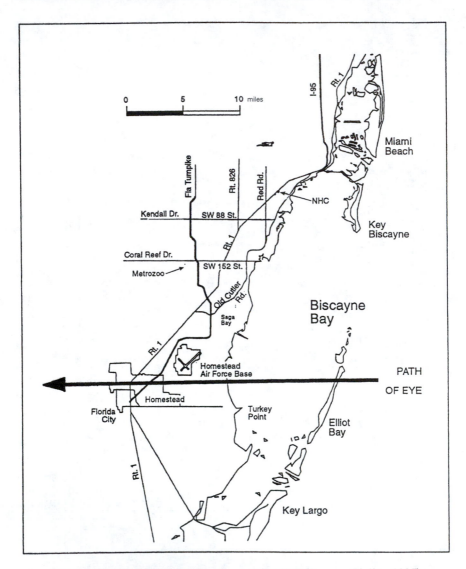

Fig. 12.5 Path of hurricane Andrew at landfall (Source: Pielke, 1995)

12.5). Downtown Miami, approximately 25 miles from this location, suffered relatively minor damage, mostly from blown-out windows. The surge was only 5–8 feet, in comparison with the 13–18 feet typically associated with category 4 storms. Moreover, the storm's rapid forward speed (approximately 20 mph) kept rainfall totals low. Andrew produced about 2–7 inches of rain in locations throughout south Florida and caused only minor inland flooding.

As mentioned above, it was the wind that caused most damage in south Dade County. At the time of landfall, sustained winds of 140 mph and gusts of approximately 200 mph were recorded. The hardest-hit cities were Homestead and Florida City, both farming and working-class retirement communities about 25 miles south-west of Miami. Virtually every building in Homestead was damaged or destroyed. Many unincorporated communities in the sprawling suburban and exurban area between Miami and Homestead also were hard hit. Middle- and upper-income areas of Kendall and Cutler Ridge were some of the most affected. Hurricane Andrew severely disrupted the southern portion of Dade County. There, as elsewhere, it shattered the daily lives of residents (Smith and Belgrave, 1995). Like its predecessor in 1926, Andrew profoundly jarred Miamians into the realization that their city could be devastated by a hurricane. Electric power and water were unavailable for days, and in some cases weeks. Damaged homes were laid bare to the late August heat and humidity of south Florida. Looting of seemingly abandoned properties became a problem in some areas. Initial emergency responses at all governmental levels appeared slow and chaotic. Not until almost a week after the hurricane struck did a coordinated governmental response take shape.

A sense of disorientation was perhaps the most distressing impact for local residents (Smith and Belgrave, 1995). Street signs, vegetation, and other landmarks were destroyed or were buried under piles of debris. Many people became lost in what was often a landscape of total devastation (Kleinberg, 1992). The level of psychological stress faced by those who experienced the storm and its damage was found to be extensive (see special issues of the *Journal of the American Academy of Child and Adolescent Psychiatry* and numerous articles in recent issues of *Psychosomatic Medicine*). Response and recovery were made worse by the spectre of criminal activity.

Soon after Andrew died away, it became evident that levels of damage were only approximately related to hurricane wind speeds. Some areas suffered heavier damage than others that had higher winds. Micro-bursts and other highly localized wind features probably explain some of these differences. Shoddy construction and failure to enforce building codes explained the rest of the damage pattern (fig. 12.6) (Leen et al., 1993; Pielke, 1995). The latter problems were widespread throughout the region, and the case of mobile homes deserves special mention. When built according to approved codes, mobile homes remain vulnerable to high winds; when not built to codes, the potential for disaster is great.

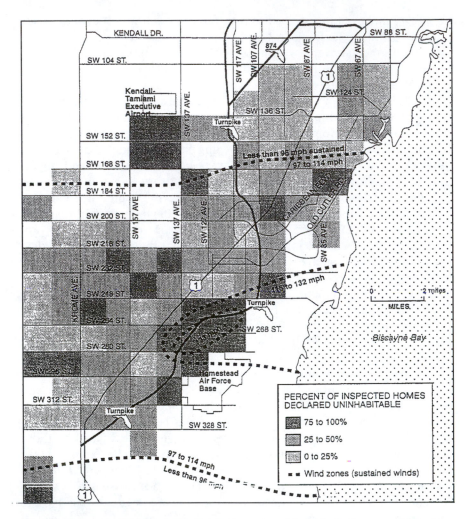

Fig. 12.6 Areas of highest wind speeds and residential damage resulting from hurricane Andrew (Source: Pielke, 1995, adapted from *Miami Herald*)

Hurricane Andrew destroyed 97 per cent of the mobile homes in Dade County (Silverstein, 1995). Some houses were blown apart by winds and sudden pressure changes, while others were swept off their moorings and sent crashing into solid objects.

Price gouging and confidence tricksters were also prevalent following hurricane Andrew. During the first several weeks, basic household and building supplies were in very short supply. The area became infested with charlatan building contractors who either never performed repairs they were paid for or did inadequate work.

Other hazards

Acute events

Rain-driven floods are associated with storms and hurricanes, but rainfall totals in Miami vary widely among events and many storms pass without serious repercussions. Flooding can also occur after prolonged wet periods and remain a problem for weeks or months thereafter. Average annual precipitation in this region ranges from 46.8 inches (119 cm) to 61.8 inches (157 cm) (US Department of Commerce, National Oceanic and Atmospheric Administration, 1987). In specific years it has been as low as 33.8 inches (86 cm) or as high as 88.1 inches (224 cm) (Sculley, 1986).

Other meteorological events that occur in the region have a much lower potential for damage. Tornadoes occur but do not pose a very high risk, and only two major tornado episodes have been recorded in Miami. Lightning also is a common hazard, given the frequency of thunderstorms; but damage and health threats do not approach those of hurricanes. Freezes also cause problems, particularly for local citrus, vegetable, and nursery growers. In central Florida, hard freezes occur every 10–20 years, but are much less frequent in Greater Miami. Freezes in central Florida have spillover effects on south Florida because displaced growers seek out new locations there that are less susceptible to frost. Throughout the twentieth century, citrus growers have been relocating ever southward in order to escape the potential for freezes (Winsberg, 1991). This movement helps fuel conflicts between farming interests and land developers in metropolitan Miami.

Chronic events

Metropolitan Miami is subject to a range of chronic hazards as well as acute ones. The most significant is drought, and droughts of varying severity occur in the region approximately every five to six years. Lack of rainfall is often not the main cause. Human alteration of the Everglades ecosystem and large increases in freshwater demands by agricultural users and urban residents have made the issue of water supply a major public policy matter. Since federal actions and regulations require that minimum water flow into the Everglades National Park be maintained at all times, conflicts among users are exacerbated. Tensions among the three main user groups (i.e. the Park, farmers, and home-owners) increase when there is either too little water (drought) or too much water (wet period) in the local hydrological system. Increasing population and the expansion of settled areas continue to make the potential for conflict more likely.

Changes in the natural hydrological system and the water-supply system have fostered additional hazards, particularly muck or grass fires and salt-water intrusions. Land drainage and water removal have lowered the water table throughout the south-east Florida–Everglades region, allowing the damp ground to dry out and increasing vulnerability to wildfires. Such fires were particularly disastrous during the 1930s in newly drained lands south of Lake Okeechobee (Bottcher and Izuno, 1994; Snyder and Davidson, 1994). Some of those fires burned for months. Property damage was minimal but many inches of fertile topsoil were burned, thereby threatening the long-term viability of agriculture in the area. Muck fires also have occurred closer to Miami, and stories of smoke over the city have been frequent in local newspapers. Increased withdrawals of groundwater have encouraged salt-water intrusions, especially in well fields close to the coast.

Levels of technological hazards such as air pollution are relatively low in Greater Miami. Little heavy industry is present in the region, and that which exists is isolated from populated areas. Water pollution is beginning to increase. Pollution of the Everglades system by mercury, nitrates, and phosphates mostly results from agricultural activities in the Everglades Agricultural Area. Miami, too, has been the site of serious biological hazards. Before the advent of pesticides and drainage programmes, the south Florida region was regularly affected by outbreaks of malaria and other water-related diseases. Miami currently also has one of the highest rates of HIV infection (AIDS) in the United States.

Vulnerability to hazards before hurricane Andrew

Before Andew struck, people and property in Greater Miami were already vulnerable to environmental hazards for a number of reasons. First, disaster-response institutions and personnel had been largely untested by a major event for a long time. Secondly, since the previous disaster the area's population had grown enormously and urbanization had intensified. Thirdly, the local economy was in a fragile state, with government spending being cut back in the wake of a neo-conservative political revolution and agriculture already hard pressed by competing demands for land and water. Finally, local society was both highly fractured and contentious, as well as filled with subpopulations whose experiences of major local disasters and whose response capabilities were both limited (e.g. the aged, non-English-speaking immigrants).

To casual observers, however, Miami residents seemed ready and able to ensure their own safety during a major disaster. Evacuation routes and shelters had been identified and marked; the public was informed about other appropriate precautions, including emergency supplies and protective measures. But, as events proved, most Miamians were not ready for the psychological, social, and economic devastation and disruption that are the mark of a major disaster. Instead they were preoccupied with what they perceived as more pressing problems.

During the early 1990s many residents, political leaders, and social commentators felt that metropolitan Miami was moving swiftly towards a state of crisis. The city of Miami and its surrounding area were described as a troubled paradise in which signs of societal stress were omnipresent. An already loose social fabric was beginning to be torn by internal tensions and problems. Chief among these were income and racial divisions. African-Americans, Hispanics, and Whites had glaringly different average incomes and contrasting rates of poverty and unemployment. African-Americans and Whites occupied opposite ends of the socio-economic spectrum, with Hispanics in the middle. Moreover, the African-American community's position was deteriorating relative to that of Hispanics, who were rapidly emerging as a major political force. Miami residents elected the first Cuban-American mayor in 1985. Frustration and desperation in the African-American community had helped fuel race riots in 1980, 1982, and 1989.

Rising crime was another obvious problem. It had several contributory factors: the legacy of cocaine drug wars in the late 1970s and early 1980s; a North American economic recession in the early 1980s (coupled with serious problems in the economies of many Latin American countries); the 1980 Marielito boatlift, which brought several thousand Cuban criminals to Miami; a widening income gap among different ethnic groups; and a broad range of other contemporary urban ills (Dunham and Werner, 1991). Well-publicized criminal attacks on tourists provided graphic evidence of Miami's social problems and severely stung the local tourist industry by driving away visitors.

The declining appeal of Miami for tourists was accentuated by increasing recognition that nearby beaches and waters were becoming polluted. This was just one of several troubling signs of environmental deterioration. Events such as the precipitous decline of sea-grass beds in Florida Bay were seen as signals of an ominous future. The Everglades were drying and dying, urban sprawl was swallowing up farmland and open space, and groundwater supplies were being overused and contaminated. Proposed solutions were controversial and caused sharp conflicts among the various interest groups. Ensuing debates further fractured regional politics along lines such as: old-timers versus newcomers; some ethnic and racial groups versus others; growth interests versus environmental conservation interests; agricultural versus urban interests; coastal versus non-coastal interests; and Native Americans versus all others.

Extra-regional issues also continued to play a major role in Miami. The future of Cuba's communist regime hung over city politics not least because the large local Cuban community were strong vocal opponents. Political change in Cuba, through either violent or peaceful means, was ardently sought, and local residents were well aware that this might have major implications for the future of Greater Miami. As events of the early 1990s seemed to bring the prospect of a new Cuban government closer, Florida's Governor, Bob Martinez, initiated a study commission (1990) to determine the potential impacts on south Florida. The weakness of Latin American economies added to Miami's problems. Although most countries had recovered from the depths of the early 1980s' recession, their economies continued to languish, which, in turn, acted as a drag on the economy of metropolitan Miami. Growth continued but at a sluggish pace.

In summary, Miami was tense and troubled in the summer of 1992. The perceived snubbing of Nelson Mandela by the city's Hispanic leaders because of his favourable stance towards the Cuban government had sharpened tensions between the African-American community and the Hispanic community. The manslaughter trial of a Colombian-American police officer charged with shooting an African-American man seemed to exacerbate the divisons. The White community also felt increasingly threatened by growth of the immigrant community in general and of the Cuban community in particular. Moreover, the regional economy was stagnant and the city of Miami was reeling from the loss of four of its largest and most prestigious employers, including Eastern Airlines and Pan American Airlines. Finally, south Florida had just experienced one of its worst droughts.

Thus was the scene on the eve of hurricane Andrew. Greater Miami had experienced fast-paced social change for several decades and was riven with serious problems. In a postscript to their seminal book about Miami, *City on the Edge*, Portes and Stepick (1993) assert that Andrew may have thrown the rate of change into "high gear." Now, four years after the event, that early evaluation can be critically assessed.

Coalition responses to hurricane Andrew

After hurricane Andrew moved away, its impacts became grist for the pre-existing interest group coalitions in Greater Miami, which sought to respond to the new realities that it had unleashed. This section looks at responses made by the three basic types of coalition.

Ethnic- and race-based coalitions

Demographic changes that were already under way sharply accelerated in the wake of hurricane Andrew. A spatial realignment of ethnic and racial groups in metropolitan Miami had been taking place before the storm arrived. Throughout the 1980s, Dade County became more Hispanic. Whites, and to a lesser extent African-Americans, had increasingly moved out of the more urbanized parts of Dade County. Many Whites resettled in Broward County to the north, while others relocated to predominantly White, unincorporated, middle- and upper-class enclaves in southern and western Dade County, particularly Kendall, Kendall Lakes, and Cutler Ridge (fig. 12.7). As a result, southern Dade County was a patchwork of different social groups in 1992. Large neighbourhoods of low-, middle-, and upper-income residents were present along with small pockets of intense poverty and wealth (fig. 12.8). A majority of the population was White, but sizeable numbers of African-American and Hispanic residents also lived in the area affected by Andrew (figs. 12.9 and 12.10). However, the largest African-American and Hispanic communities lay further north. Mobile home residents were clustered at the southern end of Dade County, in and around the cities of Homestead and Florida City (fig. 12.11). These were primarily occupied either by migrant farm labourers and their dependants or by less affluent retirees. Like much of the rest of metropolitan Miami, social and community bonds in south Dade were not strong. Most residents were newcomers who thought of other places as home, whether in Havana, Cuba, Long Island, New York, or Sonora, Mexico. Differences of income, ethnicity, race, and birthplace dampened interactions among the various subpopulations.

During a few hours, Andrew changed forever this suburban and exurban landscape and the social relations of its residents. In the initial chaotic aftermath, all residents and groups felt they were victims and that they shared a common plight (Kleinberg, 1992). Faulty construction and building code violations occurred in both wealthy and poor neighbourhoods. Older and smaller homes, owned by poorer residents, sometimes withstood the winds better than newer, more expensive ones (Pielke, 1995). Initially, as is often the case in disasters, disparate strands of the community came together. Commenting on the change, one victim stated that, "after the hurricane, there was this message – unity. Before there was a lot of segregation: among friends, ethnic groups. Immediately afterward, no one seemed to care – black, white, Spanish, Latin, or what." But the so-called "therapeutic community" did not last long. One week after the storm, people who were left homeless streamed out of south Dade County, sometimes seeking refuge beyond the county entirely. This movement was fuelled by recognition that it would – at best

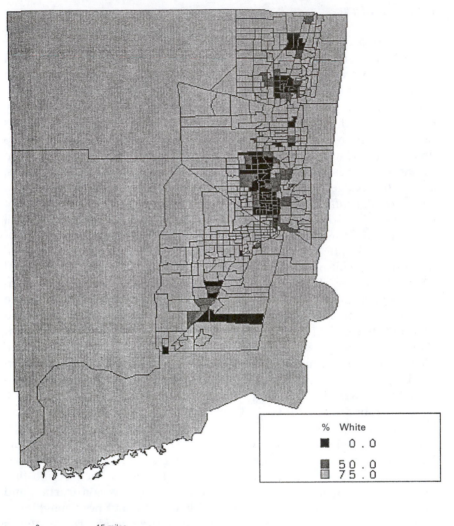

0 _____ 15 miles

Fig. 12.7 White population in metropolitan Miami, 1990 (% by Census tract) (Source: author and US Census)

– take many years to restore everyday life. However unwillingly, hurricane Andrew had provided many residents with an opportunity to move.

Residents of damaged homes and neighbourhoods belonged to three categories. Some had considered moving in the past and immediately took advantage of the devastation to do so. Others resisted moving at first and then eventually acquiesced. Still others fiercely resisted moving and

Percent By Census Troct

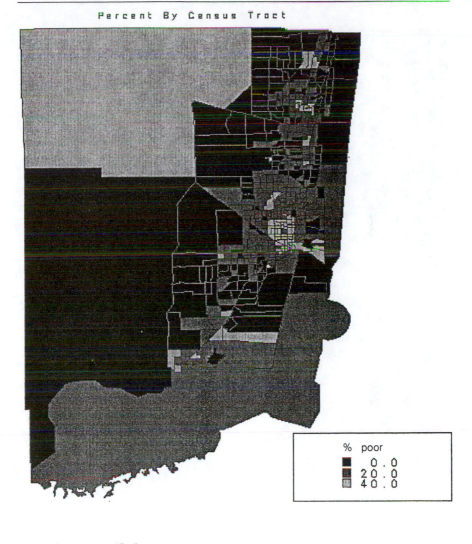

0   15 miles

Fig. 12.8 Persons below the poverty level in metropolitan Miami, 1990 (% by Census tract) (Source: author and US Census)

remained at their original locations, occupying damaged homes for extended periods or, more frequently, staying in trailers brought to the sites (Bell, 1994; Lambert et al., 1994; Smith and Belgrave, 1995; Smith and McCarty, 1996). Over the first several months, tens of thousands of people resettled out of the area; estimates typically exceed 40,000. Most did not go far and remained within commuting distance of existing jobs

% Black
■   0.0
▨ 50.0
▨ 75.0

0   15 miles

Fig. 12.9 African-American population in metropolitan Miami, 1990 (% by Census tract) (Source: author and US Census)

in Dade County. Many others moved to south-west Broward County (Bader, 1993; Smith and McCarty, 1996; Winsberg, 1994).

Early indications suggest that most migrants belonged to middle- and upper-income groups. The loss of their income to Dade County could total as much as US$500 million (Winsberg, 1996). Other evidence seems to confirm anecdotes that many of those who moved out of the county

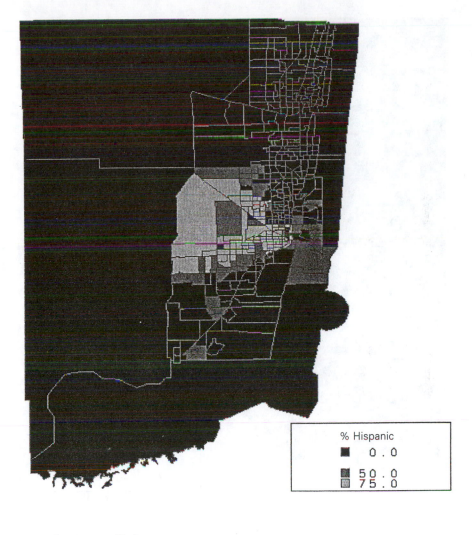

0 ———— 15 miles

Fig. 12.10 Hispanic population in metropolitan Miami, 1990 (% by Census tract) (Source: author and US Census)

were White (Winsberg, 1994). A local sociology professor who observed the migration in September 1992 assessed it as follows:

South Dade (the area hardest hit by hurricane Andrew) has been one of the few remaining areas of the county with an "Anglo" population majority. It has also been one of the few areas in Dade with affordable suburban housing. Other areas

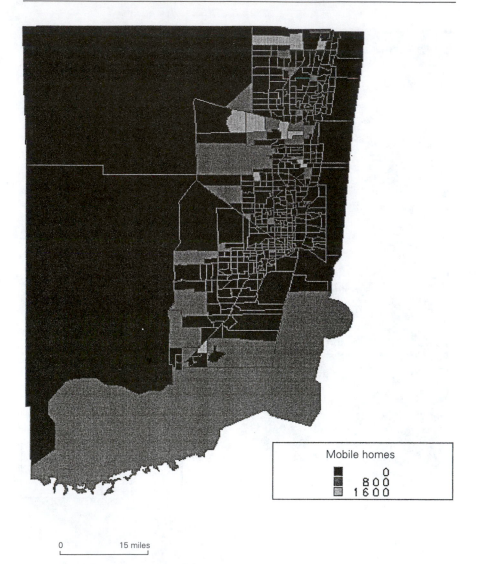

Fig. 12.11 Number of mobile homes in metropolitan Miami, 1990 (by Census tract) (Source: author and US Census)

have experienced fast suburban growth ... But those are predominantly Hispanic ... For "Anglos" choosing to leave the hurricane-stricken zones, the areas that will prove attractive, in terms of housing prices and ethnicity, are not in Dade. (Perez, 1992, quoted in Portes and Stepick, 1993, p. 225)

Increased migration out of Dade County provides an excellent example of how the storm reinforced existing social dynamics. Areas such as south-west Broward County were already growing quickly because of improved highway access, low interest rates, and the amenity appeal of a less urban location. Local residents commonly agreed that the hurricane had helped to push land development "years ahead of schedule." The overall result of storm-assisted migration and demographic reorganization has been increased spatial segregation of racial and ethnic groups and – to a lesser degree – increased economic segregation. Hurricane Andrew seems to have accelerated the Hispanicization of Dade County by enabling the movement of middle- and upper-income Whites to Broward County. Remaining non-Hispanic residents (i.e. African-American and White) are in general poorer than those who are leaving.

The post-storm migration has also begun to erase some of the social differences between Dade County and Broward County. While southern Dade County became more Hispanic and poorer, and therefore more like the rest of Dade County, migration to Broward County has tied the two counties closer together. The new residents of Broward, in general, maintain many more economic and social ties with Dade than do the rest of Broward County residents. This is particularly true of wealthier Cubans, who began to leave Dade County for Broward during this same period (Ramirez, 1994). Their business and social contacts remain firmly tied to Dade County (Sanchez and Viscarra, 1995).

Residents of the metropolitan region were linked together in other ways. Although the initially intense feeling of unity faded after several weeks, a lingering commitment to address and solve common problems has remained. In the early 1990s, many felt that Dade County and Broward County were drifting apart and that Dade County would become predominantly Hispanic and Broward County would remain predominantly White. Although underlying sentiments may not have changed, residents in each county recognize that their futures are inseparably linked (Portes and Stepick, 1993). The situation is summed up by the contrasting comments of Walter Revell, a business executive who is also a member of a Dade County reconstruction committee, and Chris Bezruki, assistant city manager of Homestead.

Is Dade [County] stronger, more united now? Revell thinks not – the size of the population and multitude of competing interests standing as barriers. Bezruki, on the other hand, is seeing a difference in his community. "We have had to believe in ourselves," he says. "And now we're showing some real concrete signs that, yes, it isn't a risk to invest here, that we will continue to build a stronger community." (Walsh, 1993).

Placed-based coalitions

The destruction of everyday life brought by hurricane Andrew fostered the development of various place-based coalitions. These appeared at three spatial scales: neighbourhoods or localities (e.g. Kendall, Homestead); subregion (e.g. south-west Broward County, Redland district of south Dade County); and region (e.g. urban zone, agricultural zone).

Differences at the local scale were perhaps most profound. Two years after the storm, some communities were better positioned for growth than before the hurricane, whereas others were worse off. For example, the city of Homestead came to regard hurricane Andrew as providing an opportunity for improvement through modernization and recapitalization (Walsh, 1993). Extensive damage in Homestead wiped out about 43 per cent of the city's tax base (Bell, 1994). By 1994, the population was still only three-quarters that of pre-Andrew days. Yet "St. Andrew" as it was referred to by many residents also brought more than US$200 million in state and federal grants, contributions, and insurance money to the city. Residents who stayed behind to watch Homestead rebuilt saw this reinvestment turn a farm town into a new bedroom community (Bell, 1994; Lambert et al., 1994). City leaders and residents also viewed the recovery period as an opportune time to change Homestead's character by initiating a private home-ownership programme that eventually converted the community from predominantly renter-occupied (60 per cent) to predominantly owner-occupied (60 per cent) (Walsh, 1993).

Other communities did not bounce back so successfully. Some of southern Dade County's poorer neighbourhoods declined even further after the hurricane (Lambert et al., 1994). Many communities suffered a delayed economic impact when moderate- and higher-income residents took their insurance reimbursements and moved out of town. A higher percentage of the remaining residents were likely to be uninsured or underinsured (Coletti, 1992b), which meant that disproportionately less money flowed back to the poorer victims, and host communities had less to spend on recovery. Hurricane Andrew also heightened the conflict between agricultural interests and development interests in southern Dade County, particularly in the area of Redland (Dunlop, 1995). Redland faced development pressures long before Andrew; but the disaster brought matters to a head by presenting farmers who had lost hundreds of acres of lime, mango, and avocado groves with three choices: "Wait six years for new trees to produce fruit; develop a diversity of crops; or sell out to the fastest and last cash crop – subdivision housing. The first two take time and diligence; the third requires only the fancy footwork of getting a zoning change" (Dunlop, 1995, p. 47). The hurricane also opened a window of opportunity for local coalitions and groups interested in preserving the rural farming landscape. Citing increased land-development pressure in the aftermath of Andrew, these groups have tried to further their own agenda of controlling urban sprawl and initiating a long-term plan for the protection of Redland farmland (Dunlop, 1995).

Why did some communities make successful recoveries while others did not? There are several possible answers. The skills of local policy entrepreneurs are one likely reason. Sometimes such skills and other resources were imported from outside the affected communities. Local and national volunteer groups worked closely with community officials and residents. For example, the blue-ribbon local committee, "We Will Rebuild," was put together at the prompting of President Bush, who visited the site of the disaster twice in the first several months. One of the objectives of this group was to promote themes of unity and community solidarity. More research needs to be done in order to determine if other groups were ethnically based (e.g. Cuban), income based (e.g. focused on helping poorer individuals or communities), or needs based (e.g. help those in the greatest need first).

Production-based coalitions

Hurricane Andrew had a profound effect on economic activity in metropolitan Miami and on coalitions associated with specific economic sectors. As predicted by many economists (Coletti, 1992b), reconstruction brought an immediate economic upswing to the previously sagging regional economy. Damage was so great that relief and insurance payments injected large amounts of new cash (West and Lenze, 1994). About 71 per cent of the US$22.65 billion worth of damage resulted from losses to residential structures (including mobile homes) and their contents. Commercial enterprises sustained roughly US$2.2 billion worth of damage. Agricultural losses amounted to approximately US$0.5 billion dollars. Remaining losses were attributable to impacts on government and utilities as well as the destruction of cars, boats, and airplanes. It was estimated that US$15.44 billion of the damage was insured, and that the total amount of estimated reconstruction expenditures would be US$17.31 billion (US$10.36 billion to the repair and replacement of damaged structures; US$2.02 billion to repair non-structural items, e.g. cars and boats; and US$4.94 billion to purchase new items).

An upturn in the local economy began to be realized within a few months of the storm. By the end of 1992, the unemployment rate had started to fall from its post-storm peak of 30,000 (Coletti, 1992b), and more than US$2.38 billion had already been spent on reconstruction (West and Lenze, 1994). Reconstruction-driven economic recovery was fully under way in 1993 (Hersch, 1993a; West and Lenze, 1994), with estimated expenditures of almost US$8.0 billion and the creation of more than 28,000 jobs (table 12.6). By mid-1994, however, the recovery boom started to fade and a more permanent post-Andrew economy began to emerge. Trends toward growth in international trade and finance and stagnation in tourism and farming, which were already present in the

Table 12.6 Estimated economic impact of hurricane Andrew

	1992	1993	1994	1995	1996
Direct jobs impact of reconstruction expenditures ('000 jobs)[a]					
Construction	2.751	16.879	18.590	21.564	0.396
Services	0.362	5.271	4.607	0.193	0.000
Trade	0.828	6.715	4.707	0.531	0.000
Total	3.941	28.865	27.904	22.288	0.396
Modified direct jobs impact ('000 jobs)[b]					
Construction	2.751	16.879	18.590	21.564	0.396
Services	−3.511	1.921	4.607	0.193	0.000
Trade	−2.767	2.670	4.707	0.531	0.000
Government	1.952	−0.352	−1.884	−1.884	−1.884
Total	−1.575	21.118	26.020	20.404	−1.448
Direct income impacts (US$ million)					
Non-agricultural wages					
and salaries	362.9	940.6	236.2	27.9	0.0
Military income	−26.8	−109.2	−112.7	−116.5	−120.3
Farm income	−187.4	−152.2	−102.6	−76.1	−44.0
Transfer payments	268.0	403.3	0.0	0.0	0.0
Dividends/interest/rent	−2,149.4	0.0	0.0	0.0	0.0
Total	−1,732.4	1,082.5	20.9	−164.7	−164.3

Source: West and Lenze (1994).
a. The "direct" impact of population dispersion into other parts of Florida is not well defined. The simulated jobs impact of this phenomenon was (in thousands of jobs): 0.0 in 1992, 6.0 in 1993, 7.5 in 1995, and 1.3 in 1996.
b. Modifications include jobs impacts of government expenditure on emergency and restoration and per diem expenditures of temporarily transferred personnel, plus estimated direct job losses.
Note: Although results are summarized here in broad industrial categories, detailed sectors of impact were used in their derivation.

early 1990s, reasserted themselves and in some cases accelerated slightly (e.g. loss of farmland in the Redland district).

Although the overall character of economic activity did not change significantly in metropolitan Miami, hurricane Andrew left some clear marks on its internal operations. The business community's perception of its own vulnerability to natural disasters is a case in point. The single most important component of Andrew's legacy was destruction of capital (e.g. loss of investments and production capacity). The amount of economic damage resulting from Andrew was staggering. Uninsured homeowners, renters, and businesses were devastated and found it difficult to recover their losses. Many had their life savings wiped out. Numerous small businesses, without significant cash reserves, went bankrupt.

Claims made by those that were insured caused a different type of impact. The local and national insurance companies that operated in Miami were overwhelmed. At least 8 firms went bankrupt, another 23 announced their intention to leave the state, and remaining firms are still in the throes of industry-wide restructuring (Hagy, 1993; Longman, 1994; McKinnon, 1995). Indeed, together with California wildfires and earthquakes in Northridge and Kobe, hurricane Andrew revealed serious flaws in the global insurance industry. In Florida, the industry generally underestimated the potential for claims from a major storm. Most industry representatives used hurricane Hugo, which struck the South Carolina coast in 1989, as a benchmark. That storm caused approximately US$4.2 billion of insured losses and, before Hugo, no hurricane had resulted in claims over US$1 billion. Price competition among companies operating in Florida compounded the problem because policy rates had not kept pace with the value of the insured property, and many companies did not have enough surplus to cover outstanding claims. Not only were the rates low in comparison with property values, the risk of hurricanes was also underestimated. Limited hurricane activity in the preceding 30–40 years, and especially in the 1980s, led actuaries to underestimate the risks (Longman, 1994).

The ensuing insurance crisis brought many policyholders – including those who had made no claims – face to face with several unpleasant alternatives: possible cancellation of policies, fewer coverage options, and dramatically increased rates. The Florida legislature passed several laws after Andrew that were designed to lessen the insurance crisis (McCabe, 1995; Westlund, 1995), but – as of August 1996 – it remained unsolved. The state of Florida has also developed a hurricane catastrophe fund to be used to pay off uncovered claims from future disasters and it continues to seek ways of dispersing the risk burden. For example, during the summer of 1996, the Florida Insurance Commissioner brought together peers from other states in an effort to generate momentum for a broad national catastrophic insurance fund (Anon., 1996b). Many policyholders found that it was easier to accept the industry's collapse than the fact that their own capital was so vulnerable both to the vagaries of the weather and to the failure of institutions that were designed to mitigate storm impacts. The hurricane delivered two key lessons to residents, businesses, and municipal officials. First, a category 4 or 5 storm was a real threat and it could cause damage far greater than expected. Secondly, more should be done to protect capital, and insurance was only one of the mechanisms that could be used. It was recognized, for example, that stricter building codes, proper enforcement of existing codes, and new mechanisms to protect emerging local industries were all urgently needed.

Home-owners and developers were some of the first to act. Demands for stronger and better-enforced building codes were numerous (Pielke,

1995). Creation of the International Center for Hurricane Damage Research and Mitigation at Florida International University institution-alized some of the local concerns (Lambert, 1995). Although part of the Center's mission is to encourage better emergency-response plans, most of its focus is on proper construction techniques and building codes, and improved ways for dispersing and using disaster relief funds. Agriculture and tourism, the other primary site-based, fixed-capital industries, were not actively involved in hurricane-preparedness and mitigation reforms. Both industries are currently preoccupied with what they perceive as larger and more pressing threats. Agricultural groups have focused on the threat of increased foreign competition, particularly with the advent of the North American Free Trade Agreement (NAFTA), and the con-tinued loss of farmland. Tourism interest groups are more worried about negative press reports about local crimes and an enduring economic recession in Latin America.

Situation-based, mobile capital industries were much more responsive to disaster concerns. Many of the finance, insurance, and real estate businesses in metropolitan Miami upgraded disaster-response plans or developed new ones (Coletti, 1992a; Hersch, 1993b; Knab, 1992). Planned or proactive responses, and non-planned or reactive (ad hoc) responses were both engaged in by many corporations in Greater Miami (Rupp, 1995). Some information-based companies increased the protection of electronic networks and digital data. Others developed a new-found appreciation of connections between themselves and the metropolitan region. Many corporations recognized that, even though their businesses might not – like the tourism industry – be fixed in place, they are pro-foundly dependent on the local community. For example, corporate facilities might have escaped the hurricane, but they were indirectly affected when employees' homes and communities were devastated. Workers would return home each night to a disorienting and stressful home life (Smith and Belgrave, 1995) that subsequently undercut their effectiveness on the job. Employers recognized the importance of restor-ing employees' home lives as quickly as possible.

Conclusion

Hurricane Andrew had the potential to be a watershed event in metro-politan Miami, but it did not fundamentally change the nature of the city's everyday life. Post-disaster reconstruction helped the metropolitan area get past its immediate economic crisis and accelerated the existing direction of social, economic, and political change. In this way, it might be said that the hurricane helped to set the stage for the resolution of some of Miami's most profound internal conflicts. Before the storm, farms and farming were disappearing from Broward County and from parts of Dade County – now, they are fading more quickly. Already the Hispanic community, particularly the Cubans, were rapidly becoming a dominant political power in Dade County – now, this transition will probably happen more quickly.

As well as speeding these social and economic transitions, the hurricane brought the residents of the metropolitan area closer together. Local people had lived through a terrible ordeal that revealed that they were dependent on others in the city not only for recovery from the hurricane but also for future economic growth. This burgeoning collective consciousness, although still relatively weak, may encourage better and more coordinated hazard responses and mitigation planning, thereby lowering the vulnerability of the city, as a whole, to future disasters.

But certain subpopulations and locations might now be even more vulnerable to disasters. The poorest neighbourhoods of south Dade did not fully recover from hurricane Andrew, and today appear more vulnerable to hurricanes than before. More research needs to be done on why communities such as Homestead were able to respond positively to the hurricane experience, whereas others were pushed to the edge of decline.

Did hurricane Andrew signal a transition in the vulnerability of metropolitan Miami to hurricanes? In many ways, it is still too early to tell because recovery is still going on (Provenzo and Fradd, 1995). It certainly demonstrated that capital investments are vulnerable to disasters. Both the material damage and the resulting insurance crisis underscore this conclusion. While not recommending any reduction in efforts to safeguard human safety, it seems clear that more attention needs to be directed to the reduction of property losses. Researchers should devote increased efforts to understanding the vulnerability of capital to disasters, to systems for protecting capital, and to measures that would enable individuals and groups to recover from their losses. The need is especially pressing in mega-cities, where a major disaster could easily cause more than US$50 billion of damage. This scale of loss could devastate the economies of most countries, and would significantly affect the economy of wealthier countries such as Japan and the United States.

Note

1. This section summarizes information from the following sources: Blake (1980); Boswell (1991); Derr (1989); Gannon (1996); Smiley (1974); Tebeau (1971).

REFERENCES

Anon. 1996a. "Cuban-born Commissioner is elected mayor of Miami." *New York Times*, 25 July, p. A7.

―――― 1996b. "Florida eyeing creation of regional disaster fund." *Tallahassee Democrat*, 10 August, p. B4.

Bader, N. 1993. "Old maxim: new roads = growth." *Florida Trend* 36(5): 56–59.

Bell, M. 1994. "For some, nightmare won't stop. For a few people, crooked contractors have left ruins even worse than the hurricane's legacy." *Orlando Sentinel*, 22 August, p. A1.

Blaikie, P. and H. Brookfield. 1987. *Land degradation and society*. London: Methuen.

Blaikie, P., T. Cannon, I. Davis, and B. Wisner. 1994. *At risk: Natural hazards, people's vulnerability, and disasters*. New York: Routledge.

Blake, N. M. 1980. *Land into water – water into land: A history of water management in Florida*. Tallahassee, FL: University Press of Florida.

Boswell, T. D., ed. 1991. *South Florida: The winds of change*. Miami, FL: Association of American Geographers.

Bottcher, A. B. and F. T. Izuno, eds. 1994. *Everglades Agricultural Area (EAA): Water, soil, crop and environmental management*. Tallahassee, FL: University Press of Florida.

Cartano, D. G. 1991. "The drug industry in south Florida." In T. D. Boswell (ed.), *South Florida: The winds of change*. Miami, FL: Association of American Geographers, pp. 105–111.

Chapman, A. E. 1991. "History of south Florida." In T. D. Boswell (ed.), *South Florida: The winds of change*. Miami, FL: Association of American Geographers, pp. 31–42.

Clark, W. A. V. and M. McNicholas. 1995. "Re-examining economic and social polarisation in a multi-ethnic metropolitan area: The case of Los Angeles." *Area* 28(1): 56–63.

Clarke, S. E. 1990. "'Precious' place: The local growth machine in an era of global restructuring." *Urban Geography* 11(2): 185–193.

Cohen, M. A., B. A. Ruble, J. S. Tulchin, A. M. Garland, eds. 1996. *Preparing for the urban future*. Washington, D.C.: Woodrow Wilson Center Press.

Coletti, R. J. 1992a. "Show your game face." *Florida Trend* 35(7): 32–34.

―――― 1992b. "Swift recovery, stronger economy." *Florida Trend* 35(6): 32–35.

Cooke, R. U. 1984. *Geomorphological hazards in Los Angeles*. London: Allen & Unwin.

Cox, K. R. and A. Mair. 1988. "Locality and community in the politics of local economic development." *Annals of the Association of American Geographers* 78(2): 307–325.

Craig, A. K. 1991. "The physical environment of south Florida." In T. D. Boswell (ed.), *South Florida: The winds of change*. Miami, FL: Association of American Geographers, pp. 1–16.

Cutter, S. L. 1993. *Living with risk*. New York: Edward Arnold.

Davis, M. 1995. "Los Angeles after the storm: The dialectic of ordinary disaster." *Antipode* 27(3): 221–241.

Derr, M. 1989. *Some kind of paradise: A chronicle of man and the land in Florida*. New York: William Morrow.

Didion, Joan. 1987. *Miami*. New York: Simon & Schuster.

Doig, S. K. 1996. "Storm warnings: 88 years in hurricane alley." *Miami Herald*, 5 September, p. 2b.

Dow, K. 1996. "The dynamics of oil spill vulnerability in the Straits of Malacca." Unpublished dissertation, Clark University, Worcester, MA.

Dunham, R. G. and R. A. Werner. 1991. "The trends and geography of crime in metropolitan Dade County." In T. D. Boswell (ed.), *South Florida: The winds of change*. Miami, FL: Association of American Geographers, pp. 97–104.

Dunlop, B. 1995. "Rescuing the Redlands (the work to preserve the rural landscape began as a hurricane recovery effort called the New-South-Dade charette)." *Landscape Architecture* 85(4): 46–48.

Ezcurra, E. and M. Mazari-Hiriart. 1996. "Are mega-cities viable? A cautionary tale from Mexico City." *Environment* 38(1): 6–15, 26–35.

Fernald, E. A. and D. J. Patton. 1984. "Water resource." *Atlas of Florida*. Tallahassee, FL: Florida State University.

Florida Department of Business and Professional Regulation. 1994. *Master file statistics: Public lodging and food service establishments, fiscal year 1994–1995*. Division of Hotels and Restaurants.

Gannon, M., ed. 1996. *The new history of Florida*. Tallahassee, FL: University Press of Florida.

Hagy, J. R. 1993. "Andrew's sucker punch." *Florida Trend* 36(3): 3033.

Haughton, G. and C. Hunter. 1994. *Sustainable cities*. Regional Studies Association, Regional Policy and Development Series. Bristol, PA: Jessica Kingsley.

Henry, J. A., K. M. Portier, and J. Coyne. 1994. *The climate and weather of Florida*. Sarasota, FL: Pineapple Press.

Hersch, V. 1993a. "Andrew's upside: A 'booming' economy." *Florida Trend* 35(12): 45–50.

———— 1993b. "Is your disaster plan in place?" *Florida Trend* 35(8): 67–69.

Hewitt, K., ed. 1984. *Interpretations of calamity*. Boston: Allen & Unwin.

Jonas, A. E. G. 1992. "Urban growth coalitions and urban development policy: Postwar growth and the politics of annexation in metropolitan Columbus." *Urban Geography* 12(3): 197–225.

———— 1993. "A place for politics in urban theory: The organization and strategies of urban coalitions." *Urban Geography* 13(3): 280–290.

Kasperson, R. 1994. "Global environmental hazards: Political issues in societal responses." In G. J. Demko and W. B. Wood (eds.), *Reordering the world: Geopolitical perspectives on the 21st century*. Boulder, CO: Westview, pp. 141–166.

Kleinberg, H. 1992. *The Florida hurricane & disaster 1992*. Miami, FL: Centennial Press.

Knab, K. 1992. "Andrew's wrath and the bankers who fought back." *Florida Banker* 6(5): 7–11.

Knoke, D. 1990. *Organizing for collective action: The political economies of associations*. New York: Aldine de Gruyter.

Knox, P. L. and P. J. Taylor, eds. 1995. *World cities in a world-system*. Cambridge: Cambridge University Press.

Lambert, M. 1995. "Hurricane Research Center planned for FIU." *Sun-Sentinel*, 17 March, p. A1.

Lambert, M., S. Borenstein, and J. Maines. 1994. "Homes and lives rebuilt two years after Andrew." *Sun Sentinel*, 21 August, p. B1.

Leen, J., S. K. Doig, L. F. Soto, and D. Linefrock. 1993. "Failure of design and discipline." In L. S. Tait (ed.), *Lessons of hurricane Andrew*. Excerpts from the 15th Annual National Hurricane Conference. Washington, D.C.: Federal Emergency Management Agency, pp. 39–44.

Linden, E. 1996. "The exploding cities of the developing world." *Foreign Affairs* 75(1): 52–65.

Liverman, D. M. 1990a. "Vulnerability to global environmental change." In R. E. Kasperson, K. Dow, D. Golding, J. X. Kasperson (eds.), *Understanding global environmental change: The contributions of risk analysis and management*. A report on an International Workshop, Clark University. Worcester, MA: The Earth Transformed Program, Clark University, pp. 27–44.

——— 1990b. "Drought in Mexico: Climate, agriculture, technology, and land tenure in Sonora and Puebla." *Annals of the Association of American Geographers* 80(1): 49–72.

Logan, J. R. and H. Molotch. 1987. *Urban fortunes: The political economy of place*. Berkeley: University of California Press.

Longman, P. 1994. "The politics of wind." *Florida Trend* 37(4): 30–40.

McCabe, R. 1995. "Insurance crisis long-winded. Andrew gives rise to another disaster that refuses to go away." *Sun-Sentinel*, 8 January, p. F1.

McKinnon, J. 1995. "Politics of wind (cont.)." *Florida Trend* 38(11): 14–16.

Merchant, C. 1989. *Ecological revolutions: Nature, gender, and science in New England*. Chapel Hill, NC: University of North Carolina Press.

Mitchell, J. K. 1995. "Coping with natural hazards and disasters in mega-cities: Perspectives on the twenty-first century." *GeoJournal* 37(3): 303–311.

Mohl, R. 1982a. "Forty years of economic change in the Miami area." *Florida Environmental & Urban Issues*, July.

——— 1982b. "Race, ethnicity, and urban politics in the Miami metropolitan area." *Florida Environmental & Urban Issues*, April.

——— 1995. "The patterns of race relations in Miami since the 1920s." In D. R. Colburn and J. L. Landers (eds.), *The African American heritage of Florida*. Tallahassee, FL: University Presses of Florida, pp. 326–366.

Mollenkopf, J. H. 1983. *The contested city*. Princeton, NJ: Princeton University Press.

Mollenkopf, J. and M. Castells, eds. 1991. *Dual city: Restructuring New York*. New York: Russell Sage Foundation.

Molotch, H. 1976. "The city as a growth machine." *American Journal of Sociology* 82: 309–330.

Monmonier, Mark. 1997. *Cartographies of danger: Mapping hazards in America*. Chicago: University of Chicago Press.

Moore, D. 1994. *To the golden cities*. New York: Free Press.

Nash, J. 1976. *Darkest hours*. Chicago: Nelson-Hall.

Nijman, J. 1996. "Breaking the rules: Miami in the urban hierarchy." *Urban Geography* 17(1): 5–22.

Parks, A. M. 1986. "Introduction." In L. F Reardon, *The Florida hurricane & disaster 1926*. Miami, FL: Centennial Press; originally published in 1926.

Perez, L. 1992. "Hurricane has severely tilted community demographics." *Miami Herald*, September: M4.

Pielke, R. A. 1995. *Hurricane Andrew in south Florida: Mesoscale weather and societal responses*. Environmental & Societal Impacts Group, National Center for Atmospheric Research.

Portes, A. and A. Stepick. 1993. *City on the edge: The transformation of Miami*. Berkeley: University of California Press.

Provenzo, E. F. and S. H. Fradd. 1995. *Hurricane Andrew, the public schools and the rebuilding of community*. Albany: State of New York University Press.

Ramirez, D. 1994. "Broward attracts Cubans: Ex-Dade residents move in hopes of a better life." *Sun-Sentinel*, 11 November, p. 1A.

Roberts, R. and Emel, J. 1993. "Uneven development and the Tragedy of the Commons: Competing images for nature–society analysis." *Economic Geography* 68(3): 249–271.

Rupp, W. T. 1995. "Toward a process model of corporate social performance in response to natural disasters: An analysis of corporate America's response to hurricane Andrew." Unpublished manuscript, Ph.D. dissertation, University of Georgia.

Sanchez, J. I. and D. M. Viscarra. 1995. "Are Hispanic employees unwilling to relocate? The case of South Florida." *International Journal of Intercultural Relations* 19(1): 45–58.

Sassen, S. 1993. "Rebuilding the global city: Economy, ethnicity, and space." *Social Justice* 20(3–4): 32–50.

——— 1994. *Cities in a world economy*. Thousand Oaks: Pine Forge Press.

Sculley, S. P. 1986. *Frequency analysis of SFWMD rainfall*. Technical Publication 86–6. West Palm Beach, FL: South Florida Water Management District.

Silverstein, S. 1995. "How safe are mobile homes?" *Los Angeles Times*, 20 August, p. A1.

Smiley, N. 1974. *Yesterday's Miami*. Miami, FL: E. A. Seemann Publishing.

Smith, K. J. and L. L. Belgrave. 1995. "Reconstruction of everyday life experiencing hurricane-Andrew." *Journal of Contemporary Ethnography* 24(3): 244–269.

Smith, S. K. and C. McCarty. 1996. "Demographic effects of natural disasters: A case study of hurricane Andrew." *Demography* 33(2): 265–275.

Snyder, G. H. and J. M. Davidson. 1994. "Everglades agriculture: Past, present, and future." In S. M. Davis and J. C. Ogden (eds.), *Everglades: The ecosystem and its restoration*. Delray Beach, FL: St. Lucie Press, pp. 85–116.

Soja, E. 1995. "Postmodern urbanization: The six restructurings of Los Angeles." In S. Watson and K. Gibson (eds.), *Postmodern cities and spaces*. Cambridge, MA: Blackwell, pp. 125–137.

Solecki, W. D. and S. Michaels. 1994. "Looking through the post-disaster policy window." *Environmental Management* 18: 587–595.

Tait, L. S., ed. 1993. *Lessons of hurricane Andrew*. Excerpts from the 15th Annual National Hurricane Conference. Washington, D.C.: Federal Emergency Management Agency.

Tebeau, C. W. 1971. *A history of Florida*. Coral Gables, FL: University of Miami Press.

Timmerman, P. 1981. *Vulnerability, resilience, and the collapse of society*. Environmental Monograph No. 1. Toronto: Institute for Environmental Studies, University of Toronto.

US Department of Commerce, National Oceanic and Atmospheric Administration. 1987. *Climatological Data No. 91, Annual summary. Florida 1987.* Asheville, NC: National Climatic Data Center.

Walker, R. T., W. D. Solecki, and C. Harwell. 1997. "Land use dynamics and ecological transition: The case of south Florida." *Urban Ecology* 1: 43–57.

Walsh, M. 1993. "Rebuilding by the script." *Florida Trend* 36(4): 70–71.

West, C. T. and D. G. Lenze. 1994. "Modeling the regional impact of natural disaster and recovery. A general framework and an application to hurricane-Andrew." *International Regional Science Review* 17(2): 121–150.

Westlund, R. 1995. "Homeowners' insurance: Can Florida weather the storm?" *Florida Realtor* 70(10): 16–19.

Winsberg, M. D. 1990. *Florida weather*. Orlando: University of Central Florida Press.

——— 1991. "South Florida agriculture." In T. D. Boswell (ed.), *South Florida: The winds of change*. Miami, FL: Association of American Geographers, pp. 17–30.

——— 1994. "Urban population redistribution under the impact of foreign immigration and, more recently, natural disaster: The case of Miami." *Urban Geography* 15(5): 487–494.

——— 1996. "Measuring Andrew's aftermath." *American Demographics* 18(5): 22.

World Resources Institute. 1996. *World Resources 1996–97: The urban environment*. New York: Oxford University Press.

13

Findings and conclusions

James K. Mitchell

"Urbanization holds out the bright promise of an unequaled future and the grave threat of unparalleled disaster."

(Wally N'Dow, Secretary-General, Habitat II; quoted in United Nations Commission on Human Settlements, 1996, p. xxi)

The preceding case-studies provide convincing testimony about the importance of environmental hazards as mega-city problems. It is now time to bring together the main findings and to discuss their implications for researchers, managers, and others who will share responsibility for guiding the development of mega-cities in the twenty-first century. In the pages that follow, emphasis will be placed on the role of natural hazards because they raise some of the most difficult intellectual and policy issues and because they have been largely neglected by the global community of urban experts. This does not imply that other kinds of hazard are lacking in importance, for the evidence shows that many types of urban environmental threat are worsening and all are becoming increasingly intertwined. However, a natural hazards perspective brings the issue of limits to human agency clearly into view at the same time as it illuminates the action-forcing nature of disjunctive *events*. In discussions of technological and social hazards, these matters are often obscured by expansive interpretations of human agency and by the privileging of *structures* (especially social structures) as causal factors.

Herein it will be argued that mega-city hazards are profuse, burdensome, symbolically potent, incompletely understood, and addressed

473

by public policies that typically make use of just a few types of possible adjustment. They are also closely bound up with many other urban issues and have acted as catalysts for change in a wide range of policy contexts. Despite noteworthy successes in some places, the bulk of existing mega-city hazard-management systems are either barely holding their own or seriously deficient. In the future, all such systems will be grievously challenged by the changing nature of hazards themselves and by the profound transformations of large cities that are now occurring through-out the world. The consequences of misjudging natural hazards in mega-cities of the twenty-first century are likely to be worse than anything yet experienced.

Hazards in 10 mega-cities: A review

The 10 mega-cities examined in this volume are a diverse group. Not only do they differ in terms of culture, population, area, growth rates, eco-nomic status, and political functions (see chap. 1); they are also subject to varied mixes of hazards. Consider, for example, Tokyo and Sydney. Though both are affluent industrialized cities of roughly similar age, their hazard profiles are quite different. Tokyo has a history of catastrophic destruction and its residents are much concerned with staving off future disasters. Here earthquakes, snowstorms, and typhoon-driven floods threaten one of the world's chief business centres – a densely built-up city that is distinguished by remarkably little social pathology. Sydney, in contrast, has historically been free from truly catastrophic natural dis-asters but is increasingly beset by technological risks and social tensions. Bushfires and flash-floods now affect what is arguably the world's most dispersed mega-city – a place whose residents pay only selective and intermittent attention to environmental hazards. Equally compelling contrasts mark the hazard profiles of Dhaka and Lima, Seoul and Los Angeles, London and San Francisco, Mexico City and Miami. Such a wide spectrum of settings and experiences provides a rich array of knowledge about the ways in which metropolitan societies have sought to come to terms with hazardous environments.

Not only are the mega-cities diverse; they are beset by complex hazard problems:
- the agents of hazard are many and the mixes varied;
- relationships between hazards and competing urban problems vary widely among different cities;
- urban hazards issues and interest groups are volatile, especially in the largest cities;

- human-modified natural processes of urban areas are imperfectly understood;
- although they are typically viewed as "public policy problems," mega-city hazards are also underlain by contentious scientific and philosophical issues about the nature of change in urban environments and about the role of nature in human-constructed settings.

In the crucible of the mega-city, those ingredients interact and are transformed, giving rise to distinctive patterns of hazard.

Mega-city hazards are profuse

Environmental hazards are present in profusion among all of the case-study cities. Every one is affected by at least two or three different kinds of natural hazard and some (e.g. Tokyo, Mexico City) by as many as seven or eight. Floods, earthquakes, and windstorms are the most common damaging phenomena; different pairs of these three constitute major risks in all of the places studied (e.g. floods and windstorms in Dhaka; earthquakes and floods in Los Angeles; windstorms and earthquakes in Tokyo). Other risks that have triggered significant disasters include (in descending order): (1) slope failures; (2) droughts/water shortages; (3) wildfires, subsidence, and fog; (4) tsunamis; (5) volcanoes; and (6) snow. Fire has been the most common and destructive human-made urban hazard, especially fire as a consequence of external or internal warfare, though natural factors also play a role. Among the technological hazards, air pollution is perhaps the most widespread, the most frequently recognized, and the most generally destructive (UNEP/WHO, 1994), while terrorism and other violent crimes are important social hazards in most mega-cities.

Mega-city hazards are important

For many urban dwellers, either natural disasters are historical curiosities that, it is hoped, will not be repeated or they are improbable fantasies that it is believed will never leave the movie screen. But the record of actual events is both recent and very real. Hazards are not only present, they are undeniably important phenomena both inside the case-study cities and beyond their boundaries.

Losses of life and property are one index of importance. During the twentieth century, human fatalities attributable to single extreme events have been as high as tens to hundreds of thousands in large cities (e.g. Tokyo in 1923; Mexico City in 1985). Moreover, the potential for catastrophic urban death tolls is growing rapidly. But the **material and eco-**

nomic impacts of environmental hazards in big cities loom even larger. For example, more than 40 per cent of Tokyo was completely destroyed by earthquakes and fires on two occasions in the past 75 years. Recent record-setting bills for hurricane damage in Miami (US$30 billion) and wildfire in Oakland (US$2 billion) and Kobe (over US$100 billion) testify to the potential for even larger economic costs. Of course, economic loss totals tend to be biased in favour of richer cities – which have more expensive properties at stake. But even modest-seeming losses can be devastating to mega-cities of the third world, as evidenced by the experience of Mexico City. Though by no means the most expensive earthquake to affect a large city, the one that struck Mexico City in 1985 came at a time when the country was facing severe economic problems. Resources were diverted to recovery in the city, thereby hampering national development for a number of years.

Between 1985 and 1995 the myth that modern cities have successfully conquered the hazards of Nature was refuted by a series of massive losses (see table 1.1 above). During this period, earthquakes, hurricanes, wildfires, and other natural extremes inflicted human casualties and record property damage on half a dozen major metropolitan areas in the United States, Mexico, and Japan. In several of these the economic losses reached proportions that had serious implications for national budgets and the stability of the global reinsurance system.

The recent record of **technological hazards** and disasters in mega-cities is more ambiguous. Chronic technological hazards, especially those associated with locally generated environmental pollutants, are, of course, almost synonymous with large cities and have attracted a great deal of attention from analysts in recent years. Some acute technological disasters also stand out in mega-cities, including explosions of gas fuel tanks in Mexico City (1984) and of sewer lines in Guadalajara (1992), as well as Bhopal's catastrophic chemical leak (1984). Large cities have been hit by massive and prolonged electricity power failures (e.g. New York in 1965 and 1977; London in 1987), which sometimes spawned secondary disasters. However, as far as the mega-cities in this sample are concerned, acute technological disasters seem to have been less deadly and less damaging than natural disasters. Much the same might be said of mega-city epidemics and other **biological hazards**, although major scares such as outbreaks of hepatitis in Lima and plague in Surat (India) suggest that worse is possible. Of course, like the rest of the world, mega-cities are at risk from technological and biological threats that occur on continental and global scales, such as the Chernobyl nuclear accident, the AIDS pandemic, and the spectre of global climate change.

If the mega-city record of technological and biological hazards is ambiguous, the same cannot be said about urban **social hazards**. In par-

Table 13.1 Cities with major incidents of social violence since 1989

Civil/internal war or urban terrorism		Riots or street protests by civilian population	External warfare
Baku	Madrid	Belgrade	Baghdad
Beijing	Manchester	Bombay	
Bogota	Mogadishu	Calcutta	
Buenos Aires	Monrovia	*Dhaka*	
Cairo	Moscow	Jakarta	
Colombo	New York	*Los Angeles*	
Kabul	Paris	Yangon	
Karachi	Phnom Penh		
Kinshasha	Port au Prince		
Lahore	Tibilisi		
Lima	*Tokyo*		
London			

Note: Case-study cities in italic.

ticular, there is growing concern about the vulnerability of large cities to acts of terrorism, internal warfare, and rancorous public protest. Table 13.1 identifies some of the larger cities that have recently experienced these kinds of event. Big-city facilities have long been prime targets of inter-state warfare, internal coups, and popular anti-government uprisings, but they are now also becoming battlegrounds in a whole new generation of problems. These include: conflicts among ethnic, religious, and other groups recently freed from the constraints of Cold War politics; as well as deprived or alienated members of a growing urban underclass. Access to powerful weapons of mass destruction and the availability of soft targets such as easily disrupted electronic information technologies and urban infrastructures are contributory factors. Thus far, public buildings, business offices, shopping centres, and transportation facilities have been the most common targets, but there are fears that basic life-support systems (e.g. food, water, and energy supplies) could easily be disrupted – or hazardous facilities deliberately destroyed – with catastrophic results.

Losses are not the only measure of importance. The malign ubiquity of extreme events is manifest in other ways. Among the case-study cities are several in which nearly every occupant was affected negatively by extreme events for extended periods. The experience of Dhaka in 1987 and 1988, when three-quarters of the city was covered by floodwaters for many weeks, is a case in point. It is not so much that some Dhaka residents died, many lost their possessions, and others contracted waterborne diseases, or that space satellite images of the submerged city quickly conveyed its plight to an international audience of concerned scientists and curious laypeople. Rather, it is that the floods also became yardsticks for measuring other experiences in the lives of victims and managers alike. In other words, they become **marker events** for resident populations.

Very often such natural disasters have ushered in major changes of built environments and far-reaching institutional reforms. Lima was almost entirely reconstructed after the catastrophic earthquake of 1746. The pivotal significance of the 1906 earthquake for the city of San Francisco is too well known to warrant further comment, but more recent disasters have also reconfigured the landscape of other Bay Area communities. For example, the Loma Prieta earthquake (1989) radically changed both the Marina District of San Francisco and the lowlands of Oakland, while the firestorm of 1992 extensively remade upscale neighbourhoods in the hills above the city of Oakland. The urban fabric of modern Tokyo is largely a product of rebuilding after the 1923 Kanto earthquake and the fire bombing raids of the Second World War. Miami hurricanes of the 1940s gave a strong push to massive water-management schemes that encouraged metropolitan expansion, and the hurricane of 1992 significantly accelerated the Hispanicization of Dade County's population. In London, the inner city's physical layout owes much to the Great Fire (1666)[1] and the economic decline of the port was sharply assisted by the 1940s' blitz. More recently, a catastrophic 1953 east coast storm surge paved the way for construction of the Thames flood barrier, which, in turn, helped to secure investments in the rejuvenated Docklands development area of east London. Mexico City's 1985 earthquake provided opportunities for grass-roots citizen organizations to take on responsibilities for relief and reconstruction, which had formerly been the exclusive preserve of government agencies, and to use the experience as a base for challenging government authority in other realms of public policy.

Some of the changes that took place in the wake of natural disasters had effects well beyond the affected cities. For example, the 1933 Long Beach (Los Angeles) earthquake triggered state legislation (The Field Act) that subsequently revolutionized school building standards and siting criteria throughout California, as well as inspiring similar changes elsewhere. The Kanto earthquake promoted the development of country-wide civil defence programmes and the creation of a national disaster day (1 September), which is marked throughout Japan by emergency-management exercises and simulations that involve tens of thousands of citizens.

Unfortunately, changes that occurred in the wake of major disasters were not always beneficial. Rubble from the Tokyo fire of 1657 was used to create filled land in the Koto Delta, which is now one of the most troublesome earthquake- and flood-prone locations in Tokyo. After the Second World War, many of Tokyo's bombed-out neighbourhoods were filled with cheaply constructed wooden housing that remains a perennial problem for today's hazard managers. Fiscal policies adopted by the Dhaka Municipal Committee of 1840, in an effort to address a crisis of sanitation and economic decline, helped push residents out of the old city into adjacent flood-prone areas; the new polders that have recently been constructed near Dhaka may also increase the city's catastrophe potential while providing needed space for expansion. The much-afflicted Marina District of San Francisco is founded on rubble from the 1906 earthquake and from a subsequent Exposition that was held to mark the city's resurrection. Aqueducts that were intended to reduce the earthquake vulnerability of San Francisco's water-supply system cross the very faults that will likely rupture during future quakes. Drainage works undertaken during the nineteenth century in response to historic flooding have accelerated soil compaction and subsidence in much of Mexico City. The lesson is unmistakable: when a big disaster affects a major metropolitan area its consequences can be permanent and profound – both positively and negatively – for the city itself and for the larger society.

On occasion, natural disasters may turn into something more than just marker events – they may become *metaphors that place environmental hazard near the centre of a city's identity*. The conjunction of natural and human disaster impacts in Los Angeles in 1992 (i.e. earthquakes in the flatlands, wildfires in the hills, and riots in the ghettos) not only inflicted record-setting economic losses but reinforced the city's civic reputation as a place of environmental and social insensitivity that continually flirts with apocalyptic demise. Athough Los Angeles may be the most flamboyant example of a mega-city that has embraced disaster as a central component of urban identity, it is not the only one. Foundational disaster imagery and myth-making are also present to varying degrees in Tokyo, Seoul, San Francisco, Dhaka, Mexico City, Lima, and perhaps elsewhere. It is not possible here fully to assess the implications of such characterizations for urban populations, but they should not be neglected in the formulation of future urban strategies and policies – especially those that have to do with hazard management (see the final sections of this chapter).

The management of mega-city hazards is truncated

Certain mega-cities (e.g. Greater San Francisco and Tokyo) possess civic cultures that are both sensitive to hazards issues and active in mounting appropriate responses, but most city governments and institutions are far less committed to the containment and reduction of hazards as primary goals of public policy. Typically, the gap that separates awareness of risks from responses to mega-city hazards is wide and is filled with mediating factors that tend to retard the adoption of improved public policies. *Existing public policies strongly favour professionalized warning, evacuation, and emergency-management programmes for a wide range of acute threats backed up by separate sophisticated engineering technologies for different chronic risks.* London's movable flood barrier, Dhaka's embankments and polders, Seoul's flood pumping stations, and Los Angeles' sediment catchment basins are typical examples of mega-city flood adjustments. Hazard insurance and hazard-resistant building practices also are sometimes available in the more affluent mega-cities. In most places the emphasis is on expert-driven systems and formal procedures carried out by public institutions, although a high degree of co-operation by well-informed lay citizens is often expected. Lay residents of large cities also participate in informal, traditional, or folk measures, such as networks for providing mutual aid.

Many improvements to the formal public adjustments are possible, including the upgrading of emergency services and the installation of hazard-warning and evacuation technologies in cities that do not yet possess them, as well as the development of appropriate methodologies for assessing hazards and incorporating risk-management strategies into public budgets, plans, statutes, and other regulatory devices. However, even in relatively well-provisioned mega-cities of Japan, North America, and Europe, the areal and demographic coverage of formal public sector hazard-management programmes is incomplete, and the extent to which they address the premier hazard concerns of resident populations is often uncertain.

What else that might be done remains missing from the preferred range of management alternatives? Broadly speaking, the *neglected approaches involve non-expert systems, informal procedures, non-structural technologies, private sector institutions, and actions taken by individuals, families, neighbourhood groups, firms, and similar entities.* Among others, these include measures that: (1) encourage hazard-sensitive decisions about site selection, land management, and facility operations; (2) control the installation and replacement of infrastructure; (3) relieve institutional and social inequities that shift hazard burdens onto certain (already disadvantaged) groups; (4) buttress local grass-roots capacities for hazard management; and (5) promote less environmentally stressful non-structural hazard-mitigation technologies. In addition, there is a *lack of initiatives that jointly address different kinds of hazard*, a *slowness to integrate hazards management with other problem-solving urban programmes*, and a *failure to investigate other roles that hazard plays in the lives of urban residents.*

It is not that the latter approaches are entirely absent in mega-cities. For example, local hazard-management programmes staffed by lay-persons have made useful contributions to hazard reduction in Mexico

City and Los Angeles. Neighbourhood hazard volunteer organizations have been pioneered in Tokyo, along with local stores that sell emergency supplies and community centres that stock low-techology emergency-management equipment. Internet pages now carry lists of profit-making companies that make a living from serving the hazard-protection needs of residents in some US cities, and some commercial organizations have established trade shows of hazard-management technology that is suitable for private as well as public sector adoption. However, these kinds of facilities and activities by no means exhaust the list of alternatives that are possible and they today account for only a small minority of existing responses.

Much is still unknown about mega-city hazards

Of the four basic components of hazard (i.e. risk, exposure, vulnerability, response), information about risks and responses in mega-cities is generally more plentiful and accurate than is information about exposure and vulnerability. Likewise, the information base for mega-cities in more developed countries (MDCs) is typically better than for mega-cities in less developed countries (LDCs). But the relationship between information availability and public policy is not a simple matter and is continually changing under the impress of scientific investigations, the experience of disasters, and fluctuations in public anxiety about different types of hazard.

Information about risks and responses is relatively good in the mega-cities of MDCs, at least with respect to long-established natural hazards. For example, the general probability of earthquakes and of areas susceptible to destructive ground motion is well known in places such as Tokyo, San Francisco, and Los Angeles, though the details are far from complete. Large faults are well documented but many important small faults go undetected until earthquakes occur, and geologists are only now beginning to explore the complex interference ground motion patterns that appear in lowland basins when seismic waves are reflected off surrounding mountains. Likewise, the information necessary to construct reliable earthquake-resistant high-rise buildings is readily available, but much less is known about fail-safe construction practices for highway overpasses, underground infrastructure systems, and other lifelines (Platt, 1991). Compared with risks and responses, exposure and vulnerability are highly volatile components of urban hazard that sometimes respond quickly to demographic and land-use changes (Montz and Gruntfest, 1986). Prospects for identifying changes in exposure have improved with the advent of remote sensing and GIS (Geographic Information System) technologies, although most MDC mega-cities are only now beginning to apply these tools and their potential is far from being realized (Geo-Hazards, 1994; Gruber and Haefner, 1995; Oppenheimer, 1994). Socio-spatial patterns of earthquake vulnerability are poorly known in virtually all MDC mega-cities.

Compared with the MDCs, *many mega-cities in LDCs often lack very basic information about hazards.* For example, data about flooding in Seoul and in Dhaka are sparse. Despite the fact that floods have claimed hundreds of lives in Seoul in recent years, the flood risk has not been assessed by public agencies or research organizations. The case-study of Seoul's flood risk reported in this volume is believed to be the first effort of its kind in the city's history and certainly the first to provide even rudimentary quantitative documentation of the relationship between land development practices and the severity of flooding. It is also clear that the growing flood hazards of Dhaka have not been analysed for purposes of disaster mitigation. International agencies and the national government have devoted much more attention to rural flooding in Bangladesh, although the city of Dhaka now contains around 8 million people and rapid infilling of open space in low-lying neighbourhoods is feeding a major potential for urban flood disaster. Likewise, the case-study of social and economic vulnerability to earthquakes in Mexico City represents the first known attempt to address this kind of knowledge gap in a mega-city of the developing world.

Information about informal groups and non-governmental organizations that are engaged in urban hazard management is also woefully deficient. Many such groups are national or international in scope (e.g. religious, fraternal, charitable, and public service organizations); some are indigenously urban (e.g. neighbourhood associations, urban cooperatives, local union welfare groups), others have been transferred from rural areas by migrants to the cities (e.g. family and locality kinship networks); and some are emergent groups that are still developing *in situ*, either in response to specific hazards or for other purposes. Given the growing importance of the informal economic sector in many large cities, especially throughout the LDCs, and corresponding shrinkages in the role of the state, especially in MDCs, such groups may become more important as hazard managers in the future. Of course, other outcomes are also possible. For example, cities have often acted as agents of "modernization," encouraging the replacement of informal organizations by formal ones and lay competence by professionalization. The struggle to create more effective hazard management in mega-cities may be as much a contest between outdated professional institutions and emergent ones as between experts and laypeople or among citizen groups that are differently empowered. These are matters that are not well understood at present.

The contrasts between scientific information about hazards in mega-cities of poor countries and rich countries are real enough but it would be a mistake to exaggerate them. *The comparative information advantage of MDC mega-cities is eroding* both because of improved scientific infra-structures in some LDCs and because of institutional barriers that impede further advances in the application of scientific information in many MDC mega-cities. For example, in Latin America scientific infor-mation about earthquake risks, micro-zonation, and structural vulner-ability is improving rapidly, partly as a result of research initiatives stimulated by the International Decade for Natural Disaster Reduction (GeoHazards, 1994). Conversely, MDC mega-cities are often hampered by public institutions that lack the authority to apply available hazards information at the metropolitan scale. Some US cities, including San Francisco and Los Angeles, have developed metropolitan hazards-research and -management organizations but many others (e.g. Miami, New Orleans, Houston) do not possess them.

In certain important respects there are no significant information gaps between MDC and LDC mega-cities. Indeed, *all mega-cities start from similar bases of ignorance when it comes to hazard "surprises."* These unprecedented events are, by definition, poorly known everywhere. A great deal of additional research will be necessary before it is possible to assess the effect of surprises associated with climate change, new indus-trial chemicals, or hybrid hazards that combine natural, technological, and social risks. Likewise, there is a general lack of information about the relationship between hazards and other urban issues in all kinds of large cities. The complex nature of contemporary mega-cities virtually ensures that hazard policies and programmes rarely stand alone as claimants of public support. Competition among different urban constituencies with quite different agendas is the rule, and public policies inevitably involve compromises and trade-offs among responses to different issues. By the same token, the complexity of mega-cities affords scope for the creation of coalitions and collaborative ventures among unusual partners. Some contributors to this volume (e.g. Solecki in chap. 12) hint at the possibil-ities that changing coalitions might portend, but the general research lit-erature is thus far largely silent about such matters. In summary, lack of information about mega-city hazards is clearly a serious problem, but it is often overshadowed by the volatility of the urban crucible, which con-tinually recatalyses even the known components of hazard.

Mega-city hazards are changing

Although information gaps prevent a definitive assessment, it is very clear that mega-city hazards are in transition at the close of the twentieth

Table 13.2 Trends of mega-city hazards

* Mixes of natural, technological, and social hazards are increasingly common
* Risks are changing slowly
* Loci of hazard are shifting markedly
* Differentially vulnerable groups are becoming polarized and segregated
* Public support for hazard-management initiatives may be faltering
* Overlaps among hazard and urbanization issues offer opportunities for
  managerial intervention

century.[2] Everywhere, the composition of hazards is changing; the management of hazards is changing; and the ways people think about hazards are changing (Mitchell, 1995). Nowhere is this more true than in the world's largest cities. At the same time, the global population is undergoing a vast demographic shift towards a predominantly urban world and the process of urbanization is itself in flux. Some of the more important changes are summarized in table 13.2.

Changes of interactivity

Urban hazards and disasters are becoming an interactive mix of natural, technological, and social events. In the past, these events were analysed as separate types of phenomena, but that approach may no longer be desirable or possible. Most potential victims – and many hazard-management agencies – do not make such distinctions; for laypersons, urban hazard is usually a composite and unstable class of events. Experience is carried over from one type of event to another and coping measures frequently straddle several types of hazard.

Interactivity occurs in different ways. Most commonly there is a *burgeoning of hybrid hazards* composed of different mixes of natural, technological, and social risks.

Recent representative examples from the United States include: fiery floods in the Houston metropolitan area of Texas; the incorporation of toxic chemicals into Mississippi flood detritus in St. Louis; and problems in Chicago's central business district involving the collapse of outworn canal walls, inundation of derelict underground rail tunnels, and subsequent electrical power failures. More of these kinds of event are likely to occur in the future, thereby raising serious difficulties for reductionist analyses that treat the components of complex hazards as if they were separate phenomena. In any event, the sheer numbers and diversity of mega-city hazards mean that populations at risk customarily must simultaneously juggle with the prospects of many different kinds of threat (see chap. 1). Some of what urban residents know about hazards comes from direct experience and visual inspection, but many of the cues are indirect and are derived from images of risk and hazard projected by mass media sources for purposes that are not connected with hazard management (e.g. partisan politics, advertising, entertainment). The hybridization of mega-city hazards is not solely an objective process that involves overlaps between risks that are both active and real; it is also the subjective conflation of experienced and imagined events and the reinvention of past hazards to fit the explanations of the present.

Interactivity need not require direct physical involvement among different kinds of event; it can occur indirectly, as when *experience gained with one type of hazard informs adjustments to another*. The widening cascade of effects that flow from urban terrorist campaigns provides a good example of this form of interactivity. Terrorist incidents have raised the hazard awareness of residents in places such as London, New York, and Paris in complex ways. For example, bombings of New York's World Trade Center (26 February 1993), the Federal Building in Oklahoma City (19 April 1995), and the Olympic Park in Atlanta (27 July 1996) have not only led to tighter security precautions at major public facilities throughout the United States but raised the general visibility of emergency-management agencies. Thus, in successive State of the Union speeches (1995, 1996) made by the US President, the Federal Emergency Management Agency has become both a lauded exemplar of compassion and efficiency in government and a pivot of political campaigns to encourage citizens to take increased responsibility for their own safety. Anxiety about the general vulnerability of densely populated urban areas was also fuelled by poison gas attacks carried out by the Aum Shinrikyo cult in Japan (20 March 1995). In London, Irish Republican Army bombings and fires in the Underground railway system have had repercussions for public safety measures that resulted in increased security against a wide range of natural and human extremes (e.g. heightened readiness among public safety services, upgraded facilities and techniques for treating traumatic injuries, more stringent building standards). They have also affected the business property insurance market and have encouraged owners and tenants of risk-susceptible buildings to give public safety a prominent place among their priorities. With urban terrorism a continuing problem in mega-cities from Colombo to Lima, the prospects for increased adoption of overlapping or mutually supportive adjustments to natural, technological, and social threats have probably increased.

Changes of risk

Risk (as measured by the probability of experiencing an extreme event) is among the more stable components of urban hazard. That is not to say that the urban physical environment remains unchanged. In many cities, geological substrata are sinking, river beds are rising in response to accelerated sedimentation, natural flood-detention systems are being converted to other uses, and vegetated slopes are being cleared for housing developments. For example, subsidence in response to underground water withdrawals is a continuing problem in Mexico City. In Lima, the bed of the Rimac River is slowly rising as sediment accumulates in the channel; seepage from the elevated river is weakening foundations and walls of nearby buildings, which are predominantly built of dried mud. The disappearance of open spaces to provide building lots for a voracious housing market is held to be directly connected with increased runoff and flood losses in Seoul. Infilling of floodwater-detention ponds in Dhaka is also believed to be cutting the time between low water and peak discharge during major floods and is thought to be increasing the damage potential of lesser floods. Moreover, the spectre of global climate change and sealevel rise threatens a significant number of large cities. However, despite their continuing importance, changes in physical risks probably pose fewer problems and fewer uncertainties for urban populations than do changes in the other three components of hazard.

Changes of exposure

Large changes in exposure and vulnerability are perhaps the most striking of the case-study findings. In virtually all of the mega-cities that were analysed, the locus of hazard is shifting – often rapidly. For example, despite popular beliefs that associate the city of San Francisco with catastrophic earthquakes, other communities on the eastern fringes of the mega-city now have much larger potentials for disaster. In Tokyo, the inner and outer suburbs are increasingly at risk because of both rapid expansion of the city's periphery and lack of attention to hazard-sensitive design in the newer developments. By contrast, the central city and adjacent neighbourhoods have been the focus of major investments in emergency preparedness, hazard mitigation, and other disaster-reduction alternatives. Unlike North American cities, which appear to be dying from the centre towards the periphery – with parallel increases in exposure and vulnerability to hazard – Tokyo appears to be renewing itself and upgrading its disaster resilience from the centre outwards. In Greater London, a shortage of buildable land since the Second World War has encouraged occupance of previously open floodplains everywhere except in a protected green belt and despite land-use planning controls. At the same time there has been a movement of the locus of economic investment and reinvestment downstream (eastward) along the Thames Valley. This is making London increasingly more susceptible to tidal flooding, especially if predictions about sealevel rise (connected with global climate change) are confirmed by subsequent events.

Changes of vulnerability

Increasingly, *polarization and spatial segregation of groups that have different degrees of vulnerability to disaster are becoming the norms for most cities*. In almost every case it is clear that poor and rich urban neighbourhoods are diverging with respect to the types of hazard they face and the degrees of risk they are exposed to. Contrasting types of hazard in rich and poor areas are most obvious in the mega-cities of developing countries (e.g. Mexico City, Lima, Dhaka), where the sheer volume of recently arrived poor migrants from rural areas and their association with hazardous marginal lands are also prominent features. Available information does not permit a definitive assessment of polarization and hazard segregation in Tokyo and Seoul, but there too evidence suggests that "safety gaps" are opening up between different social groups. Polarization and segregation with respect to income and risk are also increasingly visible in European and North American mega-cities. The contrast between poor earthquake-prone inner-city populations in Los Angeles and affluent suburbs at risk from wildfires and slope failures has received much attention from analysts both in this book (chap. 11) and elsewhere (Davis, 1995; Kirp, 1997). Similar but less well-known contrasts characterize Greater Miami, where poor south-side neighbourhoods bore the brunt of hurricane Andrew whereas the richer north side escaped relatively unscathed. Indeed, as far as the hazard consequences of polarization and segregation are concerned, it can be argued that the cities of the developed world are beginning to resemble those of the developing world.

Changes in efficacy of hazard management

Responses to hazard include measures adopted by formal public and private institutions and adjustments adopted by private individuals. Most of the case-study authors report *increasing difficulty in developing and sustaining public support for hazard-management initiatives*. In mega-cities, patterns of response to hazard are complex and varied. Some cities (e.g. Tokyo) have adopted multiple reinforcing public adjustments, involving sophisticated technologies, backed up by high levels of training and self-reliance on the part of civilian populations that together are designed to address all parts of the disaster cycle (i.e. preparedness, emergency management, recovery, mitigation). Others possess no effective formal programmes for natural-hazard management (e.g. Dhaka).

Most occupy intermediate positions on the spectrum of response. However, hazard reduction appears to be losing ground almost everywhere in recent years, even in the best-protected cities of Japan, Europe, and North America. For example, the pace of hazard-related legislation and associated administrative actions has slowed noticeably in Tokyo since the 1970s; so too has the construction of large "hazard-proof" building complexes. Whether the disastrous effects of the 1995 Hanshin earthquake will undermine public confidence in existing Japanese approaches to hazard management or bring about renewed investments in alternative measures remains to be seen. In London, there has been a decided weakening of government support for planning and development controls, including those that affect the use of hazard-prone land and funding for emergency services. Despite Sydney's long-established formal commitment to principles of comprehensive planning and significant investment in institutions that respond to floods, bushfires, and other sudden emergencies, the city seems unable to grapple with the chronic hazards associated with long-term environmental deterioration.

Counterposed against the very real problems of mounting actions that are exclusively focused on hazards are some promising opportunities for *jointly attacking hazards and other overlapping urban issues.* In such cases, shared constituencies help to keep hazard reduction high on the public agenda. Moreover, it is when hazard issues overlap with other issues that analysts have opportunities to examine the link between hazard systems and the larger problematiques of which they are a part.

The momentum to change public policy or to find the resources necessary to implement existing policy is usually short lived in most large cities. It is often present for a limited period during a major disaster but wanes quickly thereafter. Although disaster recovery and improved preparedness can stand alone as public issues during these periods because they command the support of many victims or other affected groups, it is often difficult to sustain public interest and involvement in disaster policymaking for very long at other times. However, when hazard issues overlap with other issues there is considerable potential for mobilizing a large and continuing joint constituency. The post-hurricane rebuilding of Greater Miami was effectively assisted by the joining together of separate groups concerned about poverty, immigration, migrant labour, crime, national defence, the peacetime role of military forces, mass transit, building code enforcement, intergovernmental relations, and tourism, among other matters (Mitchell, 1995). Though the broad interest group coalitions that were born during the emergency have not persisted, the hurricane's differential impact on these groups has created the potential for new cooperative alignments in local politics that might be exploited by the managers of hazards and urban communities (see chap. 12 in this volume). If disaster-reduction efforts are to succeed, urban leaders and managers need to have a better understanding of the potential for these kinds of intersecting issues to emerge and the probability that some combinations may be especially felicitous for the adoption of innovative disaster-reduction measures. The sustainability of cities in the face of disaster is as much a function of enhancing institutional and behavioural capacities to deal with uncertainty as it is an outcome of material investment in hazard-management technologies and physical infrastructure.

Spatial parameters of mega-city hazards

The spatial parameters of mega-cities have been singled out for attention for two main reasons. First, the internal and external spatial organization of cities – including their populations, physical environments, and activities – is one of the fundamental themes of urban research. Secondly, the "goodness of fit" between people and environments, which is a central problem of hazards research, has a strong spatial component. In other words, issues of space and place are particularly germane to the analysis of hazards and of urbanization. Some of these matters have already been touched on above (e.g. the shifting locus of mega-city hazard), but they will be revisited here in order to highlight a whole family of changes that are explicitly spatial in nature. The discussion that follows examines mega-city hazards at a number of spatial scales from global to neighbourhood.

Global and national linkages of mega-city hazards

On 12 May 1997, a well-defined tornado was captured on television as it passed the downtown district of Miami without causing significant damage. The image of a powerful vortex spinning behind glass and steel high-rises that are the product of Miami's recent emergence as an international centre of finance for Latin America vividly underlines one aspect of the connections between the global economic system and the hazardousness of mega-cities. In this case, the outcome may have been positive from a hazard perspective. The influx of new investment helped to rebuild a business district that is now more resistant to extreme winds. What of other effects elsewhere?

First, *the global economic system has a definite but uneven influence on mega-city hazards*. Investments and economic functions are being redistributed among and within countries, thereby stimulating changes in national and local levels of exposure and vulnerability to hazards. For example, loss of investment in Los Angeles leads to the closure of firms that contribute to – and are victimized by – environmental hazards. But it also means reductions in local public and private resources for managing hazards and speeding recovery from disasters. Conversely, London in recent years has attracted a disproportionate share of new investments in urban hazard management precisely because it contains so many globally important financial services that are vulnerable to disruption. Such divergent experiences illustrate the difficulties of assessing net effects of global economic changes on mega-city hazards. Mexico City's recent economic history reinforces the point. In the period between 1940 and 1980, during the era when import-substitution economic policies were in vogue in Mexico, large numbers of new businesses established themselves in the city's inner districts. But recently, with the advent of export-oriented policies, many of Mexico's larger, more productive, and more profitable firms have migrated elsewhere, especially to the strip of land adjacent to the US border. The city still confronts a legacy of serious natural hazards that were much aggravated by recent meteoric urban growth, but it does so without the involvement of some of the large-scale economic enterprises that it had previously monopolized. These examples suggest that global economic forces help to change the calculus of hazard in mega-cities but do not produce any one predominant outcome.

A second way in which external influences can affect the management of mega-city hazards is via the *extension of national government authority over local public policy*. Here again the evidence is difficult to interpret both because local politics often have a disproportionately large effect on national policy-making in primate cities that are also national capitals and also because the constitutional division of responsibilities for hazard management between different levels of government varies among countries. For example, in the past, national government policies largely shaped the development of places such as Seoul, Lima, and Dhaka, but elsewhere metropolitan government institutions such as the Greater London Council pursued independent orientations. With the re-emergence of laissez-faire philosophies of national government during the 1980s, the latter role became less evident in London, although the departure in May 1997 of a British political regime that was sympathetic to free market ideology may signal a possible reversal of this trend. Even in Japan, which remains very much a centrally directed bureaucratic state, the city (prefecture) of Tokyo has a high degree of autonomy, as evidenced by its proliferation of specially tailored metropolitan hazard-management laws and planning initiatives. In short, although the evidence is mixed, mega-city hazard-management policies and programmes are just as likely to be different from those pursued by national governments as they are to be similar.

A third way in which external influence comes to bear on mega-city hazards is through the *operation of broad cultural notions about the societal role of environmental hazards.* From a public policy standpoint, the main issue here is whether the myths are appropriate for the management of mega-city hazards. Many seem to be out of phase with urban realities. For example, the Australian national myth of a colonizing frontier people who have contested and subjugated a harsh land may not be well suited to the relatively benign precincts of Sydney. There the most pressing environmental problems are now more likely to arise through human mismanagement than because of natural risks. Hazards are largely missing from the dominant British environmental myth, which runs to bucolic images of well-cultivated fields, neat gardens, and an undemanding climate. Yet London remains susceptible to very serious flood and storm hazards. Finding and reinforcing appropriate hazard myths for mega-cities deserve a significant place on the agenda of urban management in the next decade.

Urban hinterlands supply yet another set of external influences on mega-city hazards. For Puente (chap. 9 in this volume), a mega-city is more or less vulnerable according to the degree of its dependence on its hinterland. For example, *rural disasters are a potent generator of migrants to urban areas.* Dhaka receives a seasonal influx of rural flood victims, many of whom remain to swell the city's permanent population and increase its long-term vulnerability. The migration of rural victims from rural Peru to Lima under pressure from Shining Path guerrillas, or from Haiti to Miami in response to civil unrest, are other examples. Elsewhere the stimulus might be drought or famine, perhaps combined with civil war (e.g. Ouagadougou or Khartoum). The preceding examples are of cities in LDCs, but the principle extends to all mega-cities. In the case of London or New York, hazard hinterlands are much larger. Both of these cities are bases for governmental and non-governmental organizations that have global responsibilities for disaster relief and both have been subject to terrorist incidents that reflect geographically distant conflicts. The size of the hazard hinterland increases as does the city's centrality to the global urban network.

Another set of spatial properties derives from the position of mega-cities within systems of cities. *Mega-cities tend to be at or near the top of national urban hierarchies, where they generally have superior access to national resources.* This enhances their resilience against disasters. For example, national policy for disaster relief was liberalized when Britain's largest city – London – was badly affected by a severe storm in 1987 (Mitchell et al., 1989). Miami, the eleventh-largest US metropolitan area, did not receive nearly such favourable treatment from national policy makers in the wake of a much more destructive storm – hurricane Andrew (1992). *Mega-cities also lie at different distances from the core of the global economic system. This affects their capacities to mobilize inter-*

national aid in the wake of disasters and to secure other forms of external support for hazard-mitigation programmes. The general spillover effects of responses to security threats that were taken by private financial institutions in London have already been noted. Similar changes occurred in the San Francisco Bay Area's Silicon Valley after the Loma Prieta earthquake. On the other hand, the extent to which international businesses have upgraded adjustments to hazard in cities such as Dhaka and Lima is not known but is suspected to have been much less.

Urban form as a factor of mega-city hazard

Now let us turn to the *mega-city as an urban spatial form*. In this respect there are great variations among the case-study mega-cities. Some are relatively compact, densely populated, and intensely developed places (e.g. Lima), whereas others sprawl in all directions at low density (e.g. Sydney). Some pivot around one or two central nodes (e.g. Mexico City), whereas others have multiple urban nuclei (e.g. Los Angeles). Some are set on more or less flat land with few major barriers to surface movement (e.g. Miami); others are punctured by rivers, coastal embayments, steep slopes, and similar impediments (e.g. San Francisco Bay Area). Some are served by well-integrated transportation grids with many alternative routes, modes, and interchange points (e.g. London); others squeeze traffic into a handful of paralysing bottlenecks (e.g. Seoul). The hazard consequences of these different layouts are considerable.

Each urban form offers a different mix of advantages and disadvantages to hazard managers and potential victims. For example, compared with compact cities such as Dhaka, low-density ones such as Los Angeles are serviced by complex infrastructure networks (e.g. electricity, water, gas) that may be disrupted more often and more easily by a wider range of locally prevalent hazards. But dispersed urban forms lend themselves to emergency-management systems that optimize vehicular mobility and the transfer of relief aid into affected neighbourhoods from external sources. Problems can arise when movement is impeded; winding tree-lined roads in the low-density hilly suburbs of Oakland proved to be deathtraps for escaping motorists and barriers to incoming emergency vehicles during a recent wildfire! Conversely, Dhaka's heavily built-up inner city, served by narrow, congested streets and often improvisational infrastructures, raises nearly insurmountable access problems for "rapid response" teams that must be dispatched from elsewhere and offers advantages to neighbourhood-based emergency systems.[3] Shortages of open space in densely built-up cities put severe strains on emergency evacuation systems that use such places as accessible assembly points where temporary shelters can be erected and neighbourhood feeding stations established. Diseconomies of scale certainly come into play sooner for large hazard engineering projects in low-density cities than in compact mega-cities, but this does not mean that such projects are precluded. The cost of building large protective structures (e.g. dykes) around lightly occupied neighbourhoods is daunting, but this has not prevented even poor cities such as Dhaka from investing in such measures when all existing urban lands are already developed at high densities. The point here is not that the advantages and disadvantages of either type of city form are permanent; rather it is necessary to design hazard-management systems that are best adjusted to the changing needs of specific mega-cities. It is also necessary to be careful about applying policy guidelines for urban development that do not take much account of hazard criteria. For example, as Parker notes in chapter 7, London does not conform to the model of urban development preferred by the European Commission (i.e. high-density, mixed-use, compact). The reasoning may be sound from the perspective of amenity preservation and the efficient delivery of services, but the hazard implications of such urban forms are far from clear.

Internal urban dynamics

Core–periphery relations in the city, the differentiation of neighbourhoods, and the sequential occupation of urban districts by different land uses and populations have frequently been of interest to urban analysts. Information about environmental hazards throws new light on all three topics. Perhaps the most interesting finding relates to patterns of upgrading and renewal of urban areas in response to concerns about natural hazard. Particularly distinctive is the migration of vulnerability in Tokyo. There the zone of structural vulnerability to hazards is moving outwards toward the suburbs and the zone of social vulnerability is differentiating inwards toward particular marginal populations that frequent places with special hazard characteristics. Though less clear cut, a similar process seems to have been occurring in Miami before hurricane Andrew devastated the south side of the metropolitan area and accelerated the transfer of more resilient middle-class populations to north-side locations. Variants may be in progress in the San Francisco Bay Area and in Los Angeles, where the exposure and vulnerability of buildings to a suite of natural hazards are also very high along the outer urban margins and where there has been less investment in upgrading the hazard resistance of neighbourhoods near the urban cores.

Mention of urban margins and marginal populations recalls the marginalization theory of hazard ecology (Wisner, chap. 11 in this volume). This suggests that marginal places (i.e. peripheral, hazardous, or otherwise undesirable) and marginal people (i.e. disadvantaged groups) often coincide, usually because social forces push those without resources to the literal and figurative "edges" of community life. It is this that leads Wisner to assert that, in so far as Los Angeles is concerned, "there are worse things than earthquakes." However, evidence from the full range of mega-cities seems less clear. Sometimes areas that lie at the geographic heart of a great city, and are much in demand by many potential users, are also highly hazardous (e.g. the filled lands of Tokyo and the former lake beds of downtown Mexico City). Sometimes, too, the outer periphery of a mega-city is exposed to great physical hazard but is also much sought after for homesites because it possesses offsetting amenities or other values (e.g. the San Francisco Bay Area's Oakland Hills; the new polder developments of Dhaka). Yet there is no denying that the dry lakebed neighbourhoods of Mexico City's north-east side or the *pueblos jovenes* in peripheral districts of Lima are occupied by poor people and often subject to serious environmental hazards. Given the implicit logic of marginalization as a theory of hazard, these cases deserve close analysis. The various kinds of margins need to be more clearly distinguished[4] and the relationships between them and conditions of social advantage or disadvantage spelled out.

The space–time continuum

"Compression of the space–time continuum" has been identified as a distinguishing feature of worldwide societal change in the late twentieth century. This term connotes both: (1) a speeding up of the circulation of information, capital, and people; and (2) a reduction of distinctions between different places. These processes are believed to be occurring more or less everywhere but are said to be most evident in urban areas. The record of hazards and hazard responses in this sample of mega-cities provides considerable support for the first of these notions but much less for the second.

Urban patterns of exposure and vulnerability are sensitive barometers of speeded-up circulation and they are the dimensions of mega-city hazard that are changing fastest and most dramatically. New spatial patterns and widening gaps between population subgroups are detectable nearly everywhere. Hazard effects attributable to the accelerated circulation of capital and people are more apparent than those attributable to speeded-up information. Sometimes the results are positive – that is, they lead to improved protection against environmental hazards for workers and residents (e.g. in the financial districts of London and Miami); sometimes they are negative, as in the experience of cities that lost taxable investments that paid for safety services and provided residents with the financial resources to buffer hazard (e.g. parts of Los Angeles). But there are signs that exchanges of information between MDC and LDC mega-cities may be working to reduce hazards in some places (e.g. Latin American cities). At present such exchanges mainly involve the physical movement of experts and students between information-sending and information-receiving countries by means of international conferences, courses for foreign students, and consultant visits abroad. The first Internet Conference on Urban Hazards was held in 1996 under the sponsorship of the International Decade for Natural Disaster Reduction (Geneva) and signalled the beginning of faster exchange of ideas and knowledge.

Conversely, the hazard landscapes of mega-cities do not appear to be becoming more alike. Residents confront a kaleidoscope of choices and constraints produced by layered combinations of: physical risks; human-modified environmental processes; the forms and dynamics of the built urban environment and its social ecology; and existing and emerging patterns of human exposure, vulnerability, and response to hazards, among others. For managers of urban hazard, the central problem remains one of adjustment between environments and peoples that started out both diverse and changeable and have only become more so in recent decades.

Summary of case-study findings

The picture of mega-city hazards that emerges from the case-studies is complex and susceptible to different interpretations. Its main features include: (1) the diversity of risks that confront urban populations and growing interactivity among those risks; (2) the extent to which previous urban disasters (especially natural ones) have had deep and long-lasting repercussions on built environments and societal institutions as well as more obvious immediate human effects; (3) the build-up of catastrophe potential in mega-cities; (4) the narrowness of existing urban hazard-management policies and programmes; (5) important gaps in scientific information; and (6) reorganization of the urban ecology of environmental hazard, most notably reflected in shifting patterns of exposure and vulnerability. In short, according to the case-study evidence, the environmental hazards of large urban areas are already highly important and they are changing in ways that will increase their significance during the twenty-first century. Urban managers would do well to pay attention to these trends and to include hazard management among their priorities.

In the meantime, there are a few simple steps that can be taken to pave the way toward improved management of environmental hazards in mega-cities. There is a pressing need for "horizontal" exchanges of information about hazards and hazard-management policies among individual mega-cities. Such exchanges offer valuable opportunities for learning about the effectiveness of different policy tools in different hazard contexts. For example, London and Seoul might compare the use of green belts as hazard-management devices; Tokyo and Miami might exchange data about the taxation of peri-urban agricultural land; Mexico City and Seoul might compare the impacts of programmes for decentralizing the economic and political functions of capital cities; Sydney and Los Angeles might focus on the nexus of transportation–air pollution–emergency-management problems that affects both cities; Dhaka and Lima could share experiences about links between squatters, housing policies, and hazardous sites.

Historical data on previous hazard adjustments and data about present hazard-management strategies adopted by individuals and small firms are scarce in most mega-cities. We need carefully to sort out adjustments that failed because they were fundamentally overwhelmed or mismatched with the new conditions created by recent urbanization from those that atrophied from lack of awareness by public officials and new urban residents, or from lack of investment, or from other causes that might still be rectified or redressed. In an era when private sector initiatives are being encouraged by public policies (e.g. the World Bank's new urban-management programme), promising innovations at the grass roots also need to be identified and nurtured. Researchers can begin both activities now.

The potential of theoretical constructs such as "urban metabolism" and "urban carrying capacity" is largely unexplored from a hazards-management perspective. To date, these notions have been chiefly applied to the exchange of resources and wastes between urban areas and surrounding territories and to the environmental impact of urban resource development projects. But they might profitably be extended to include the impact of hazards on the carrying capacity of mega-city infrastructures. Hazards that impair the conduits on which metabolism depends or that reduce the carrying capacity of urban infrastructures, even on a temporary basis, might be taken into account by urban sustainable-development programmes (Berke, 1995; MacLaren, 1996).

New departures

Beyond the immediate findings lie other observations that open up new questions and suggest new directions for future research. Broadly speaking these fall into two categories, with pairs of contrasting themes: (1) the effects of hazards on mega-cities and of mega-cities on hazards; and (2) the ways in which hazard mitigation might be added to the agenda of mega-city managers while hazards managers might take on board some of the broader concerns of mega-city management.

Quite apart from their obvious impacts – destructive and otherwise – natural hazards, natural disasters, and other natural extremes affect mega-cities and the people who live there in three distinctive ways that have received inadequate scrutiny by researchers. First, they are important agents of *diversification* in urban settings. Secondly, they provide vivid examples of *contingencies* that affect human life. Thirdly, they are fertile sources of *metaphors and myths* about urban existence.

It is often alleged – and not without cause – that large urban areas are increasingly structured by global economic and political forces that encourage uniformity of landscapes and lived experiences. Whether the process is unilinear or bifurcated – towards one type of urban entity or towards "more developed" and "less developed" variants – similarity appears to be growing. Natural hazards are a potent antidote to this process, for they segment urban space into areas of differing risk that are continually being modified and shifted as people search out and apply a growing roster of adjustments that change net risks and alter levels of exposure and vulnerability.

Likewise, an entirely praiseworthy concern for alleviating the *everyday* problems of big cities blinds many urban analysts to the vast importance of unexpected, infrequent, and otherwise *exceptional events* that compel leaders and residents to change focus and redirect their energies toward different ends. Natural hazards and disasters are prime examples of a class of change factors that frustrate long-term schemes of planning and management that are based on assumptions of continuity. Of course, local government has often been considered an institutionalized form of crisis management, mostly characterized by muddling through. But the numbers and significance of events that continually bubble to the top of the urban crucible are such that more than crisis-response is now necessary. The public and private sectors of metropolitan areas should learn to take disjunctive events into account systematically and deliberately, not just as inconvenient disruptions of "normalcy." In other words, mega-cities need to equip themselves with more than preparedness systems, emergency operations centres, and disaster-response capabilities. The more contingencies occur in cities, the more cities will have to develop flexible responses across the full spectrum of governmental responsibilities. Broadly construed, hazard mitigation – in all its forms and for a broad range of events – should become a continuing basic part of urban governance.

Finally, the ultimate importance of environmental hazards for big-city populations may be their symbolic value as fertile sources of metaphors and myths about appropriate human behaviour in an uncertain universe.

Too often, myths about great cities and disasters have tended toward the heroic and the apocalyptic. They have crystallized into unchanging behavioural guides and public attitudes that may have been historically useful but are now ill suited to the fast-breaking multiple crises of the postmodern world. The phoenix myths of Chicago or Seoul or San Francisco have outlived their former purposes and now stand as half-truths at a time when more subtle lessons about the ongoing nature of human adjustments to environmental constraints might offer appropriate pointers toward the long-term sustainability of urban areas.

The effects of large-scale urbanization on natural hazards are manifold, but three new avenues are particularly worthy of further investigation:
1. the changing (spatial) ecology of urban hazard;
2. the role of indirect information about hazard that is presented in the urban landscape and the urban experience; and
3. opportunities for raising the low visibility of hazard as a topic of urban governance.

The bulk of previous research on human adjustments to natural hazard has tended to adopt an all or nothing approach toward the role of group processes. Hazard adjustments are regarded either as preponderantly the product of individual decision-making or as primarily the creations of powerful social forces that are brought to bear by the dominant on the dominated. Neither of these views accords with the reality of life in a complex urban environment where both factors operate and the entire setting is quite evidently undergoing far-reaching change. So it is necessary to employ investigative tools that are more sensitive to the joint effects of individual and group processes. Given the ample evidence that the socio-spatial organization of mega-cities is in flux, it is worthwhile to invest more effort in *mapping the various elements of hazards*. Patterns of risk, exposure, vulnerability, and response are known to be changing, but the changes are poorly documented. If long-term strategies for hazard mitigation are to be effective, it will be necessary for them to be periodically readjusted to take account of changing land use and occupancy, new technologies, emerging social issues that bear upon hazard management, and other factors. Better understanding of urban hazard ecology is a prerequisite for this process.

Mega-cities can be thought of as complex information-transfer nodes that display and transmit a vast array of signs, signals, and messages. In such settings, direct messages about hazard (e.g. hazard maps, warnings, informational pamphlets, public service announcements by aid agencies) are apt to be diluted or lost before they reach the intended recipients. At the same time, the urban environment is rich in indirect information about hazard that may be more effective in influencing decisions and behaviour than the formally targeted messages. Sources of indirect information include, among others: events observed in the street or evidence of previous occurrences; knowledge acquired in schools, libraries, museums, or other educational establishments; news reports picked up from the mass media; warning notices and symbols; advertising signs and broadcasts; consumer product information; entertainment programmes and performances; gossip among family members, friends, and neighbours; rumours passed along by others; acquired folk beliefs, etc. Although there has been a large amount of research on the perception of specific risks and on judgements about certain sets of risks, little attention has been focused on links between the broad array of indirect influences on risk judgement in cities and assessments of particular natural hazards or on the interplay of indirect information and direct information pertaining to the full range of mega-city risks. The *contextual embeddedness of mega-city hazards warrants greater attention* by the research community.

A third aspect of urbanization that deserves further exploration as an influence on hazard management is the *emerging structure of interest group competition in city governance*. As recognized by the United Nations Commission on Human Settlements (1996) and other organizations, public policy in contemporary cities arises through struggle among interest groups that focus on one or more of the following subjects: economic growth, environmental protection, and social justice (Campbell, 1996). To a large extent, economic interest groups exist not just because of a desire for profit but also because of previous experience with economic stagnation and fears that such conditions could occur again. Likewise, social justice interest groups exist not just because of a commitment to the ideal of social harmony but also because of experience with social injustice and fears of worse to come. However, urban environmental protection interests are much less united about the goals of environmental protection and they are rarely concerned about hazards that arise in Nature; instead their focus is on stemming the deleterious effects of human mismanagement of the physical world. In other words, natural hazards are largely unrepresented in the urban governance models of contemporary analysts. The tendency to marginalize the physical environment as a variable in urban affairs is already marked in many cities, and the low salience of natural hazards in city politics is disturbing and requires further analysis.

This brings us to the second pair of issues, namely, the insertion of hazard considerations into urban management and the broadening of urban hazards management to take account of other goals. If the central problem of contemporary urban management is to resolve conflicts among economic, environmental, and justice interests, the central problem of contemporary hazards management is to shift the emphasis of public policy from reaction to anticipation – from dealing with emergencies after they occur to addressing the problems that "cause" them. What happens when these agendas are combined?

A very likely possibility is that tensions will be generated among existing urban policy interest groups. For example, judged by previous experience, economic interests may favour policies that improve preparedness and emergency-management alternatives that reduce foreseeable disruptions to business operations and markets but may be less willing to invest in mitigation measures whose pay-off lies far in the future. Environmental interest groups may prefer the inverse priorities. So little is known about the hazard perspectives of interest groups concerned with urban social justice that a wide range of orientations is possible; they might favour either, neither, or both alternatives or they might promote quite different ones. It is premature to fix on any single distribution of preferences, but the likelihood of *incompatibilities among the principal interest groups* is significant. Again there is a need for additional research, especially with respect to the stance of justice interest groups toward issues of urban hazard.

Environmental hazards have been, are, and will continue to be important problems for the world's mega-cities. Indeed, their visibility is likely to increase over the next several decades. It will be important to monitor how well urban societies respond to the challenges that they pose, because hazards and disasters possess special significance as public issues. To a significant degree the success of urban sustainable-development initiatives will be determined by the ability of mega-city leaders and their diverse constituencies to join in metropolitan-wide systems of governance and management that are responsive to disjunctive changes as well as to continuing problems. Nor will it be possible to exempt mega-cities from the broader global agenda of sustainable development, which requires humans to balance the demands that we make upon the physical environment with the limits that the environment sets on the possible. As agents of disjunctive change and dramatic constraints on human activities, natural hazards and disasters set demanding tests for the emerging institutions of urban living.

Notes

1. New street patterns and new construction materials both appeared in the wake of the fire. The changes would have been more profound if Charles II had been both rich enough and powerful enough to impose a formal reconstruction programme on the city. As it was, massive Baroque-style reconstruction schemes proposed by Sir Christopher Wren and others were never carried out (Knox, 1993, p. 92).
2. Parts of this section have been summarized in Mitchell (1995).

3. This does not mean that both practices are in fact avoided. The "poldering" of new suburban neighbourhoods behind large embankments in Dhaka is one obvious example of the former. Because of urban congestion, Calcutta's official ambulance services are mainly dispatched from their bases to carry non-critical patients to hospitals, whereas victims whose lives are at stake are generally brought to medical facilities by passing motorists.

4. Among others, margins might include: (1) the physical edges of the city, namely, the outer edge, which is in contact with rural districts; inner edges adjacent to physical boundaries such as coasts, rivers, and steep slopes; (2) time–distance edges that connote remoteness from opportunities for work, necessary facilities, and the like; (3) perceptual edges such as boundaries between social areas occupied by different – perhaps hostile – groups and places located beside zones of exclusion such as high-crime districts or red-light districts; and (4) zones of physical difficulty such as areas exposed to natural risks and hazards.

REFERENCES

Berke, Philip R. 1995. "Natural-hazard reduction and sustainable development: A global assessment." *Journal of Planning Literature* 9(4): 370–382.

Campbell, Scott. 1996. "Green cities, growing cities, just cities? Urban planning and the contradictions of sustainable development." *Journal of the American Planning Association* 62(3): 296–312.

Davis, Mike. 1995. "The case for letting Malibu burn." *Environmental History Review* Summer: 1–36.

GeoHazards. 1994. *The Quito, Ecuador earthquake risk management project: An overview*. San Francisco: GeoHazards International.

Gruber, Urs and Harold Haefner. 1995. "Avalanche hazard mapping with satellite data and a digital elevation model." *Applied Geography* 15(2): 99–114.

Kirp, David L. 1997. "There goes the neighborhood: After the Berkeley fire, an architectural disaster." *Harper's Magazine* 294 (March): 45–53.

Knox, Paul. 1993. "Cities of Europe." In Stanley D. Brunn and Jack F. Williams, *Cities of the world: World urban development*. New York: HarperCollins, pp. 88–150.

MacLaren, Virginia W. 1996. "Urban sustainability reporting." *Journal of the American Planning Association* 62(2): 184–202.

Mitchell, James K. 1995. "Coping with natural hazards and disasters in U.S. mega-cities: Perspectives on the twenty-first century." *GeoJournal* 37(3): 303–312.

Mitchell, James K., Neal Devine, and Kathleen Jagger. 1989. "A contextual model of natural hazards." *Geographical Review* 79(4): 391–409

Montz, Burrell and Eve Gruntfest. 1986. "Changes in American urban floodplain occupancy since 1958: The experiences of nine cities." *Applied Geography* 6(4): 325–338.

Oppenheimer, Clive. 1994. "Discussion meeting on natural hazard assessment and mitigation: The unique role of remote sensing, The Royal Society, London, 8–9 March, 1994." *Disasters* 18(3): 294–297.

Platt, Rutherford H. 1991. "Lifelines: An emergency management priority for the United States in the 1990s." *Disasters* 15(2): 172–176.

United Nations Commission on Human Settlements. 1996. *An urbanizing world: Global report on human settlements 1996*. New York: Oxford University Press.

UNEP/WHO (United Nations Environment Programme/World Health Organization). 1994. "Air pollution in the world's mega-cities." *Environment* 36(2): 4–13, 25–37.

Postscript: The role of hazards in urban policy at the millennium

James K. Mitchell

Habitat II, the Second United Nations Conference on Human Settlements (Istanbul, 3–14 June 1996), provides a valuable benchmark of contemporary thinking about urban policy by researchers, managers, and political leaders throughout the world. During the past 25 years, the United Nations has sponsored a series of international conferences on pressing public issues, ranging from population and development to women's role in social change. Many of these gatherings have functioned as highly publicized forums for formulating policies that were subsequently adopted by international institutions and national governments. Befitting its status as the last major global policy meeting of the twentieth century, Habitat II was intended to function as a capstone to the entire conference series.[1] Though rural settlements were not excluded from the agenda, the consequences of increasing urbanization were the focus of concern (UNCHS, 1996).

Students of urban hazard watched the Istanbul meeting with considerable interest because they anticipated that it would reflect substantial recognition of the growing importance of environmental risks and disasters in cities, especially the very large cities that are now coming to dominate the global settlement pattern. This was a reasonable expectation because the toll of mega-city disaster losses had spiralled dramatically upward during the previous decade and several international organizations had already taken up the issue. In the 1980s, agencies with responsibilities for disaster relief (e.g. the League of Red Cross and Red

503

Crescent Societies; the UN High Commissioner for Refugees) drew attention to growing urban hazards and the urban effects of rural hazards (Wijksman and Timberlake, 1984; El-Hinnawi, 1985). Now other international agencies, for which urban hazards had not historically been a central concern, added them to their agendas. These included the United Nations Environment Programme, the World Health Organization, and the World Bank (World Health Organization, 1992; UNEP/WHO, 1994; Kreimer and Munasinghe, 1992; Serageldin et al., 1995; World Resources Institute, 1996, p. 144). Furthermore, natural disasters were recognized as an important global issue in their own right. Among the UN meetings that led up to Habitat II was the World Conference on Natural Disaster Reduction (Yokohama, 23–27 May 1994), which highlighted urban hazards (Ichikawa, 1995). The accompanying International Decade for Natural Disaster Reduction has given rise to a variety of significant initiatives for mitigating the natural and technological hazards of large cities (IDNDR, 1996). Sensing the growing importance of urban disasters, the United Nations Commission on Human Settlements (UNCHS) adopted a resolution that directly addressed the role of natural disasters in urban areas and outlined a substantial programme of action (UNCHS, 1995).

Had the spirit of the UNCHS resolution carried over into Habitat II there would be little need for concern about the salience of hazards on the global urban management agenda.[2] Unfortunately, expectations that Habitat II would produce a firm global commitment to hazard-sensitive urban development were largely disappointed. Hazards, disasters, and associated topics were barely mentioned in conference texts and then only in contexts that were marginal to the main thrusts of the meeting.[3] Housing losses due to natural and man-made disasters *were* chosen as one of 29 indicators used to construct a global urban database, involving 235 cities in 110 countries (Appendix 3). But hazard-reduction goals were almost invisible among over 600 so-called "best practices" of urban management that were identified in a global inventory solicited by conference staff. For example, just 2 of the top 105 "best practices" of urban management identified during Habitat II involved responses to environmental hazards or disasters, and only one of these affected a mega-city (Cairo) (Appendix 4). Fewer than one-third of the 34 natural disaster projects included in a larger database of over 600 "best practices" were located in large cities (Appendices 5 and 6).

In light of the evidence provided in this book, the apparent failure to highlight burgeoning problems of urban hazards and disasters – especially natural disasters – is disturbing. It is disturbing not just because an important problem is being ignored by an influential constituency but also because failure properly to address urban environmental hazards may signal flaws in the conceptualization and operationalization of urban sustainability as a basis for managing large cities.[4]

The question that mega-city hazards pose to policy makers goes right to the heart of sustainability. How, if at all, can large and rapidly changing cities be made sustainable in the teeth of potentially devastating events that are also highly uncertain? Given the centrality of sustainable development as a guide to policy-making for all aspects of the human environment, the contention that it does not – as currently construed – adequately take account of environmental hazards is a serious challenge. A detailed argument in support of that claim is beyond the scope of this volume, but it is appropriate to introduce some important pieces of supporting evidence.

First, urban sustainability is a concept that is contested between advocates of so-called "Green" and "Brown" agendas; hazards play different roles in these agendas and are affected by different kinds of policy responses (Satterthwaite, 1996; World Resources Institute, 1996). The Green agenda gives pride of place to hazards that are linked with anthropogenic degradation of the physical environment (e.g. resource exhaustion, erosion, pollution) (Beatley, 1995; Mitchell and Ericksen, 1992). The Brown agenda highlights hazards in less developed countries that are linked to poverty and inadequate urban services (Main and Williams, 1994; McGranahan and Songsore, 1994). Acute geological, meteorological, and hydrological hazards are not excluded from consideration, but other types of human-constructed hazards that affect the poor on a daily basis are heavily emphasized. Surprises (i.e. unprecedented hazards), especially those that affect more affluent cities, receive little attention. Even if combined, these two agendas do not provide a comprehensive basis for addressing the hazard-management problems of large cities.

Secondly, differences between hazard mitigation and sustainable development ensure that important parts of each subject remain outside the frame of reference of the other. In other words, safety (a prime consideration in hazards management) does not necessarily equal sustainability, and contingencies (of which hazards and disasters are good examples) may require different responses than enduring problems (Mitchell, 1992; Berke, 1995). The truth is that large and complex cities require expansive management initiatives that can simultaneously address incommensurable goals. Mega-cities must be prepared to cope with unexpected or unfamiliar events as well as long-term problems; acute natural hazards as well as chronic crises of environmental degradation. Along with the evidence about trends in urban hazard that has been presented in this volume, the disjunctive events of recent history clearly support this claim (Hobsbawm, 1996).

To ignore the role of environmental hazards in cities is to deny important lessons of urban history. To discount the importance of natural hazards in contemporary mega-cities is to leave their populations exposed to worsening risks. To assume that sustainable urban development can be achieved without attention to problems of contingency – of which natural hazards are a pre-eminent example – is to court frustration and failure.

Notes

1. The conference was an enormous enterprise involving approximately 25,000 representatives, delegates, and other participants from governments and non-governmental organizations in almost every country. After a period of lengthy – sometimes heated – argument and intense negotiations, the conference issued a Declaration, an Agenda for further work, and a Global Action Plan.
2. Resolution 15/11 of the UN Commission on Human Settlements was adopted at its fifteenth session in Nairobi (25 April–1 May 1995). It summarizes various concerns that indicate a need for sustainable urban development and identifies measures for achieving that goal (see UNCHS, 1995, pp. 423–424). These are summarized as follows. Natural hazards and disasters are 2 of the 10 itemized concerns, namely: (1) "Natural disasters are an outcome of the interaction between natural hazards and vulnerable conditions which cause severe losses to people and their environments and they usually require outside intervention and assistance at national and international levels in additional to individual and communal responses"; and (2) "The challenge of comprehensive disaster mitigation programmes in urban areas is to continue general economic development and provide jobs, shelter and basic amenities while addressing the environmental and equity problems which are the real causes of vulnerability to natural hazards." Eight specific measures for improving the mitigation of natural disasters in urban areas are proposed: (a) setting up institutional structures that will ensure that natural disaster mitigation becomes an integral part of sustainable settlements development; (b) building national collective memories of disasters and responses to them; (c) improving access to safe building sites for poor people; (d) encouraging the siting of new settlements in safe areas; (e) identification of hazardous sites and conversion of them to productive uses in order to preclude illegal occupation; (f) reduction of threats associated with existing hazardous sites; (g) development of hazard-resistant housing; and (h) provision of technical assistance for hazard management to technicians, professionals, and administrators.
3. In Istanbul the main focus was on issues of urbanization, especially in large cities of developing countries. The most serious problems were identified as: inadequate financial resources; lack of employment opportunities; spreading homelessness; and the expansion of squatter settlements. The special concerns of refugees, migrants, and street children received particular attention. Much of the debate revolved around rights to adequate housing; reproductive health care; the future of the UN Centre for Human Settlements; and the inclusion of references to various social reforms affecting women, disadvantaged groups, and environmental health and justice. Recommendations for action reflected these topics. They included: (1) elimination of sexual exploitation of young women and children; (2) gender-disaggregated data collection; (3) lead poisoning prevention; (4) measures to take account of the social and environmental impact of policies; (5) a strong commitment to the economic empowerment of women, including references to the right to inheritance and flexible collateral conditions for credit; (6) affirmation of the right to an adequate standard of living for all people and their families; (7) language on environmental justice and environmental health; and (8) a reaffirmation of the call from the Beijing Conference on Women for control and regulation of multinational corporations and an appeal to the private sector to invest in communities (various issues of the *Earth*

Negotiations Bulletin, 1996, an independent reporting service that provides daily coverage of official UN negotiations for environment and development agreements wherever they take place in the world; it is published by the International Institute for Sustainable Development and supported by governments, United Nations agencies, and private foundations).

4. The concept of sustainable development was the guiding principle around which much of Habitat II was organized and it is often viewed as a vehicle for shifting the emphasis of hazard-management policies from relief to mitigation.

REFERENCES

Beatley, Timothy. 1995. "Planning and sustainability: The elements of a new (improved?) paradigm." *Journal of Planning Literature* 9(4): 382–395.

Berke, Philip R. 1995. "Natural-hazard reduction and sustainable development: A global assessment." *Journal of Planning Literature* 9(4): 370–382.

El-Hinnawi, Essam. 1985. *Environmental refugees*. Nairobi: United Nations Environment Programme.

Hobsbawm, Eric. 1996. *The age of extremes: A history of the world, 1914–1991*. New York: Vintage Books.

Ichikawa, Atsushi. 1995. "Coping with urban disasters." *OECD Observer* 197: 15–16.

IDNDR (International Decade for Natural Disaster Reduction). 1996. *Cities at risk: Making cities safer ... before disaster strikes*. Supplement to *Stop Disasters* No. 28. Geneva: IDNDR.

Kreimer, Alcira and Mohan Munasinghe, eds. 1992. *Environmental management and urban vulnerability*. World Bank Discussion Paper 168. Washington, D.C.: World Bank, pp. 51–76.

Main, Hamish and Stephen Wyn Williams, eds. 1994. *Environment and housing in third world cities*. Chichester and New York: John Wiley.

McGranahan, Gordon and Jacob Songsore. 1994. "Wealth, health, and the urban household: Weighing environmental burdens in Accra, Jakarta, and Sao Paulo." *Environment* 36(6): 4–11, 40–45.

Mitchell, J. K. 1992. "Natural hazards and sustainable development." Presentation to the 17th annual Natural Hazards Research and Applications Workshop, Boulder, 15–16 July; summarized in Berke (1995).

Mitchell, J. K. and N. J. Ericksen. 1992. "Effects of climate change on weather-related disasters." In I. M. Mintzer (ed.), *Confronting climate change*. Cambridge: Cambridge University Press, for Stockholm Environment Institute, pp. 141–151.

Satterthwaite, David. 1996. "Revisiting urban habitats." *Environment* 38(9): 25–28.

Serageldin, I., M. A. Cohen, and K. C. Sivaramakrishnan, eds. 1995. *The human face of the urban environment: Proceedings of the second annual World Bank conference on environmentally sustainable development*. Washington, D.C.: World Bank.

UNCHS (United Nations Commission on Human Settlements). 1995. Resolution 15/11. Adopted by the United Nations Commission on Human Settlements, 25 April–1 May, Nairobi. As reported in UNCHS (1996), pp. 423–424.

——— 1996 *An urbanizing world: Global report on human settlements 1996*. New York: Oxford University Press.

UNEP/WHO (United Nations Environment Programme/World Health Organization). 1994. "Air pollution in the world's mega-cities." *Environment* 36(2): 4–13, 25–37.

Wijksman, Anders and Lloyd Timberlake. 1984. *Natural disasters: Acts of God or acts of Man?* London: International Institute for Environment and Development.

World Health Organization. 1992. *Our planet, our health*. Geneva: Report of the Commission on Health and Environment.

World Resources Institute. 1996. *World resources 1996–97*. Special issue on "The Urban Environment". Washington, D.C.: WRI, UNEP, UNDP, World Bank.

Appendices

1. What is the population of the world's largest cities?

It is impossible to be sure of the exact population of the world's largest cities. Among others: definitions of cities vary; municipal boundaries vary; the existence, frequency, and accuracy of urban censuses vary; and rates of population change vary. The following table provides the most authoritative estimates of urban population for 14 of the largest cities.

City	Population, 1994 (millions)	Per capita GNP, 1991 (US$)
Tokyo	26.8	26,824
São Paulo	16.4	2,920
New York	16.1	22,356
Mexico City	15.6	2,971
Shanghai	15.1	364
Bombay	15.1	330
Los Angeles	12.4	22,356
Beijing	12.4	364
Calcutta	11.7	330
Seoul	11.6	6,277
Jakarta	11.5	592
Buenos Aires	10.0	3,966
Tianjin	10.7	364
Osaka	10.6	26,824

Source: United Nations, *World Urbanization Project. The 1994 Revision*, New York, 1995.

2. Known pre-twentieth-century urban disasters that killed more than 10,000 city residents

Year	City	Nature of disaster and death toll
365	Alexandria (Egypt)	Tsunami killed "many thousands"
526	Antioch (Syria)	Earthquake killed about 250,000
1041	Tabriz (Iran)	Earthquake killed 40,000
1138	Kirovabad (Tadjikistan)	Earthquake killed 130,000 in and around the city
1169	Catania (Italy)	Volcanic eruption killed 15,000 – most of city's population
1627–29	Mexico City	Floods killed about 20% of city's 127,000 people
1642	Kaifeng (China)	Deliberate breaching of a dyke on the Hwang Ho River killed most of the city's 200,000–300,000 inhabitants
1693	Naples (Italy)	Earthquake killed over 90,000 of about 200,000 residents
1721	Tabriz (Iran)	Earthquake killed about half the population of 150,000
1746	Lima (Peru)	Earthquake killed many of the city's 40,000 people
1755	Lisbon (Portugal)	Earthquake and tsunami killed 10,000–60,000 of the city's estimated 300,000 people
1773	Guatemala City	Earthquake killed many of city's 30,000 people
1797	Quito (Ecuador)	Earthquake killed 40,000 in and around this city of 30,000
1822	Aleppo (Syria)	Earthquake killed 100,000 out of 150,000 inhabitants
1824	St. Petersburg (Russia)	Ice jam floods killed 10,000
1853	Shiraz (Iran)	Earthquake killed 12,000 of 22,000 inhabitants
1864	Calcutta (India)	Cyclone killed "tens of thousands"
1881	Haiphong (Vietnam)	Typhoon killed 300,000 in and around the city
1882	Bombay (India)	Cyclone killed 100,000 people in and around this city of 800,000+ residents

Note: Many other cities suffered extensive destruction of property but relatively few deaths and injuries, especially as a result of catastrophic urban fires. The Great Fire of London (1666), the Chicago fire (1871), and the San Francisco (post-earthquake) fire (1906) are well-known examples. For additional US examples, see Christine Meisner Rosen, *The limits of power: Great fires and the process of city growth in America*, New York: Cambridge University Press, 1986.

3. Habitat II: Urban indicators that reflect the effects of environmental (natural, technological, social) hazards and disasters

Participants in the Habitat II conference process identified 27 indicators that are useful for making comparisons among conditions of urban living throughout the world. Indicator data from 235 cities in 110 countries were collected during 1995–1996. The following three indicators directly or indirectly measure some aspect of environmental hazard:

Indicator 19: Housing destroyed. Defined as the proportion of housing stock destroyed per thousand by natural or man-made disasters over the past 10 years.
Indicator 10: Median price of water, scarce season.
Indicator 6: Crime rates. Defined as reported murders and reported thefts per thousand population annually.

4. Habital II: Top 105 best practices of urban development that involve disaster management and humanitarian investment

Title of project	Location
Urban Planning and Reconstruction of a War-torn City Centre, Beirut	Lebanon[b]
Resettlement in Northern Iraq	Iraq
Palestinian Housing Council	Palestine
Housing Settlement Project in Shanghai	China[b]
Post-Calamity Reconstruction of Anhui Province's Rural Areas[a]	*China*
Urban community development for the resettlement of Ein Helwan, Cairo[a]	*Egypt[b]*

Source: United Nations Conference on Human Settlements (Habitat II), *Annotated bibliography of best practices 100 list*, A/CONF.165/CRP.3, Nairobi, 11 April 1996.
a. Projects undertaken in response to natural disasters.
b. Projects located in large urban areas (mega-cities).

5. Habitat II: Best practices of urban development that involve the reduction or management of natural hazards and disasters

Title of project	Location
A Structural Program for Hazardous Slum Areas in Belo Horizonte	Brazil[a]
Favela-Bairro Program	Brazil
Natural Disaster Control Plan in Serro do Mar, Cubatao Region	Brazil[a]
Rehabilitation of Urban Areas – Guarapiranga Project	Brazil
Prevention and Reduction of Geological Risks in the Hills of Santos	Brazil[a]
Post-disaster Reconstruction and Rehabilitation of Rural Areas in Anhui	China
Reconstruction of Ethnic Miao Wood Houses, Zhengdou village, Guangxi	China
Applicable Technology for Rebuilding Houses Damaged by Earthquakes	Costa Rica[a]
Disaster Management	Cuba
A Comprehensive Urban Development Project for Earthquake Victims	Egypt[a]
The Construction of 100 Schools – Contribution of Egyptian citizens	Egypt[a]
Developing the Constructive Urbane Environment in Sohag	Egypt
El-Tadamon Village – Assuite	Egypt
The National Projects for Establishing New Villages in Upper Egypt	Egypt
The Role of the Ministry of Urban Communities	Egypt
Innovative Shelter Delivery Mechanisms for Earthquake Affected Villages	India

Source: Best Practices Database, Habitat II (1997). http://www.bestpractices.org.
a. Projects located in large urban areas (mega-cities).

6. Habitat II: Best practices of urban development that involve the reduction or management of technological hazards and disasters

Title of project	Location
Vienna Air Monitoring Network	Austria[a]
The Establishment of the Pedagogical Process of Traffic	Brazil
The Intersectoral Involvement of Society in Traffic Safety	Brazil
Helsinki – The Energy Efficient City	Finland[a]
Intervention in Historic Settlement of the Plaka – Conservation Revival	Greece
Solar Village 3	Greece
Tehran's Action Plans for Improving the Living Environment	Iran[a]
Promotion of Latrine Construction by Low-income Households in Maputo	Mozambique[a]
Introducing Public Transport by Trolley-Bus	Romania
Reducing Pollution and Improving Environmental Quality	Romania
Improving Living Environments Through Comprehensive Local Policy	Sweden
The City of Stockholm	Sweden[a]
Electric Buses, Application and Research	USA
Switched Onto Safety	United Kingdom
Changing Travel Behaviour and Public Attitudes to Transport in Hampshire	United Kingdom
Energy Efficient Best Practice Programme	United Kingdom
Sustainability Indicators in Merton	United Kingdom
Blueprint for Leicester – Focus on Leicester's Home Energy Strategy	United Kingdom

Source: Best Practices Database, Habitat II (1997). http://www.bestpractices.org.
a. Projects located in large urban areas (mega-cities).

Contributors

John Handmer, School of Geography and Environmental Management, Middlesex University, Queensway, Enfield EN3 4SF, United Kingdom

Saleemul Huq, Bangladesh Centre for Advanced Studies, 620 Road 10 A (New) Dhanmondi, GPO Box 3971, Dhaka 1209, Bangladesh

Kwi-Gon Kim, Department of Landscape Architecture, Seoul National University, College of Agriculture, Suwon, Korea 170

Yoshio Kumagai, Institute of Socio-Economic Planning, University of Tsukuba, 1-1-1 Tennoudai, Tsukuba, Ibaraki 305, Japan

James K. Mitchell, Department of Geography, Rutgers – The State University of New Jersey, Piscataway, NJ 08854-8045, USA

Yoshiteru Nojima, Urban Planning Department, Building Research Institute, Ministry of Construction, Tokyo, Japan

Anthony Oliver-Smith, Department of Anthropology, University of Florida, Gainesville, FL 32611, USA

Dennis J. Parker, School of Geography and Environmental Management, Middlesex University, Queensway, Enfield EN3 4SF, United Kingdom

Rutherford H. Platt, Department of Geology and Geography, University of Massachusetts, Amherst, MA 01003-0026, USA

Sergio Puente Aguilar, Centro de Estudios Demograficos y de Desarrollo Urbano, El Colegio de Mexico, Camino al Ajusco 20, Col. Pedregal de Santa Teresa, 01000, Mexico, D.F.

William D. Solecki, Department of Earth and Environmental Studies, Montclair University, Upper Montclair, NJ 07043, USA

Ben Wisner, Director of International Studies, California State University, Long Beach, CA 90840, USA

514

Index

515